Global Maritime Conference (GLOBMAR 2018)

SHS Web of Conferences Volume 58 (2018)

Sopot, Poland
19-20 April 2018

Editors:

Theo Notteboom
Chris Peeters
Helena Maria Gregorio Pina Calado
Ernest Czermanski

ISBN: 978-1-5108-7600-2

Printed from e-media with permission by:

Curran Associates, Inc.
57 Morehouse Lane
Red Hook, NY 12571

Some format issues inherent in the e-media version may also appear in this print version.

No changes have been made to the content of these proceedings. There may be changes to pagination and minor adjustments for aesthetics.

Printed by Curran Associates, Inc. (2019)

For additional information, please contact EDP Sciences – Web of Conferences
at the address below.

EDP Sciences – Web of Conferences
17, Avenue du Hoggar
Parc d'Activité de Courtabœuf
BP 112
F-91944 Les Ulis Cedex A
France

Phone: +33 (0) 1 69 18 75 75
Fax: +33 (0) 1 69 28 84 91

contact-edps@webofconferences.org

Additional copies of this publication are available from:

Curran Associates, Inc.
57 Morehouse Lane
Red Hook, NY 12571 USA
Phone: 845-758-0400
Fax: 845-758-2633
Email: curran@proceedings.com
Web: www.proceedings.com

TABLE OF CONTENTS

Introduction

The maritime economy, extended with the concept of Blue Economy, is considered to be highly important in the present world. The rapidly growing marine space and increasing numbers of resource stakeholders have given rise to a global problem of sustainable use of world oceans and seas, as well as a challenge to solve conflicts between different marine space users.

Shipping, one of the foremost industries to exploit the waters, has raised a challenging issue for all those, who have an interest in using marine environment in a commercial and non-commercial way. The variety of these ways has faced us with the problem of marine spatial planning, in the process of which all stakeholders will share in the drive for sustainability.

Sustainable shipping with its complexity (vessels, ports and terminals, shipbuilding, maritime administration) is facing the challenge of reducing the negative environmental impact, safety, security, biodiversity preservation and many other issues calling for a modification to the existing approaches to the maritime economy.

On the other hand, all maritime business is under increasing pressure to provide efficiency in economic and energetic terms, which often has the opposite effect to global trend forces. In such difficult conditions, the operation and development of the maritime transport requires smart solutions, innovations and awareness from all maritime stakeholders.

In this meaning, The Global Maritime Conference (GLOBMAR) 2018 has taken the challenge of gathering high-level experts, researchers and professionals and presenting new concepts and solutions for sustainable shipping development and took place on 19-20th April 2018 in Sopot, at the University of Gdańsk – Faculty of Economics,

The conference was organized in a cooperation of City University of Applied Sciences Bremen, Free Hanseatic Bremen, University of Szczecin, Maritime University of Szczecin, Gdynia Maritime University and Maritime Institute in Gdansk.

The Honorary Patronage over the conference was assumed by Minister of Maritime Economy and Inland Navigation, Marek Gróbarczyk, Marshal of the Pomorskie Voivodeship, Mieczysław Struk, Rector of the University of Gdańsk, prof. Jerzy Gwizdała.

The conference was supported by Business Partners: Port Gdańsk, Rezerwa Sp. z o.o. and Media Partners: Baltic Transport Journal, Harbours Review, NAMIARY and the Baltic Sea Region Interreg Programme 2014 – 2020 project BalticLINes.

The event brought together 37 speakers, including scientists and professionals in the academia, industry and government. The thematic areas of the conference were divided into 7 sessions: the role of green & sustainable shipping, global impact of shipping industry, Baltic sea region impact shipping industry, shipping in maritime spatial planning, delivering MSP, digitalization of shipping & port operations and efficiency improvement.

This Conference Proceedings consists of 33 blind peer reviewed papers in an alphabetical order, but, primarily forming four basic thematic groups:

1. Sustainable shipping
2. Port & Harbours sustainable development
3. Maritime Spatial Planning
4. Maritime Law & Administration

Finally, we want to express our huge gratitude to all participant, particularly keynote speakers, session moderators, presenters, the scientific and organizing committee, editors and reviewers. He also expressed his satisfaction with the wide-range of topics discussed during the conference and hope that the extensive networking would continue in the future, for instance Sustainable Shipping Conference 2019 in Bremen and GlobMar 2020 in Sopot.

edp sciences

Web of Conferences

Statement of Peer review

In submitting conference proceedings to *Web of Conferences*, the editors of the proceedings certify to the Publisher that

1. They adhere to its **Policy on Publishing Integrity** in order to safeguard good scientific practice in publishing.
2. All articles have been subjected to peer review administered by the proceedings editors.
3. Reviews have been conducted by expert referees, who have been requested to provide unbiased and constructive comments aimed, whenever possible, at improving the work.
4. Proceedings editors have taken all reasonable steps to ensure the quality of the materials they publish and their decision to accept or reject a paper for publication has been based only on the merits of the work and the relevance to the journal.

GLOBMAR 2018 - Global Maritime Conference

Title, date and place of the conference 19 - 20 April 2018, Sopot, Poland

Prof. Theo Notteboom
Prof. Chris Peeters
Prof. Helena Maria Gregoria Pina Calado
Proceedings editor(s): dr Ernest Czermański

Prof. dr. Chris Peeters
22/10/2018

Prof. Dr. Theo Notteboom
5/11/2018

Date and editor's signature

Deregulation of laws and border facilitations, and their impact on the increase of competitiveness of Polish seaports

Magdalena Adamowicz[1] *,

[1] University of Gdansk, PhD, Maritime Law Department, Faculty of Law and Administration, Poland

Abstract. This article attempts to explore the effects caused by the implementation of facilities for maritime entrepreneurs introduced under the act deregulating economic law (hereinafter: "Ports 24h" or port package). Moreover it was analysed whether the procedures implemented by the Customs Service and border units, and in particular the simplified procedure, one stop shop and single window, shortened the time of customs clearance in ports. It was examined whether new solutions affected seaports in Poland and helped to build their competitive advantage over the Baltic Sea. The port of Gdansk was particularly interested due to the annual increase in transhipments. There is a surprising lack of studies on customs conveniences in sea ports. The bulk of academic work focuses on the competitiveness of sea ports as seen through infrastructural investments, with no consideration given to factors such as stable, clear and reliable law which does not impede enterprises in their business.

Introduction

Access to the sea and the possession of well-developed sea ports count among the determinants of economic growth for every country. It is commonly believed that a sea port is a window to the world [1].

In Poland, just as in many other countries of the world, sea ports are a critical ingredient of and a major influence on the operation of the economy. If favorable conditions exist for sea ports to develop and operate efficiently, the entire economy is at an advantage, while any barriers or negative factors have a detrimental impact on the economy of any country. It is very important therefore to provide sea ports in Poland with opportunities for action in order not only to clear hurdles to their growth but also to enable their development and progress to the position of leaders in the Baltic area.

It was only a few years ago, in 2012, that one, in 2012, one of the world's leading consulting companies produced a report which found that Polish sea ports would have to face numerous challenges that slow down their development at present [2]. Among said barriers were issues related to VAT payments, unreasonably long customs procedures and problems with obtaining necessary documents as quickly as in the Western ports. The report estimated that 670 000 TEUs worth of traffic arriving at or departing from Poland

*Corresponding author: madamowicz@prawo.ug.edu.pl

was handled by other than Polish ports, [2] with ineffective customs, taxation and sanitation procedures at border checkpoints provided as the reasons.

Similar criticisms and postulates came as early as 2004 before Poland's accession to the European Union from various environments, including importers, exporters, freight forwarders, carriers, terminal operators and even customs authorities themselves, calling for reform in relevant legal regulations. The need for new solutions followed not only from trends in nation-wide policies intended to introduce facilitations for entrepreneurs, but also from the necessity to implement procedures consistent with solutions adopted in the EU member states and around the world. It was predicted that new solutions in this field, once implemented, would enable a substantial volume of cargo traffic to be re-directed towards Polish ports. To that end, preparations were started to set up indispensable conditions in Poland to create the Baltic logistics and transport centre for Central and Eastern Europe. For such an ambitious project to become a reality, however, a number of legal and organizational changes would have to take place, including facilitations for entrepreneurs, modernization of the existing and construction of new road, rail and port infrastructure on land and water alike. Another barrier to increasing the volume of cargo is posed by shortcomings in the IT system and an inadequate exchange of information and documents by electronic means, as well as system incompatibility among various administrative authorities and between the same authorities and entrepreneurs.

Not only will this article attempt to explore the legal effects of facilitations for entrepreneurs introduced within the framework of the 4th deregulation act, [3] the so-called "Ports 24h" packet and procedures implemented by state administrative bodies (e.g. the Customs Service), but it will also determine their impact on sea ports in Poland building their competitive edge, with special reference being made to the Sea Port of Gdańsk. Due to size constraints, the present article will omit any issues related to infrastructural investments at the ports and in their vicinity as well as broadly understood digitization, as the foregoing have been discussed extensively in the literature [4,5].

1. EU policy on facilitations in processing ships at sea ports

EU Directive 2010/65 on reporting formalities for ships , which entered into force in 2015 [6], obliged all member states to implement appropriate domestic regulations. Its goal was to simplify and harmonize administrative procedures applied in maritime transport by making electronic transmission of information standard and rationalizing reporting formalities. The Directive applies to ships entering and departing from ports situated in member states.

The Directive made provision for the so-called "single window", or a single place of storage of all data concerning sea-going vessels and cargo. The idea was that information would be reported and collected in one place to be shared with all authorized institutions involved in ship's clearance, i.e. coast guard, veterinary, sanitary and customs authorities as well as other authorized [7]. The aim of the legislation was for ships arriving at ports to transfer all the necessary data to the relevant domestic authority and be processed even before mooring. Any in-port checks by customs services or other officials would then be carried out on an ad-hoc basis only.

Another goal of the directive was to create a uniform internal market which would – despite the elimination of cross-border controls – ensure safety for inhabitants and the environment. Poland as a border state of the EU was tasked to cause its customs services to allow goods brought from outside the EU into free circulation in such a way as to comply with safety standards for all users and the environment.

To ensure standardized clearance and processing for ships at the international level and because of the global dimension of maritime transport, EU legislation, including the present directive, had to take into account the requirements of the International Maritime Organization (IMO) [8]. Therefore, the Directive obliged member states to adopt harmonized and updated forms (FAL forms) prepared by the IMO in the FAL Convention in order to facilitate traffic [9]. However, compliance with reporting requirements by completing the paper versions of FAL forms was acceptable only until 1 June 2015, with electronic documents being the only permissible form from then onwards.

Besides, the Directive obliged member states to extend cooperation between the competent authorities, such as the authorities responsible for customs, border checks, and public health and transport, to simplify and harmonize reporting formalities in the EU and rationalize their use of electronic data transmission and data exchange systems, while also removing barriers in maritime transport and creating a European maritime transport area without barriers.

2. The "Port Packet" – reasons for revision

The foregoing EU Directive on reporting formalities for ships was one of the principal reasons for introducing the changes in our national system of customs and border clearance procedures at sea ports.

Another reason for the changes stemmed from the fact that customs procedures at Polish sea ports took too long a time in comparison with their Western counterparts. Clearance procedures were subject to overly long verifications involving a great deal of red tape. Imported goods, especially food, were subject to control by specialized bodies such as the Customs Service, Veterinary Inspectorate, State Sanitary Inspectorate, Inspectorate for the Commercial Quality of Agricultural and Food Products, State Inspectorate for the Protection of Plants and Seeds. Each of these bodies was authorized to carry out mutually independent checks. There was no coordination between the activities of the individual bodies and those of the mentioned authorities. Before 2015, a delivery was subject to checks by more than one authority and so it had to undergo the same logistical operations a number of times, resulting in a long handling time and increased costs for entrepreneurs. If the authorities had acted jointly, the checks could have been carried out more quickly than when each of them was performed separately. The average waiting time at a Polish sea port prior to enacting the legislation was 30 hours and – in earlier times – 5 days, a week or even longer. In Western European ports, the same process took from 1 to 2 days, allowing importers to save time and money [10].

The main reasons for long and costly procedures at ports before enacting the legislation included:

(1) The absence of communication between the respective links in the supply chain and control bodies at ports.

(2) The absence of precise pre-arrival cargo information which could be used for risk assessment, identification of the goods and of the transacting parties.

(3) Multi-stage risk assessment.

(4) Insufficient technical and organizational measures at the disposal of port operators.

(5) The absence of mutual cooperation between the inspectorates, especially when drafting and issuing post-control decisions, due to a long response time.

Other relevant factors include: a) insufficient working time of the inspectorates and frequent delays in the issuance of decisions; b) unequal level of digitization of the inspectorates, limiting the possibility of a quick exchange of information; c) discrepancies in the internal procedures of each of the inspectorates with regard to control and issuance of decisions, d) organization and work conditions – some of the inspectorates do not have their offices at container terminals and their officials have to commute.

Therefore, a discussion sprang up around issues related to customs and general difficulties in the access to Polish sea ports as well as near-discriminatory legal provisions giving preference to German and Dutch sea ports over Polish ones. In the discussions, it was said that the largest "Polish" cargo-handling port was Hamburg rather than Szczecin, Gdynia or Gdańsk. Unsurprisingly, such issues underlie persistent complaints of Polish entrepreneurs about administrative bodies and port operators. The parties concerned together with representatives of the government signed a declaration of intent entitled "The Polish Gate – The Polish Ports", whose signatories undertook to take joint action to amend the discriminatory laws as they now stand. The idea proposed is that cargo arriving by sea should be cleared by the Customs Service and arious inspectorates within a reasonable time-limit to fast-track procedures and improve the competitive position of Polish sea ports.

Another problem is VAT settlement. The EU has in place a cashless VAT system that eliminates the need for entrepreneurs to reserve additional funds to cover VAT payments, with deferment periods of up to 90 days. Poland adopted a similar solution in 2012, but it requires posting a large deposit, which means that only 5 to 10 per cent of forwarders qualify. Therefore, it is cheaper for forwarders to choose the port in Hamburg or Rotterdam, as this does not necessitate any additional funds. Multitude of inspectorates, control institutions is Is yet another issue.

3. "Port 24h" implementation

On 1 January 2015, the Act on Facilitating Business Activities of 7 November 2014 came into force. It imposed a 24-hour time-limit for completing any formalities related to the carriage of goods by sea, counting from the time of their declaration at customs. The Customs Service was tasked to coordinate inspections at ports in such a way as to ensure that all inspection activities are carried out with the least possible inconvenience to importers and operators. Under the coordinating efforts, the Customs Service was obliged to:

1) inform the authority competent under separate regulations to carry out inspections of goods brought from third states of the arrival in the EU customs area of goods subject to obligatory inspection carried out by that authority;

2) set a time, date and place of inspection of goods brought from third states in coordination with the port operator and holder of the goods.

The 24-hour time-limit did not apply to cases requiring laboratory tests, quarantine or – in the case of livestock transports – isolation. In all other situations, the duration of inspections at the port, including those carried out by the Customs Service, of goods brought by sea from third states should not exceed 24 hours, as counted from declaration at customs, filing a complete inspection application form with the proper authorities and providing notice of the time and place of inspection until the release of goods.

Therefore, the 24-hour time-limit begins to run when all of the requirements have been met, i.e. declaration of goods, filing the paperwork and setting a time and place of inspection.

The Customs Service can carry out the task thanks to an exchange-of-data system (Inspection Coordination System – ICS) implemented to bring together other border control bodies and entrepreneurs. ICS replaces paper documents and allows progress tracking on all freight processing procedures.

Other measures implemented under the Act relate to increasing the competitiveness of Polish sea ports, including:

1. digitization of applications, i.e. it is no longer necessary to physically inspect the freight, resulting in processing times of between 15 and 40 minutes where the time-limit set by EU standards is 2 hours;
2. curtailed paperwork – documents are accepted in electronic form (e-Attachments)
3. pre-arrival clearance – freight is declared for customs clearance before its arrival at the port, which shortens processing times
4. certificate of origin – makes it easier to obtain approvals for exported goods, curbing paperwork and reducing the number of bureaucratic steps and the time needed for carriers and authorities to complete formalities (authorized exporter, registered exporter, in blanco certificates).
5. VAT – VAT deferment on terms similar to those applied in German ports, but limited to AEO holders and enterprises applying simplified procedures
6. export procedure – extends export procedure to goods, refers to EU goods exported out of EU or to special tax areas.

4. Effects of the reform

It was in 2017, two years after the new regulations had entered into force obliging the Customs Service to coordinate inspections, National Treasury Administration (NTA) was appointed. NTA was supposed to optimize and coordinate the activities of three authorities: the Customs Service, the Treasury Audit Authority and Tax Administration. NTA was tasked to actively support enterprises and systematically raise standards of service through applying and extending the range of available conveniences and shortening processing times. The reformed Customs Service operating within the new structure as part of NTA holds a respectable place the World Bank's "Doing Business 2018" ranking which describes and compares the conditions of running a business in 190 countries. Polish solutions related to the trade of goods with foreign countries with respect to customs services at border checkpoints came first in a tie against a number of countries [11] (10. The short time and low costs of export and import-related logistical process were particularly applauded).

The port of Gdańsk is undeniably an excellent example of a spectacular success – a record volume of cargo-handling operations amounting to 40.6 million tons in 2017 – only 2 years after the entry of regulations. From the example of the port of Gdańsk, it can be said that only 2 years after the entry into force of the new regulations, a spectacular success was recorded – a record volume.

Table 3. Budget receipts from goods passing through DCT

Budget receipts from goods passing through DCT	2015	2016	2017	I-III 2018
VAT	4 583	5 406		
excise duty	191	4		
customs duties	670	789		
TOTAL (mln zł)	5 444	6 199	22500	2100

Source: PGA SA based on the data of the Customs Chamber in Warsaw and the Statistical Compendium of the Customs Service

The port authority's plan for 2017 projected a throughput of 38.9 million tons, which was in itself an ambitious goal, compared to 2016 when the figure stood at 37 million tons. Such good performance was due to the growth of Deepwater Container Terminal DCT [12], which handled 1.6 million containers, 22 per cent up against the previous year's result; also, thanks to an increase in automotive transports from 14 thousand in 2016 to 74 thousand, marking a radical 412 per cent growth. General mass cargo formed nearly half (44 per cent) of all containerized freights. There was a minimal (0.05 per cent) rise in the volume of coal, which accounts for 12.5 per cent of cargo-handling operations. It is assumed that, with a good economic situation, the year 2018 could bring another record, exceeding 50 million tons of handled cargo, with DCT set to service 2 million containers. Despite the successes, there still remains a weak link in the chain, and that is a small share of food transports in the overall cargo traffic through Polish ports, most of which arrives in our country through Rotterdam.

The ports of Gdańsk and Gdynia are the gateway into Poland for 60 per cent of total imports into our country from China. A 19-percent rise in the volume of containerized cargo volume passing through Gdańsk and Gdynia, as compared with the year 2016, was achieved thanks to fast and smooth customs procedures. The container terminal at the port of Gdańsk delivers the bulk of tax contributions.

Table 1. Participation of the Port of Gdańsk in receipts from customs duties, VAT and excise duty from goods handled at the Port of Gdansk

	In the income of the state		
Year	**Inflows in millions of PLN**	**Polish revenues in millions of PLN**	**The port's share in the State's revenues**
2008	2 223	253 547	0,90%
2009	3 035	274 184	1,10%
2010	3 758	250 303	1,50%
2011	6 479	277 568	2,30%
2012	6 264	287 595	2,20%
2013	13 970	279 151	5,00%
2014	16 001	283 543	6,00%
2015	18 400	289 137	6,00%

Source: PGA SA based on the data of the Customs Chamber in Warsaw and the Statistical Compendium of the Customs Service

Table 2. Participation of the Port of Gdańsk in receipts from customs duties, VAT and excise duty from goods handled at the Port of Gdansk

	In total revenues for Poland from customs duty, VAT and excise tax		
Year	**Polish customs offices (PLN million))**	**Customs branches Port (PLN million)**	**Participation of Gdańsk**
2008	78 557	2 223	3%

2009	75 880	3 035	4%
2010	80 375	3 758	5%
2011	88 962	6 479	7%
2012	91 272	6 264	7%
2013	83 414	13 970	17%
2014	84 225	16 001	19%
2015	76 835	18 400	24%

Source: PGA SA based on the data of the Customs Chamber in Warsaw and the Statistical Compendium of the Customs Service

According to NTA data for 2017, the amount of VAT, duty and excise fees totalled PLN 22.5 billion. The statistical value of goods handled was estimated at PLN 77.5 billion. It is estimated that sea ports account for 6 per cent of the GDP.

It is also responsible for 40 per cent of state treasury revenue. In the first quarter of 2018, the cargo handled at DCT contributed PLN 2.1 billion in duties, VAT and excises.

As a result, Polish sea ports are becoming increasingly competitive against western ones in terms of the time of customs procedures. This could bring an increase in traffic through the port and inthe cargo handled. In all ports, the customs clearance is in electronic form, with most freights not even required to physically undergo inspection. This goes to show that legislative modifications and the smooth operation of public administrative bodies have a large impact on the competitive position of sea ports especially relative to ports in Germany and other Western European countries.

Every third container is cleared under the notification procedure for goods and customs clearance when the cargo is still on board the ship. Physical inspections are becoming more of a rarity where manual checks are used as a supplementary measure. Increasingly, risk analysis is based and relies heavily on electronic data.

Conclusion

The object of the article was to show that amendments to the law to deregulate and simplify procedures have contributed to the increase of containerized cargo volume in Poland, which means development and a better competitive position for Polish sea ports.

The "Ports 24h" programme, in force for 3.5 years now, has been successful and has simplified inspection procedures. The implementation of the Act on Facilitating Business Activities was acclaimed by businesses in the port industry. The clearance of cargo is supervised by the Customs Service which coordinates between all other authorities.

The amended and simplified sanitary, customs and taxation procedures have helped to procure some cargo, handled before by Western ports, mainly in Hamburg and Rotterdam.

The organizational and legal reforms have helped to significantly increase the capacity of Polish sea ports. As a result, these ports will go on to play an increasingly paramount role for the Polish economy and logistics. A further growth in the volume of handled cargo will fall into step with investments into infrastructure. Such tendencies will evidently further support and strengthen the international position of Polish sea ports and Poland in the Baltic Sea area.

References

1. M. Adamowicz, *Legal aspects of seaports in Poland*, Maritime Law **32,** 160-170 (2017) http://journals.pan.pl/dlibra/publication/107759/edition/93420/content
2. Jones Lang LaSalle Report, Polish Container Ports: New directions of development of the logistics market - September 2013, pdf ss; http://www.jll.pl/poland/pl-pl/raport
3. Polish Act on on facilitating business operations; 7 November 2014; Polish Journal of Laws 2014/ 1662.
4. M. Pluciński, *Zasady i źródła finansowanie inwestycji infrastrukturalnych w polskich portach morskich – stan obecny, kierunki ewolucji* , Logistyka **2,** 2195 – 2203 (2011).
5. M. Matczak, *Inwestycje infrastrukturalne realizowane w polskich portach morskich jako narzędzie kształtowania ich pozycji konkurencyjnej,* [in:] Polska gospodarka morska. Restrukturyzacja. Konkurencyjność. Funkcjonowanie. Rozwój, editor : H. Salmonowicz, (Wyd. Kreos, Szczecin 2010).
6. Directive 2010/65/EU of the European Parliament and of the Council of 20 October 2010 on reporting formalities for ships arriving in and/or departing from ports of the Member States and repealing Directive 2002/6/EC, Official Journal of the European Union, L 283/1, 29.10.2010.
7. P. Morales-Fusco, S. Saurí,A. M. Lekka, I. Karousos, *Assessing Customs Performance in the Mediterranean Ports. KPI Selection and Best Practices Identification as Part of the MEDNET Project*, Transportation Research Procedia, **18,** 374 – 383 (2016), http://www.sciencedirect.com/science/article/pii/S2352146516308043.
8. J. Nawrot, *Ship report ing facilitation under EU regulation. The Polish pat* ; Maritime Law **33,** 75-86 (2017) http://journals.pan.pl/dlibra/publication/118628/edition/103205/content
9. Convention on Facilitation of International Maritime Traffic (FAL Convention), adopted on 9 April 1965, as amended.
10. M. Pluciński, *The duration of border control procedures related to the handling of freight transported by means of sea-road transport chains as a component in competitiveness of the polish seaports,* Contemporary Economy **8,** 59-69 (2017), www.wspolczesnagospodarka.pl
11. The World Bank, Doing Business, Measuring Business Regulations, in categories: Trading across borders, http://www.doingbusiness.org/rankings
12. M. Adamowicz, *Deepwather Container Terminal DCT- Genesis and realization of investments,* Contemporary Economy **3**, 91-108 (2016), www.wspolczesnagospodarka.pl

SHS Web of Conferences **58**, 01002 (2018) https://doi.org/10.1051/shsconf/20185801002
GLOBMAR 2018

A preliminary evaluation of the impact of Sulphur Emissions Control Area in the Baltic Sea on air quality in port cities. Case port – the city of Gdańsk

Michalina Bielawska[1], *Oskar Czechowski*[2], *Ernest Czermański*[3], *Aneta Oniszczuk-Jastrząbek*[3], *Tomasz Owczarek*[2]

[1] Agency of Regional Air Quality Monitoring in Gdańsk Metropolitan Area
[2] Faculty of Entrepreneurship and Quality Science, Gdynia Maritime University
[3] Institute of Maritime Transport and Seaborne Trade, University of Gdańsk

Abstract. The purpose of this research is to attempt to evaluate the extent, to which technical standards related to marine fuels and thereby also sulphur dioxide (SO2) content in engine exhausts from vessels operating on the Baltic Sea have been effective in curbing the negative impact of marine shipping on air quality, in particular in port cities. Marine environment protection is governed by the provisions of the MARPOL 73/78 International Convention, which Poland ratified as a party. Different areas of concern for marine shipping have been regulated in separate Appendices to the Convention. The first step was to introduce severe restrictions on SOx emissions in view of the fact that heavy marine fuel is the lowest-quality kerosene-derived fuel with a large content of sulphur. A gradual process was put in place to reduce its content in marine fuel. As a consequence, the world's marine areas were divided into sulphur emission control areas (also known as SECA) and other areas. In Europe, these areas include the entire Baltic Sea and large portions of the North Sea. Another important technical and economic measure was to lower the limit of sulphur content in marine fuel to 0.1% in all SECA areas, with the limits remaining unchanged in the other areas. Two dates were of key importance for the investigation: 2010, when the reduction in sulphur content of marine fuels from 1.5% to 1% was mandated, and 2015, when the standard for sulphur content was dramatically lowered to 0.1%. In the first stage, the concentration of sulphur dioxide was researched as one of the factors preceding air contamination with suspended particles in the Gdańsk - Gdynia area in the period from 2005 to 2016, as investigated by four automatic reference measurement stations in the ARMAAG network (hourly data) located in the immediate vicinity of the sea (AM4, AM5, AM6 and AM8). The research concerned the arrival of high concentrations of sulphur blown in from the sea by the wind. Another key factor was the secondary data on the number of ships, in the form of monthly series, starting from 2007. The analysis was performed in stages. In the first stage, the quality of measurement and secondary data were evaluated using a unique data quality assessment method. Further on, Principal Component Analysis (PCA) models were constructed to identify spatial correlations between SO2 concentration distributions, which were used later as a basis on which to determine synthetic measures of average hourly concentrations for the entire agglomeration area. Subsequently, the impact of the SO2 source (influx from the Baltic Sea to the agglomeration areas) was gauged separately for each individual station. The PCA models constructed on the basis of hourly data corroborated the synthetic measures as correct, making it possible to identify the similarity of concentration distributions across the investigated stations. Multi-Dimensional Regression Models and Generalized Regression Models (GRM) have made it possible to identify the period, in which the concentration of sulphur dioxide dropped steadily (from 2010 to 2016), as well as the seasonal impact of variation in SO2 concentration and the number of ships. The hourly data was converted to average monthly, quarterly and annual values, depending on the mathematical model and purpose of research.

Keywords: shipping, Sulphur Regulation, ship emissions, port emissions, port – city correlation, port of Gdansk

1. Introduction

Human activity at sea generates serious consequences for the environment and society. The activity is therefore heavily regulated by international and national rules and acts. One of the most important legal acts in recent years, dedicated to the environment (with a focus on the sea) is the MARPOL 73/78 – International Convention for the Prevention of Pollution from Ships. The entire act, with a number of annexes, has been introduced to address the following concerns: oil pollution (I), noxious liquid substances (II), sewage (IV), garbage (V) and the prevention of air pollution by ships (VI). The Annex VI standards came into force in May 2005. This act established the Emission Control Areas (ECA – an area where special mandatory measures are required to control NO_2, SO_2 and particulate matter (PM), or all the three types of emissions from ships) with a global cap of 4.5% on the sulphur content in fuel and a 1.5% limit in SECA (Sulphur Emission Control Area) zones. Standards for SO_2 limits were incorporated into EU law by Directive 1999/32/EC, as amended by Directive 2005/33/EC. The Directive also set two additional SO_2 limits for EU member states: 1.5% for passenger vessels operating regular services from 2006, and 0.1% for ships at berth from 2010. Generally, ECAs were established across the whole Baltic Sea area (see Fig. 1.1) – as defined in Annex I of MARPOL (SO_2 only). The latest SO_2 limits allowing a maximum 0.1% of sulphur content in marine fuel in SECA zones came into force as of 1 of January 2015. This was predicted to mark a serious technical and economic turning point for shipping companies in the Baltic Sea Region [1].

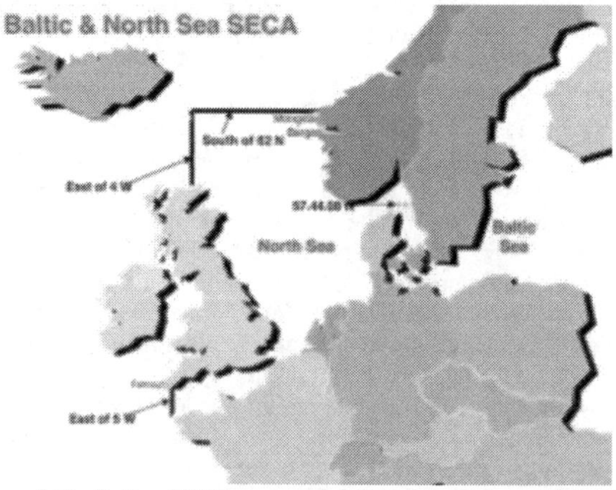

Figure 1. Baltic and North Sea SECA zone established by Annex VI MARPOL 73/78.
Source: [2]

MARPOL Annex VI means therefore a serious reduction of sulphur pollution in the SECA zones and, naturally, in coastal regions and especially ports and port cities. Due to the on-going debates, the current tendency to tighten restrictions and establish worldwide sulphur content limits on marine fuels, it seems that as of 2020, a limit of 0.5% will be enforced for global maritime transport.

This research paper is an attempt to evaluate the effectiveness of the implementation of MARPOL Annex VI in terms of limiting sulphur content in cities with large sea ports located within their boundaries. The object of this research will be to test the quality of air based on its sulphur oxide content, having regard also to such relevant factors as air temperature, wind force and direction, and vessel traffic at port.

2. Methodology

2.1. Models

In order to verify the hypotheses, models were selected for use with the adopted research procedure, depending on the measurement scales of the empirical, stochastic and exploratory variables and assumptions under investigation. Among the key models the following should be mentioned: simple models of the ANOVA family in averages compared and multidimensional principal component (PCA) models, as well as advanced generalized regression models (GRM). The obtained results allowed the implementation of all the assumed research goals. At the same time, their cognitive value is high not only in scientific but also in practical terms.

The GRM models were used to identify factors in weak measurement scales, i.e.: nominal and ordinal scale, in addition to (impact of time periods on SO_2 concentration) and in conjunction with the strong scale. GRM is a path rather than a model. It subsumes many classical models (e.g. ANOVA, MANOVA) into one simple model. It allows testing the effects and interactions. Simple models are easier to put to test again in replication and cross-validation studies, less costly to put into practice in predicting and controlling the outcome in the future and – last not least – allow using, to a larger extent than individual models taken separately, a wider range of fit statistics and diagnostic calculations, example leverages and outlier observations.

The analysis of variance (ANOVA) is used to determine whether there are any statistically significant differences between the means of two or more independent (unrelated) groups (levels of one variable in weak measurement scales) in the one-way ANOVA or two variables in the two-way ANOVA with interactions.

The one-way ANOVA compares the means between the groups we are interested in and determines whether any of those means are statistically significantly different from each other. Specifically, it tests the null hypothesis:

$$H_0: \mu_1 = \mu_2 = \mu_3 = \cdots = \mu_k$$

where μ = group mean and k = number of groups. If, however, the one-way ANOVA returns a statistically significant result, we accept the alternative hypothesis (H_1), which is that there are at least two group means that are statistically significantly different from each other [6,7].

2.2. Research procedure

The adopted research procedure explored SO_2 concentration as an indication of other concentrations. Further on, the other contaminations and their compounds, such as suspended dust, nitrogen oxides and others, will be looked at.

Of key importance were the two dates when new standards were introduced: 2010 when the permissible content limit of sulphur was lowered from 1.5% to 1% and 2015 when sulphur content was slashed dramatically to 0.1%. The first-stage investigation covered sulphur dioxide concentration (designated as SO_2 or SO2 on computer printouts) as one of the indicators of air pollution with suspended particles in the Tricity area from 2005 to 2016, as measured at one of the four automatic survey stations within the ARMAAG network (hourly data) posted in the immediate vicinity of the sea (AM4, AM5, AM6 and AM8).

Initially, the research additionally looked at the results of concentration measurements for SO2 at AM2 and AM3 stations, but these series were ignored in the subsequent stages of the research for technical reasons.

The variables were series of results collected at hourly intervals between 1 January 2005 to 31 December 2017 (113 950 items). In fact, they are continuous measurements recorded on the ARMAAG surveillance system, sectioned into hourly intervals and averaged accordingly. Besides the measurement result series for SO_2 concentrations, the research used hourly measurements of wind speed and direction for each station separately. The measurement results for SO_2 concentrations after factoring in the direction of wind from the sea were designated as "SO_2.wsea".

The data on the number of ships come from the Port of Gdańsk. These are numerical strings representing the number of ships entering and leaving the Port of Gdańsk in the years 2010 – 2017.

Depending on the purpose of the investigation and the applied mathematical model, the original daytime strings were averaged (concentrations) or combined (number of ships) to match the required monthly or annual cycles.

The research was completed in stages. In the first stage, all the series were tested for quality by two methods: a manual method carried out in operational mode by the ARMAAG foundation employees and an automatic method based on proprietary algorithms of measurement data diagnostics [3]. Only verified data were considered in further research.

The next stage involved the construction of PCA (Principal Component Analysis) models in order to identify the mutual relations in spatial distributions of SO_2 concentrations to provide a basis for determining the synthetic measures used to select the average hourly SO_2 concentrations for the entire agglomeration area.

For each year separately, PCA models were constructed for SO_2 concentration measurements at each station for all wind directions and for all SO_2 concentration measurements only for seaward winds.

For the first principal components (PC1) and second principal components (PC2) in the years 2005 – 2016, for all wind directions and seaward winds, Pearson total correction factors were calculated between the current and the previous year (designation: lag 1), as presented in Fig. 2. Starting from 2013, the correlation values are high (0.8 for PC1 and 0.5 for PC2) and stable, indicating a high degree of similarity of the phenomenon's structure. The earlier years (2009 – 2012) are characterized by dynamic change, exhibiting in full the dynamics within the relations of SO_2 concentrations at the individual stations year after year. For PC1, the correlation increase from 0.1 between 2010 and 2011, and to 0.9 between 2013 and 2012 means a complete change in the relations of SO_2 concentrations across the entire coastal area for concentrations accumulated from the seaward directions (Figure. 2 left).

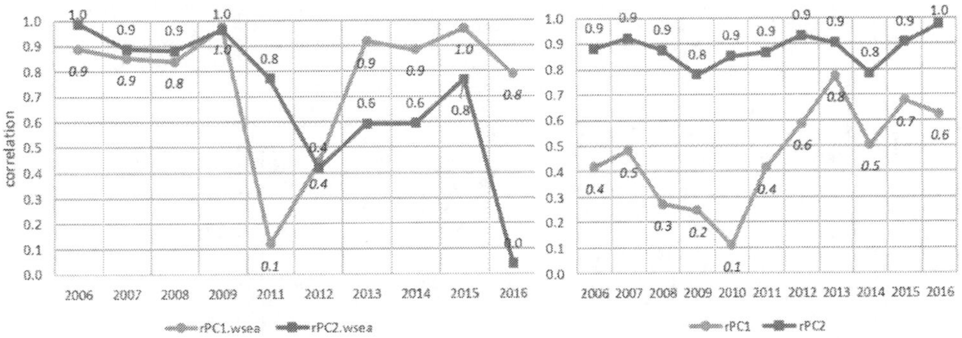

Figure 2. Correlation between PC1 and PC1 lag 1 year in PCA structure models; left: wind speed from sea, right all wind directions.
Source: The authors' own elaboration based on Agency of Regional Air Quality Monitoring in Gdańsk Metropolitan Area.

The PCA models for SO_2 concentrations confirmed that it was a correct decision to leave out the AM2 results due to the completely different type of measurement they represented (Figure 3). This was related to the fact that SO2 concentrations at the location of the AM2 station are impacted by emissions from large and troublesome industrial developments (refinery, power plant) and the densely forested surrounding area.

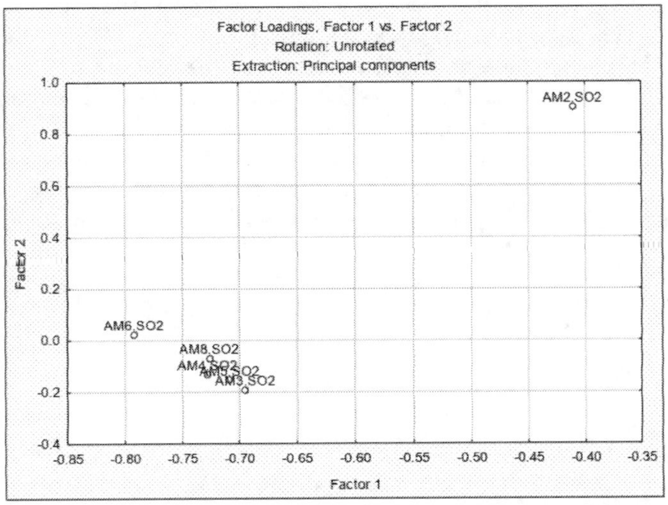

Figure 3. Model PCA results (2005-2016) from all stations and all wind directions.
Source: The authors' elaboration based on Agency of Regional Air Quality Monitoring in Gdańsk Metropolitan Area.

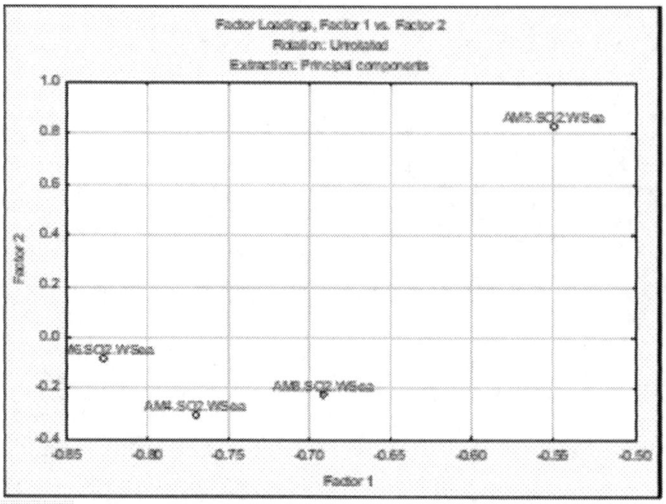

Figure 4. Model PCA results (2005-2016) from selected stations and wind directions from sea.
Source: The authors' own elaboration based on Agency of Regional Air Quality Monitoring in Gdańsk Metropolitan Area.

Ultimately, the following stations were qualified for further research: AM4, AM5, AM6 and AM8. Their mutual relations, together with the direction of the seaward wind, are presented in Figure 5. The AM5 station is located a little further inland, therefore its results differ slightly from the others'; in that sense, it forms the background for those stations situated closer to the sea. Ultimately, the PCA models for SO_2 concentrations were compared with no regard to the wind direction and models for the direction of seaward winds (Figure 6); at the same time, the models were calculated for one-hour data and any gaps were filled with averaged figures. The models thus identified exhibit a high degree of similarity, which serves as proof that the right stations have been chosen for further investigation.

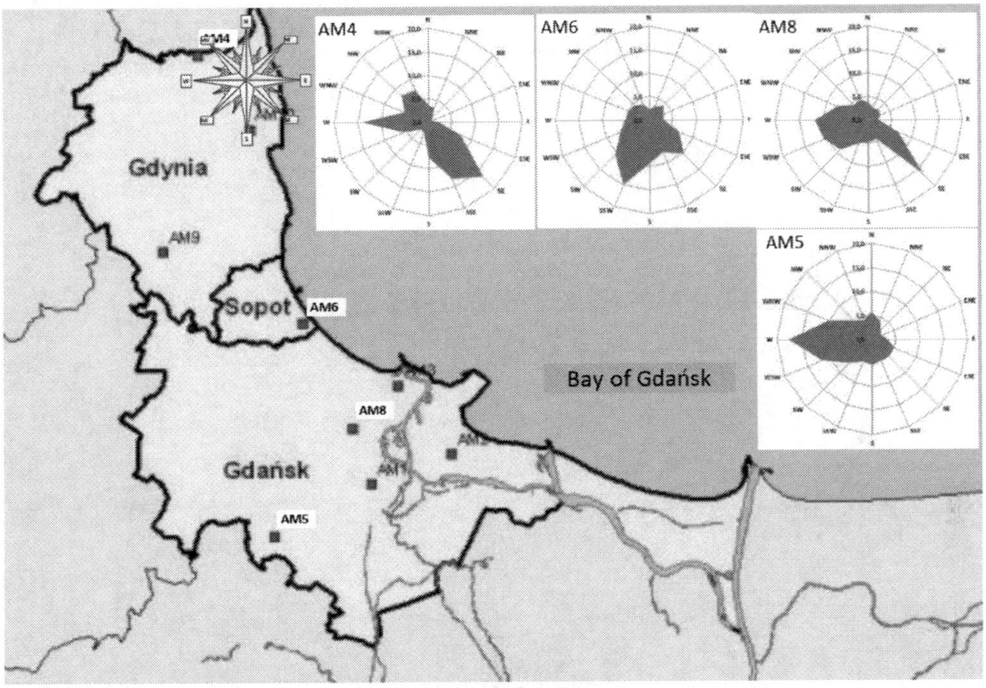

Figure 5. Wind roses (2010-2017) on selected ARMAAG stations.
Source: The authors' own elaboration based on Agency of Regional Air Quality Monitoring in Gdańsk Metropolitan Area.

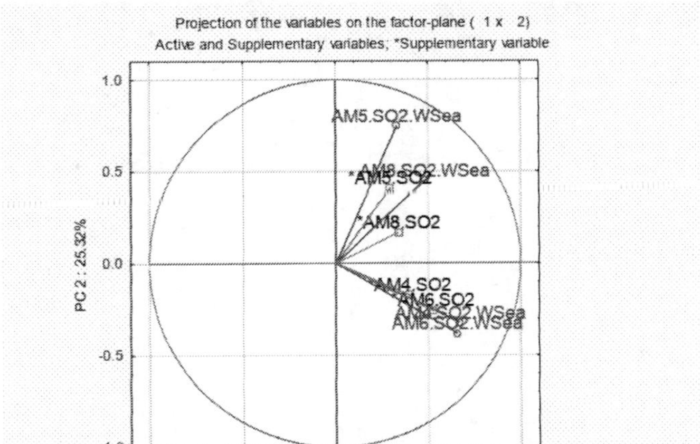

Figure 6. Comparison of PCA models: all wind directions and from the sea.
Source: The authors' own elaboration based on Agency of Regional Air Quality Monitoring in Gdańsk Metropolitan Area.

Further calculations were based on synthetic measures for the entire area and these were used to select averages for the four stations. Two measures of SO_2 concentrations were calculated for hourly results in accordance with

Table 1: for all wind directions and for seaward winds which accumulate the largest amount of SO_2 emissions from ships onto the coastal and agglomeration areas.

Table 1. Wind directions and ranges in deg in selected stations

Station	Wind direction from sea	Range in deg
AM4	From N to SE	349-360;1-146,99
AM5	From NNE to ENE	12-78,99
AM6	From N to ESE	349-360;1-123,99
AM8	From N to E	1-101,99

Source: The authors' own elaboration based on Agency of Regional Air Quality Monitoring in Gdańsk Metropolitan Area.

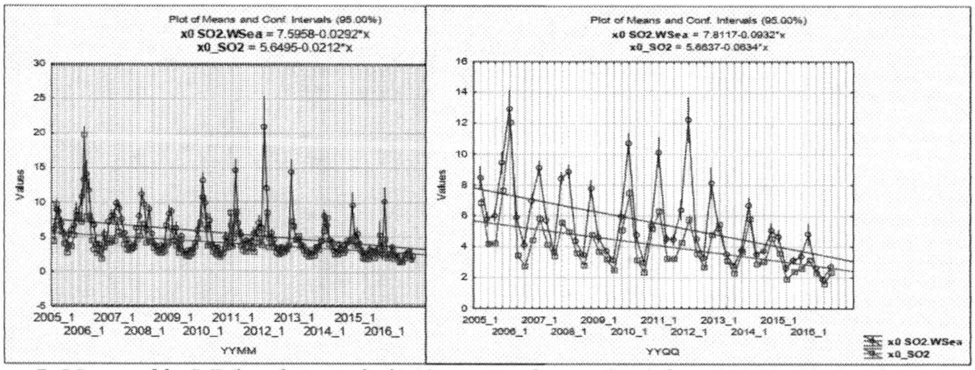

Figure 7. SO_2 monthly (MM) and quarterly (QQ) averages in year (YY) from 2005 to 2016 with OLS trend lines

Source: The authors' own elaboration based on Agency of Regional Air Quality Monitoring in Gdańsk Metropolitan Area.

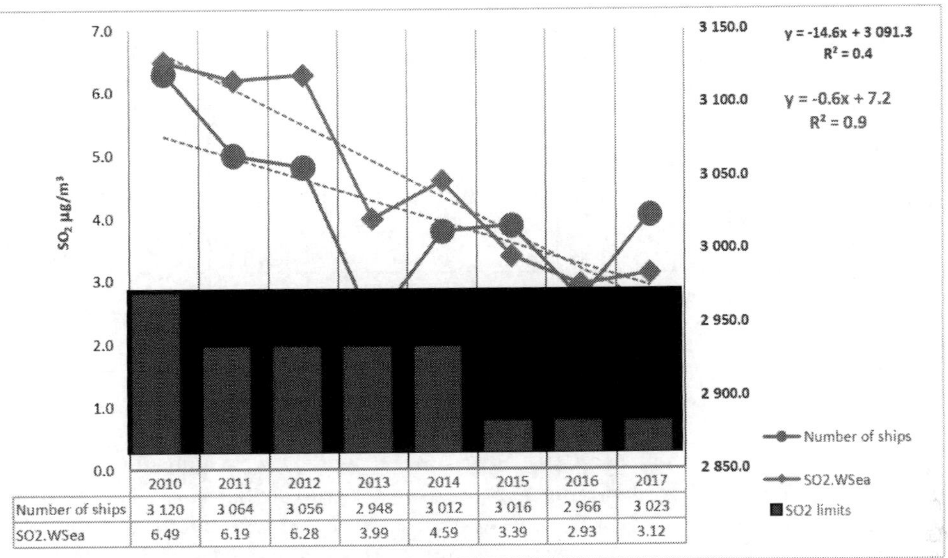

	2010	2011	2012	2013	2014	2015	2016	2017
Number of ships	3 120	3 064	3 056	2 948	3 012	3 016	2 966	3 023
SO2.WSea	6.49	6.19	6.28	3.99	4.59	3.39	2.93	3.12

Figure 8. SO_2.wsea time series. number of ships with OLS trend models identification with R^2 and SO_2 limits

Source: The authors' own elaboration based on Agency of Regional Air Quality Monitoring in Gdańsk Metropolitan Area, Maritime Office in Gdynia.

Figure presents a summary of change patterns in the average SO_2 concentrations brought in from the sea, as measured by four coastal stations, changes in the number of ships leaving the port of Gdańsk and the percentage limits on sulphur in the individual years.

As mentioned earlier on, the new, more restrictive norms for SO_2 emissions (0.1%) were introduced in early 2015. Ship owners had had over 5 years to comply with the requirement, even though an overwhelming majority of them had decided to wait until the last minute before making decisions regarding the adoption of technological strategies and solutions allowing ships to operate in compliance with the new rules. Few of them had undertaken any work on new fuels (e.g. Stena Line – methanol) or fume purification technologies (Finnlines – scrubber). Most of them decided to switch over to MGO (Marine Gas Oil). After a short period of uncertainty between 2014 and 2015, the situation has not stabilized. The research results have now shown the real efficacy of the implemented Sulphur Directive.

Despite a small hike in the number of ships (3016) compared to the corresponding figure for 2014 (3012), average SO_2 concentrations fell dramatically from 4.59 µg/m^3 in 2014 to 3.39 µg/m^3 in 2015. The following years showed the level of average SO_2 concentration accumulated from the seaward direction into Gdańsk Metropolitan Areas stabilizing at 3.00 µg/m^3 even despite the increase in the number of ships from 2966 in 2016 to 3023 in 2017.

Especially remarkable are the effects of imposing new limits on concentration on land. In 2015-2016,even despite a growing number of ship calls, the concentration level is significantly lower compared with 2014, just before the new regulations were implemented (Table 2). A similar trend for SO_2 concentrations could be observed even earlier, i.e. from 2011 (Fig. 8). At that time, however, a dramatic drop in SO_2 concentrations did not come until two years later (2013) even though the number of ships remained constant.

Table 2. Percentage changes in SO_2 concentrations and the number of ships compared to 2014

Year	SO2.Wsea average	Number of ships	SO2.Wsea percentage change to 2014	Number of ships percentage change to 2014
2014	**4.59**	**3 012**		
2015	3.39	3 016	-26.1%	0.1%
2016	2.93	2 966	-48.7%	-1.5%
2017	3.12	3 023	-50.1%	0.4%

Source: The authors' own elaboration based on Maritime Office in Gdynia.

For the relevant period of 2010 – 2016, trend models were constructed (**Błąd! Nie można odnaleźć źródła odwołania.**), fit measures were calculated (R, R2, corrected R2 and RMSE) and overall model relevance was checked (F test with p level and the relevance of each parameter based on t-Student distribution)[3]. Trend models for SO_2 concentrations both for seaward wind and a variable number of ships leaving the port are statistically relevant ($p<=0.05$). Both models indicate a falling trend.

Table 3. Pearson moment correlations; Marked bold correlations are significant at $p < .05000$ (2010-2016)

Variable	Number of ships	SO_2.WSea	SO_2limits	SO_2
Number of ships	1.0	**0.85**	0.63	0.68
SO_2.WSea	**0.85**	1.0	**0.83**	**0.92**
SO_2limits	0.63	**0.83**	1.0	**0.94**
SO_2	0.68	**0.92**	**0.94**	1.0

Include condition: Year>=2010 AND Year<=2016

Source: The authors' own elaboration based on Maritime Office in Gdynia.

Correlation analysis (
Table 3) for the period from 2010 to 2016 seems particularly interesting for the purpose of this research. SO_2 concentration limits are significantly and strongly correlated with SO_2 concentrations on land, both overall concentrations (0.94) and concentrations with the wind direction factored in (0.83). There is a strong and relevant statistical correlation between the number of ships leaving the port (0.85) with SO_2 concentrations accumulated from the sea into Gdańsk Metropolitan Area, as corroborated both by the impact of SO_2 emissions from the sea and the positive long-term effects of the new limits. The impact of SO_2 emissions from the sea on the air in the metropolitan area causes negative health effects, as confirmed by many years of research [4,5].

3. Conclusions

The following conclusions emerged in the first stage of research on the impact of new, more restrictive norms on the content of sulphur in marine fuel:

1. The impact of SO_2 concentrations emitted by fuel combustion in ships on the air in the Tricity area is noticeable and felt by the residents.
2. There has been, since 2010, a steady year-by-year decrease of average SO_2 concentration by 0.6 μg/m^3 for the whole area. For the Tricity area, this is a considerable drop ($p<=0.05$) of SO_2 concentrations emitted by fuel combustion by ships operating at sea.
3. The largest decrease of SO_2 concentrations arising from marine fuel combustion at sea has been noted since 2015. This trend is a positive effect of the new norms for sulphur content in marine fuel.
4. In 2015, despite a hike in the number of ships, from 2948 in 2014 to 3016 in 2016, the average level of SO_2 concentration fell in the entire area from 4.6 μg/m^3 in 2014 to 3.4 μg/m^3 in 2015 and to 2.9 μg/m^3 in 2016.
5. *Table 2* suggests that the new 2015 limits reduced the estimated level of SO_2 accumulated by wind into Gdańsk Metropolitan Area by around 50% compared to the same figure in 2014 (the year when the directive came into force) even despite a 0.4% increase of the number of ships calling at the port.

To sum up, it can be concluded that the entry into force of the Sulphur Directive has benefited the port city of Gdańsk through improved air quality and a reduction of SO_2 content in the air.

4. Discussion

The conducted research and the resulting outcomes indicate a high (50%) effectiveness in reducing the sulphur content in the air in port cities as a consequence of the Sulphur Directive. However, two issues should be stressed which prevent us from relating the results to the entire Baltic Sea Region, not to mention the whole globe.

First of all, the research was carried out based on data for only one city, Gdańsk. Apart from methodological correctness which was assured at every stage of the investigation, conclusions have emerged that dictate the need for analogous research to be conducted as well for other port cities situated in the Baltic Sea region or, indeed, the entire Baltic Sea. However, this requires considerable financial resources and infrastructural adjustments at the ports in question. It should also be remembered that many ports, or port cities, do not have any system in place to allow measurement of air pollution.

Secondly, the technologies used by ships calling at the port of Gdańsk are unknown, making it difficult to determine which of the technologies implemented by ship owners on the Baltic Sea have contributed to the current state of affairs and to what extent. It is said that open-loop scrubbers, for example, purify their fumes by diverting the sulphur produced in the process back into the sea waters. It is also noteworthy that the Baltic waters are especially heavily polluted, including by hydrogen sulphide. Therefore, the 50% reduction in SO_2 content in the air could have been better or worse depending on the specific technology adopted by ship owners themselves. This may in turn be determined indirectly by the port management bodies through a system of preferences and price discounts for those ships which are the most environmentally friendly.

References:

[1] E. Czermański (ed.), *Sulphur Regulation – technology solutions and economic consequences for the Baltic Sea Region shipping market*, Institute of Maritime Transport and Seaborne Trade, Sopot 2014.
[2] eceurope.net/wp-content/uploads/2007/08/seca1, 14/03/2014.
[3] P. O. Czechowski, A. Badyda, G. Majewski, *Data mining system for air quality monitoring networks,* Archives of Environmnetal Protection 39/4/p123-144, 2013 DOI: 10.2478/aep-2013-0041.
[4] A.J. Badyda, P. Dąbrowiecki, P.O. Czechowski, G. Majewski, *Risk of bronchi obstruction among non-smokers—Review of environmental factors affecting bronchoconstriction*, Respiratory physiology & neurobiology, Elsevier 2015/4/1, 209.
[5] P.O. Czechowski, A.J. Badyda, G. Majewski, A. Oniszczuk-Jastrząbek, A.K. Kraszewski, W. Rogula-Kozłowska, T. Owczarek, *Providing high-quality measurement data in analytical system of air pollution monitoring and their key importance for smart cities residents,* Annals of Warsaw University of Life Sciences–SGGW. Land Reclamation, De Gruyter Open 2017/12/1, 49, p. 241-253.
[6] D.G. Kleinbaum, L.L. Kupper, A. Nizam, K.E. Muller, *Applied Regression Analysis and Other Multivariable Methods* (Duxbury Applied) 4th Edition, 2008, ISBN 978-0495384960.
[7] G. Majewski, W. Rogula-Kozłowska, P.O. Czechowski, A. Badyda, A. Brandyk, *The impact of selected parameters on visibility: First results from a long-term campaign in Warsaw, Poland*, Atmosphere, Tom 6, Wydanie 8, Strony, p.1154-1174.

Selected aspects of corporate social responsibility in maritime freight services on the example of selected courier companies

Agnieszka Budziewicz-Guźlecka[1], *Anna* Drab-Kurowska[2]

Abstract. The aim of the article is to present the activities of corporate social responsibility in the CEP industry, with particular emphasis on maritime freight services. The article will present good practices and environmental activities proposed by the leaders of the CEP industry. An example may be the new "green services" of sea freight, which allow for reduction of CO2 emissions. The article puts forward the following research hypothesis - Raising awareness in the CSR area in the CEP industry will allow the implementation of environmental activities. In reference to the assumed goal and the hypothesis adopted in the article, the following research question was posed - what specific CSR implementation tools should be adopted by the CEP industry to actively shape environmental activities in maritime freight services? In an attempt to answer the question, the article first addressed the aspects of corporate social responsibility at CSR 3.0 level. Next, the focus was on presenting the results of the analysis of the maritime freight services offered by the leaders of the CEP industry. The following research methods were used in the article: method of critical analysis of literature, method of observing the proceedings of the entities examined, logical inference method, method of analysis and synthesis.The key element of the article is the presentation of proposals for actions aimed at intensifying activities strengthening maritime transport services through activities in the area of CSR.

1 The essence of CSR

Enterprises, including CEP market enterprises, are increasingly aware that successful operation on the market requires the creation of a cooperation network including not only customers but also employees, suppliers or intermediaries. Corporate Social Responsibility (CSR) is an essential and increasingly important topic for many organizations. So far, there is neither a general definition of this concept nor a universal concept of Corporate Social Responsibility in the literature. Most definitions describe it as a concept of integrating social and environmental issues into their business activities and their stakeholders on a voluntary basis; acting responsibly towards various social, environmental and economic pressures in order to respond to the expectations of the various stakeholders they interact with, such as employees, shareholders, investors, consumers, public authorities and non-

[1] University of Szczecin, Poland, agnieszka.budziewicz@wzieu.pl
[2] University of Szczecin, Poland, anna.drab@wzieu.pl

governmental organizations (NGOs). It is worth stressing, however, that there is a lack of one primary definition of Social Responsibility.

Business responsibility, as an ethical value consisting in incurring the consequences of the company's activities, has been discussed in the world for many years. Some researchers see CSR sources in the business ethics and regard as its beginnings works such as Leo XIII's encyclical Rerum Novarum of 1891, The Fundamentals of Business Ethics, published in the USA in 1926, and a later work - Social Responsibilities of the Businessman, published in 1953. [1]

The CSR idea evolved into a debate about business society in the 1920s. During the crisis and the Second World War, the debate fell silent and reappeared in the 1950s. The debate on CSR focused on the obligations of enterprises to society. However, defining what is the main message in the CSR concept was considered too complex and vague to be possibly indicated for the management teams of organizations. In the 1970s it was proposed that CSR could be considered as a "social reaction of an enterprise" to the expectations and requirements of society towards companies [2]. Freeman introduced the perspective of stakeholders as a way to manage organizations. He pointed out that external stakeholders should be reckoned with, but not only traditionally understood (clients, suppliers), but also non-market stakeholders, such as the media or local authorities [3]. Then, in his article, Frederick outlined a conceptual transition from the philosophical and ethical concept of CSR to the action-oriented concept of managerial corporate social responsibility in relation to social pressure [4]. Therefore, society should have social responsibilities, and then management teams should respond to social needs.

As a result, CSR has moved from ideology to reality, and many consider it absolutely necessary that organizations should play an important role in society and apply social, ethical, legal and responsible standards to their enterprises [5]. On the other hand, Young defines CSR as a strategic and long-term approach, based on the principles of social dialogue, transparent relationships and the search for solutions that are beneficial to everyone; it also means achieving lasting profit while wisely shaping relationships with all stakeholders. He also points out that it is also conducting business in a way that takes into account ethical values, law, respect for employees, society and the natural environment, as well as contributing to sustainable development through cooperation with stakeholders in order to improve the quality of life of all citizens [6]. CSR is a process aimed at taking over responsibility for the company's activities and encouraging a positive impact through its activities on the environment, consumers, employees (creating conditions for development and access to the best knowledge transfer techniques [7], communities, stakeholders and all other members of the public sphere, who may also be considered as stakeholders.

On the example of CEP market entities, it can be indicated that, it can be pointed out that CSR evolved from CSR through CSR 2.0 up to CSR 3.0. CSR 2.0 assumes that relations with the local community and other stakeholders are built on the basis of cooperation and partnership, which increases the chances of a positive solution to many global problems. In accordance with CSR 2.0, business creates value by directing its innovation and creativity to solving burning social and ecological problems on a global scale [8].

In turn, the concept of CSR 3.0 is based on a systemic approach to corporate social responsibility and inclusion of national as well as international law, economy and civil society in its activities.

The CSR 3.0 concept is characterized by the following features [9]:

- social cooperation - changes are the driving force,
- proactive management - a holistic CSR approach in all departments of the organization,
- dialogue with a network of stakeholders - strengthening the position of the company, local communities and the country,

- systemic approach to CSR by the state, law, economy, civil society - a global approach to problems and local action.

The CEP market players are involved in corporate social responsibility (CSR), mainly because they can benefit from such involvement. As a result, various NGOs have started developing CSR reporting models or frameworks, seeking inspiration in ISO 14001 (International Organization for Standardization), World Resources Institute (WRI) and Global Reporting Initiative (GRI). With regard to empirical research on CSR, three types of empirical research characterize the research in this field. The first concerns 'descriptive research', which is a report on the nature and scope of CSR with some comparisons of countries and periods. The second concerns 'explanatory research', which focuses on potential determinants of social and environmental reporting. The third is interested in the "impact of social and environmental information" on various users, mainly on market reactions. [10]

In this context, the awareness and social and ethical involvement of society have increased. Treating the environmental protection and interaction with the environment in the convention of a value chain and managing it in the aspect of creating added value is a promising concept. At the same time, various motives for undertaking environmental initiatives in the value chain are indicated [11]. CSR is defined as caring for the well-being of others and the environment in order to create value for the business. CSR manifests itself in operational strategies and practices that it develops. [12]). This idea draws attention to non-economic aspects of enterprises' activities, which should care not only for their own interest but also take into account the impact they have on the environment and the generally understood social good.

Social Responsibility can adopt the following levels, which are also noticed on the CEP market:

- CSR level imposed in some way by applicable law,
- CSR level forced on the organization by strong social pressure exerted by external stakeholders,
- A voluntary CSR level implemented by organizations that feel responsible due to their sense of place in society.

The fact that there is no understanding of the concept of CSR at the international level complicates both theoretical development of the concept and the implementation and measurement of success at the company level. This fact is particularly important considering the fundamental change in the debate on CSR: today management is not so much about whether CSR activities should take place, but rather how they should be conducted. The conducted analysis of CEP market entities has shown that the leading entities of this market can be good examples of activities carried out in the area of CSR. These entities provide sea freight services, which are the driving force of the blue economy. It should be emphasized that the blue economy is characterized by enormous potential, conducive to economic growth.

2 Trends on the market of maritime transport services in the context of sustainable development

The idea of sustainable development is a concept that aims to guide broadly understood growth and socio-economic development of the European Union countries. It is indicated that the maritime economy has a chance to be one of the engines of economic growth in Europe over the next dozen or so years. This perspective is drawn by the European Union, seeing in the blue economy the potential that will improve the economic situation of member states, including Poland. Blue growth should take place in conditions of sustainable development, which means improvement of economic results with simultaneous care for the environment, employees and local communities (http://ec.europa.eu/growth/industry/corporate-social-responsibility/). It should be pointed out that seafaring is one of the oldest forms of transport. Over the centuries, both ports and ships have evolved to increase their transport capabilities and reduce the associated costs. The largest technological leap in this area was noted in the 20th century. In particular, the digitization of services has improved port efficiency, and new technologies used in the construction of larger container vessels significantly increased the number of transported goods. [13]

The data from the 2013 Maritime Transport Review show that 80% of global trade takes place by sea ((Review of maritime transport 2013). The globalization of economic activity contributes to the increased demand for global transport services. This trend is conducive to the development of maritime transport, especially its commodity branch. This is mainly due to the lower price of intercontinental transport as well as the higher load capacity of ships compared to other modes of transport. It should be emphasized that sea transport in relation to the payload criterion offers the largest transport capacity. At the turn of 1990 - 2013, there was a 58.31% increase in the volume of transport in international maritime trade. This testifies first of all to the technological and organizational development of seaports and the modernization of ships.

In addition, it is necessary to refer to global demand forecasts for cargo transport in individual transport branches in Poland by 2030.

Table 1. Forecast of global demand for cargo transport by 2030.

Branch of transport	**Years**		
	2015	**2020**	**2030**
Transport in million tons			
Railway	211	225	261
Automobile	1 643	1 800	2075
Inland waterway	6.1	8.5	14
Piping	54	56	66
Maritime	52	60	81
Air	0.1	0.1	0.2
Transport in billion tkm			
Railway	46	50	62
Automobile	279	321	405
Inland waterway	1.4	1.6	3.2
Piping	23	23	26
Maritime	**110**	**123**	**159**
Air	0.2	0.3	0.6

Source: [14]

Sea transport is considered one of the most dynamically developing. It is estimated that in the years 2015-2030 transport will increase from about 44% to 56%, depending on how the transport is measured. This perspective is reflected in the behaviour of the CEP market entities. Sea freight services are one of the key services offered by the leading CEP market operators. The increase in interest results, inter alia, from the need to reduce transaction costs (Wilmsmeier, Hoffmann, Sanchez, 2006), as well as from the development of international trade. International business and trade are largely dependent on global shipping, which as part of the logistics chain determines the effectiveness of global trade.

Factors affecting both the development and functioning of maritime transport are [15]:
- economic conditions,
- financial conditions,
- current maritime policy,
- rules for the protection of the marine environment,
- innovativeness of transport processes.

In connection with the intensive development of transport in the field of maritime transport, in 2016 the European Commission published two executive regulations[3] introducing tools for monitoring and reporting fuel consumption as well as carbon dioxide (CO2) emissions by maritime transport. These regulations came into force on November 25, 2016. Both legal acts were issued in order to implement the law previously adopted by the European Union, which is Regulation of the European Parliament and of the Council (EU) 2015/757 of 29 April 2015 on monitoring, reporting and verification of carbon dioxide emissions from maritime transport, and amendments of Directive 2009/16 / EC. Both were also published in order to introduce uniform conditions for the implementation of legal regulations resulting from Regulation (EU) 2015/757 in all Member States. In accordance with EU climate and energy legislation, all sectors of the economy, including international shipping, should contribute to such reduction of emissions. It was estimated that total CO2 emissions related to European maritime transport (taking into account internal EU routes, cruises to the EU and cruises made from the EU territory) in 2010 amounted to about 180 Mt CO2. The EU also forecasts that CO2 emissions from European maritime transport (taking into account the same assumptions regarding internal EU routes, cruises to the EU and cruises from the EU) will reach 210 Mt in 2020 (+8% compared to 2005 level). r.), 223 Mt in 2030 (+15% in relation to the level of 2005) and 271 Mt in 2050 (+39% in relation to the level of 2005).

Undertaking activities aimed at reducing pollutions emissions is reflected in the activities for corporate social responsibility. In the activities of all entities servicing sea transport, wide-ranging actions aimed at reducing pollution emissions can be observed. Some of these entities are already well advanced in these activities, while some on the principle of "good examples" of leading entities strive to introduce ecological solutions to their operations. The article pays special attention to the entities of the CEP market.

---

[3] 1. Commission Executive Regulation (EU) 2016/1927 of 4 November 2016 on templates for monitoring plans, emission reports and compliance documents in accordance with Regulation (EU) 2015/757 of the European Parliament and of the Commission on monitoring and reporting and verification of carbon dioxide emissions from maritime transport.
2. Commission Executive Regulation (EU) 2016/1928 of 4 November 2016 on the determination of cargo carried with respect to the category of ships other than passenger ships, ro-ro vessels and container ships, in accordance with the Regulation of the European Parliament and of the Council (EU) 2015/757 on the monitoring, reporting and verification of carbon dioxide emissions from maritime transport. In December 2010, the parties to the United Nations Framework Convention on Climate Change recognized that global warming could not exceed pre-industrial temperatures by more than 2˚C. This is necessary to limit the negative impact of human activities on the climate. This long-term goal requires a reduction of global greenhouse gas emissions by 2050 by at least 50% below the 1990 level. By 2050, developed countries should reduce their emissions by 80-95%.
In July 2011, the International Maritime Organization (IMO) adopted operational and technical measures, in particular, the construction energy efficiency indicator for new vessels and the ship energy efficiency management plan (SEEMP), which were to provide an improvement in the reduction of greenhouse gas emissions.
The EU is strongly committed to achieving the objective of combating climate change by limiting global warming to 2˚C. Consequently, one of the key objectives of the Europe 2020 strategy for employment and smart, sustainable and inclusive economic growth is to reduce greenhouse gas emissions by 20% below 1990 levels by 2020 or, if specific conditions, by 30%.

3 CSR on the example of sea freight services in the CEP sector

The idea of sustainable development and ecological solutions is part of the concept of corporate social responsibility (CSR) applied more and more often by enterprises operating in various branches of the economy, also in the CEP industry.

As previously emphasized, along with the development of maritime freight services, there is an intense increase in activities for social responsibility in the CEP industry. Leading, foreign industry leaders show great interest in the subject of CSR. This interest is not limited to the traditional look at corporate social responsibility (CSR 1.0). The largest entities of the CEP market not only noticed the possibility of using the advanced CSR level but also started a series of activities consistent with the next stage of responsibility - CSR 2.0. The advanced level of corporate social responsibility in the CEP sector is reflected in strategies, good practices of enterprises and the trendsetting. An example of developing a modern CSR strategy is the DPD Group proposal - DrivingChange strategy. In the strategy announced in February 2016, the company focuses on four key areas of its business profile:

- neutral for the environment;
- friendly city logistics;
- innovative entrepreneurship;
- closer to the community.

Each of the areas proposed by the company has a defined goal and measures used to achieve it, which at the same time become good practices of the organization. An example of this is the action to reduce CO2 emissions and the measures taken for its implementation - gradual modernization of the fleet of their vehicles so that ultimately they would be maximum 5-year-old vehicles. In addition to standard practices characteristic of transport companies, e.g. route optimization, this entity is involved in solving global problems, emphasizing the importance of renewable energy sources through the purchase of carbon credits. In its operations, DPD also focuses on creating a new value for society. Practices created as part of "friendly urban logistics" are to have a positive impact on the environment (reduction of pollution in city centres, reduction of congestion), but above all to bring a social benefit. An example is creating points for sending and receiving parcels. In addition, the entity also focuses on the Predict service, thanks to which it is possible to flexibly manage the time of delivery, which eliminates the phenomenon of empty runs. The DrivingChange strategy also means stakeholder-oriented activities; in line with the CSR 2.0 assumptions, the company builds a bond with its stakeholders. These activities apply to both company employees and partners. In the case of employees, DPD Group undertakes a number of activities aimed at investing in people and designs a number of incentive programs. DPD also supports local communities. However, particular attention should be paid to the ecological model of office management adopted in the company, based on reducing water and energy consumption as well as waste recycling. It is clearly visible that DPD Group is undertaking a series of activities in the area of CSR, however, referring to maritime freight services, DPD in its CSR strategy does not clearly indicate actions targeted in this area. It is necessary to strengthen this direction, all the more that from year to year the share of maritime freight services is increasing in the activity of this operator.

Another entity in the CEP sector is FedEx. As part of a comprehensive international freight forwarding program, FedEx Trade Networks offers sea freight services connecting North America, Latin America, the Asia-Pacific region, Europe and the Middle East[4]. Taking advantage of the possibilities of regional sea terminals, wide opportunities to sail

[4] Full container load transports (FCL). Less than container load transports (LCL) - LCL Economy service, FedEx® International DirectDistribution service and LCL Direct service for urgent shipments.

out each week, and privileged relationships with many shipowners, it maximizes transport flexibility while maintaining the right level of costs; all this in the context of activities for corporate social responsibility, through the EarthSmart program (www.fedex.com). This program expresses FedEx's commitment to environmental protection. It is a set of initiatives offering specific business and environmental benefits. EarthSmart consists of three pillars:

- EarthSmart Innovations - FedEx services and assets such as trucks and facilities that meet stringent standards in the field of innovation and environmental protection. Examples of EarthSmart FedEx global solutions include electrically powered tricycles that operate in Paris, hybrid trucks and all the electric vans that FedEx operates in major urban centres such as Paris and London, as well as an automated electronic document sending tool (ETD).
- EarthSmart Outreach - by running many global and local philanthropic projects, FedEx strengthens its efforts to create cleaner, healthier and more efficient living conditions by supporting environmentally friendly transport, green urban spaces and resilient ecosystems. FedEx supports sustainable development projects through its employees, allocating their time and skills to support important issues.
- EarthSmart @ Work - programs that engage employees in environmental matters and help them make a significant contribution to sustainable environmental development.

In addition, as part of CSR activities, FedEx declares that the target in the form of fuel savings was exceeded by 30% approximately 5 years before the set date. The company has achieved such ratios thanks to a strategy based on three categories of activities: reduction of total transport costs through route optimization, an introduction of more fuel-efficient vehicles, expansion of the fleet with vehicles powered by alternative fuels.

The entity that can be considered as the leading among the entities of the CEP sector in the field of sea freight is DP DHL Group. DHL Group conducts very extensive activities in the field of good practices and pro-environmental activities. It is these company activities that are rewarded and highly rated in both CSR rankings and practices in the field of "green logistics". This is not the only plane that distinguishes the company from the point of view of social responsibility. In addition to the monitored and improved CSR strategy and reporting of its activities, the company focuses on creating a new value aimed at persuading the CEP sector to act in the area of responsible logistics. A report was prepared to accelerate the implementation of the concept of sustainable development in the CEP industry, which again points to the advanced CSR level of the organization. There were indicated 15 possibilities of applying the concept of corporate social responsibility. DHL focuses on trends aimed at introducing activities related to circulation economy, increasing the availability of transport services, as well as achieving responsible trade and production. According to the report, these changes are possible through appropriate global management, relations with stakeholders and becoming aware of the impact that logistics companies have on society, the economy and the environment. DHL also gives examples of good practices. By focusing on the circular economy, the use of vehicles with so-called "Flexible interior" is predicted. This type of solution is aimed at adjusting the load space, which is divided into two parts: with consignments for delivery and with a place for storing secondary raw materials. The size of individual parts is to change during delivery, so as to fully use the cargo space both when delivering shipments and returning to the base. Thus, DHL creates a new value, solutions that are meant to be a response to global problems. Another of them is the possibilities associated with packaging. Here, the key is not only the use of degradable materials but also a change in approach to them. One of the DHL group companies, DHL Global Forwarding was the first global company in the logistics industry that set the goal of quantitative reduction of CO_2 emissions. Thanks to the optimization of the means of transport by 2020, the company intends to reduce the CO2 emissions resulting from operations by 30% compared to the level from 2007 [16]. DHL Global Forwarding is also the first freight forwarder in the industry to launch a green service with reduced CO2 emissions for sea freight. The Green Danmar service - offered by Danmar Lines NVOCC (Non-Vessel Operating Common Carrier) - allows customers to choose a carrier that emits 5 percent less carbon dioxide than the industry average. In addition, the customer may choose to combine this service with a DHL compensation solution and transport his shipment in a carbon-free way. Thanks to this, only over the last few years DHL Global Forwarding has reduced carbon dioxide emissions in sea freight by more than 40 percent.

The interest in creating more and more good practices in these areas is in line with current trends of business development in terms of sustainable development (Turoń, Golba, Czech 2015). Although these investments may seem costly and long-term, companies realize that they contribute not only to improving the functioning of their environment but also to investing in their own organization and promotion (Turoń 2015). In turn, the occurrence of an increasing number of good practices in the organizational area proves their real impact on the organization and gives the opportunity to improve it. Corporate social responsibility also brings benefits to the company related to human resources management. Research shows that the CSR policy implemented in companies in the field of work practices is manifested in the employees' greater involvement and satisfaction with the work. The aspect of increasing the organization's innovativeness should also be pointed out. By designing their own good practices, companies often go a step further, creating a new, innovative value for society, the environment, but often also for the economy. An example of this may be activities of the DHL logistics company, which, apart from developing its own CSR strategies, also publishes reports on sustainable development or future trends. In the latest report "Logistic Trend Radar DHL 2016", the company presents a set of 26 trends that can be used in the logistics industry in the near future. Of the 26 trends, 14 are activities that focus on social and business aspects that fully take into account the principles of sustainable development. The important fact is that these are the development guidelines for the entire logistics industry, not just the publication associated with the brand's activities. DHL thus creates a new value, but also provides directions for other players in the logistics industry [17].

Summarizing activities in the area of CSR, it should be pointed out that there are clear tendencies that describe CSR areas, as shown in Figure 1.

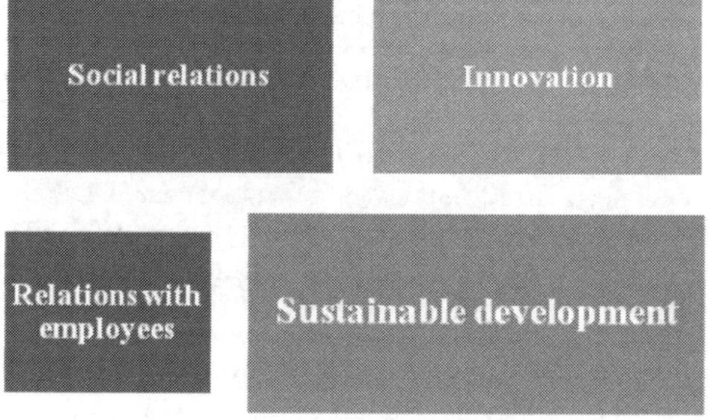

Fig. 1 CSR areas.
Source: Own study

From the point of view of the maritime economy, the most important area of CSR is sustainable development. In the case of the analyzed sector, with particular reference to sea freight, it is clearly visible that CEP market entities carry out a series of activities which, however, are not always oriented towards maritime freight services. A clear focus on shaping CSR activities for these services is necessary. Below (Figure 2), there is a model for strengthening CSR activities in the field of sea freight for the CEP industry.

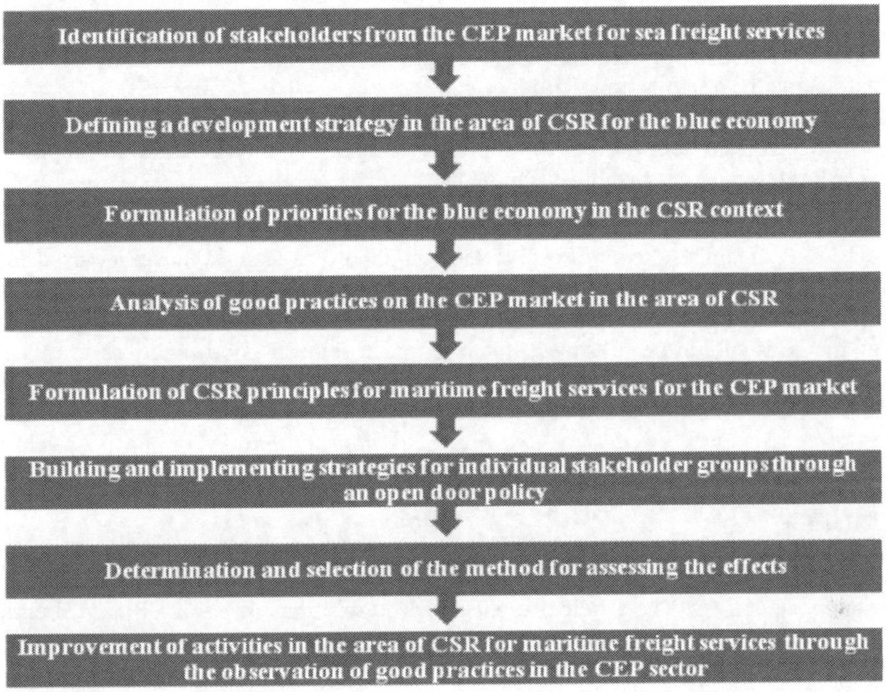

Fig. 2 Model strengthening CSR activities in the area of sea freight.
Source: own study.

The starting point should be an accurate identification of stakeholders. The analysis should concern all entities: clients, institutions, organizations, offices that may influence the company representing the CEP market as well as those who are influenced by its activities. A precise definition of the groups to which actions can be directed alongside with their needs and expectations is required. Correct identification of stakeholders, as well as dialogue with them, contribute to the more effective implementation of the objectives set in the social responsibility strategy.
The next stage should be defining the strategy in the area of CSR 3.0 for a sustainable 'blue economy'. The strategy should include:
- integration activities implemented through the active and effective involvement of all interested parties,
- clear, transparent and responsible actions. This will be conducive to taking steps also in the case of adverse unplanned effects,
- holistic, cross-sectoral and long-term activities. All decisions should be made on the basis of current and future consequences, in the context of values, costs and benefits for stakeholders,
- innovative activities taking into account the natural possibilities of the environment.

The next stage should be formulation and specification of priorities for the blue economy, with particular emphasis on CSR.
The next step should be to analyze the activities that yielded concrete, positive results, and which contain some innovation potential, durable, and that can be repetitive, applicable to similar conditions elsewhere by other entities. The provision of good CSR practice is the promotion of the organization; it is a sense of satisfaction that the effects of the work performed can serve others. Such analysis allows for the use of good practices among other CEP market organizations.
The formulation of CSR principles for sea freight services for the CEP market is another stage that details the organization's conduct.

Then, consistently strategies for all stakeholder groups should be built and implemented. An open door policy with a priority emphasis on sustainable development as well as corporate social responsibility can be used.

The next element should be the determination and selection of the method for assessing the effects. The obtained effects of a given action should exceed the costs incurred for its implementation. At the same time, what is effective must refer to the assumed goals of the action. However, it should be remembered that CSR aspects are very difficult to measure and new methods should be followed, but first and foremost, good practices in this field should be analyzed. The last but in fact, the continuous stage should be the endless improvement of activities in the area of corporate social responsibility for maritime freight services.

4 Summary

CSR is a concept that indicates the directions of development of entities in the conditions of sustainable development. The article presents specific solutions for the CEP market entities, which constitute a significant support for the blue economy by providing maritime freight services. The article analyzes the leading entities of the CEP market, presenting a very wide range of activities in the area of CSR. These entities, through their "good practices", show that CSR is not to be an additional activity, running alongside the mainstream of undertaken activities, the aim of which will only be to build an image on the basis of, for example, financing of charity campaigns or participation in undertakings defending the natural environment. Such a CSR approach would be too limited and would not reflect its main purpose. It should be emphasized that social responsibility is not a matter of an additional item in the costs incurred for the benefit of the environment or the local community. CSR principles should be inextricably interwoven with the company's basic activity, and are to be a way to conduct it. Therefore, a model was presented to strengthen CSR activities in the field of maritime freight for entities of the CEP market, which are only shaping their activity in the area of CSR.

References

1. M. Popowska, *CSR a zarządzanie wiekiem w przedsiębiorstwie,* „Przedsiębiorstwo we współczesnej gospodarce – teoria i praktyka", **3**, 29, Wyd. Politechniki Gdańskiej (2012)
2. R.W Ackerman,., R.A Bauer,. *Corporate Social Responsiveness,* Reston VA: Reston Publishing Company (1976)
3. R.E. Freeman, *Strategic management – A stakeholder approach*, Marshfield, MA: Pitman Publishing (1984)
4. W. C. Frederick, *From CSR1 to CSR2*: The maturing of business and society though', *Business & Society*, **33(2),** 150-164 (1994)

5. A. Lindgreen, V. Swaen, *Corporate Citizenship: Let not Relationship Marketing Escape the Management Toolbox*, *Corporate Reputation Review* **7**(4), 346–363 (2004)
6. S. Young, Etyczny kapitalizm. Jak na powrót połączyć prywatny interes z dobrem publicznym, Metamorfoza (2005)
7. A. Budziewicz-Guźlecka, A. Drab-Kurowska,. Zarządzanie wiedzą w dobie nowej gospodarki. ZN US, EPU **35**, 39-51 (2009)
8. W. Visser, The Age of Responsibility: CSR 2.0 and the New DNA of Business. Journal of Business Systems, Governance and Ethics, **5** (3) (2010)
9. K. Turoń, P. Czech,. *Doskonalenie funkcjonowania przedsiębiorstwa poprzez społeczną odpowiedzialność biznesu na przykładzie firm branży Transport-Spedycja-Logistyka*. Rynek-Społeczeństwo-Kultura, nr **1(23)** (2017)
10. C. Reverte, Determinants of corporate social responsibility disclosure ratings by Spanish listed firms. JBE, **88(2)**, 351-366 (2009)
11. L. Preuss), The Green Multipler, Polgrave Macmillan (2005)
12. A. Glavas, K. Kelley, *The effects of perceived corporate social responsibility on employees. Bus. Ethics Q.* **24,** 165–202 (2014)
13. W.Y Yap., J.S.L Lam., 80 million-twenty-foot-equivalent-unit container port Sustainability issues in port and coastal development, **71**, 13–25, O & CM (2013)
14. J. Burnewicz, *Prognozy zapotrzebowania na usługi transportowe w Polsce do 2020 roku* [in] Uwarunkowania systemu transportowego Polski. Red. B. Liberadzki, L. Mindur, ITE, 132 (2007)
15. T. Nowosielski *Determinanty rozwoju kontenerowej żeglugi morskiej*, , ZN UG. Si M ITiHM, **9**, 51-71 (2012)
16. (http://www.dpdhl.com/en/responsibility/ environmental-protection.html).
17. (http://www.dhl.com/en/about_us/logistics_insights/dhl_trend_research/trendradar.html#.WrwjUNRuaG4)

Overview of MARPOL ANNEX VI regulations for prevention of air pollution from marine diesel engines

Leo Čampara[1], *Nermin* Hasanspahić[2], and *Srđan* Vujičić[3]

[1]PhD student, University of Dubrovnik; Maritime Department, Ćira Carića 4, 20000 Dubrovnik, Croatia,
[2]PhD student, University of Dubrovnik; Maritime Department, Ćira Carića 4, 20000 Dubrovnik, Croatia,
[3]PhD student, University of Dubrovnik; Maritime Department, Ćira Carića 4, 20000 Dubrovnik, Croatia,

Abstract. Increased international awareness of growing global air pollution issues resulted in adoption of measures necessary to reduce ships' emissions into the atmosphere. The MARPOL convention with 1997 Protocol amendments included a new Annex VI which for the first time regulates harmful compounds emitted from ships on international voyages. The Parties of the 1997 Protocol have incorporated the MARPOL convention regulations within their own national legislation. , An International Air Pollution Prevention (IAPP) certificate is issued to ships flying their flag by a recognized organization/classification society. The IAPP certificate serves as proof of a ship's compliance with these regulations, thus assuring a ships' seaworthiness. This paper summarises the most important provisions of Annex VI as applied to air pollutants from marine diesel engines, namely nitrogen and sulphur oxides, particulate matter and carbon dioxide. Recent studies proved that once adopted, Annex VI regulations have positive effect on the reduction of air pollution from ships. The Annex VI regulations present a major challenge and obligation to stakeholders in the shipping industry. Marine diesel engines are the most common choice to propel merchant ships compared with competitive alternatives. Engine manufacturers are particularly affected, as these regulations have necessitated the development of cleaner marine diesel engines and additional innovative technological solutions.

1 Introduction

Over the last three decades, large industrial plants and road vehicles have been regarded as the major sources of air pollutants, while ships were unjustifiably considered too mobile, too far from the coast and insufficiently important to be given significant attention as sources of air pollution. While harmful emissions from land-based sources have gradually been regulated and reduced, the continuous increase in maritime traffic, which presently accounts for more than 90% of total world trade, has led to a greater contribution to global air pollution [1]. This has resulted in an increasing awareness of ships' negative environmental impact which has forced the shipping industry to considerably reduce emissions with the aim of preserving the sea, the environment in general and the health of all living beings [1].

Although most of the harmful exhaust emissions from ships engaged in international voyages end up being widely distributed throughout all seas, a number of research conducted by non-governmental environmental organizations have shown that ships are still the largest individual source of acidity and eutrophication of the European land environment. Also, due to the density and frequency of maritime traffic, ships significantly contribute to the elevated level of particles hazardous to health and to the destruction of the ozone layer. To illustrate this, annual emissions from internationally trading ships into the seas surrounding Europe[1] were estimated to contain 3.3 million tons of nitrogen oxides (NOx), 2.3 million tonnes of sulphur oxides (SO_x), and 250,000 tons of harmful particulate matter (PM) [2]. Ships' SOx and NOx emissions is estimated to increase by around 40-50% between the years 2000 and 2020. If true, Europe's emissions from ships involved in international shipping will reach or even exceed the total emissions from land based sources in all 27 EU member states by the year 2020 [2]. However, according to the latest estimation by the European Monitoring and Evaluation Programme (EMEP), the annual NOx reduction in the Baltic Sea area will be 22,000 tons as result of the newly IMO designated Baltic and North Sea NOx Emission Control Areas (ECA) [3]. A number of recent studies also estimated that international shipping accounts for 14-31%, 4-10% and 2- 3% of the total worldwide global emissions of NOx, SOx, and CO_2, respectively. These estimates take into account emissions increasing by a factor of four as the number of ships might triple by 2050 [4]. The IMO third greenhouse gas emission study foresees that despite a predicted increase in ships' average efficiency of 40%, and in the absence of any further regulations, CO_2 emissions from international shipping could increase between 50-250% by 2050[5]. These figures depend on future economic growth and energy developments.

[1] The Baltic Sea, the North Sea, the North-East Atlantic, the Mediterranean and the Black Sea.

The exhausts of propulsion and auxiliary marine diesel engines are primarily responsible for a ships' air pollution. Approximately 90% of large merchant ships, *i.e.* tankers, bulk carriers, container ships, etc., are propelled by large two stroke slow speed diesel engines, while all of them have at least one diesel generator installed on board [6]. Although very reliable, efficient, relatively easy and economical to maintain, they must currently comply with the increasingly stringent regulations on emissions of exhaust gasses pollutants [7].

This paper provides a brief overview of the most important and currently enforced International Maritime organisation[2] (IMO) regulations as applied to the prevention and limitation of harmful substances contained in the exhaust gases of marine diesel engines, as stipulated in the MARPOL Convention [3] Annex VI [8, 9, 10]. These regulations have a direct impact on the operational management of ships, particularly those engaged in international voyages. This paper mainly focuses on regulations concerning emissions of NOx, SOx and PM which are considered to be the major air pollutants from ships, including carbon dioxide (CO_2) which is the main contributor of excessive greenhouse gas formation. Although not a focus of this paper, it is important to note that in addition to aforementioned regulations on marine diesel engines emissions, Annex VI provisions also regulate ozone depleting substances (ODS), volatile organic compounds (VOC), marine fuel oils (MFO) and shipboard incinerators operation, all equally important in controlling air pollution from ships [8, 10].

MARPOL Annex VI provisions, described further in the paper, provide guidance for adopting regional and national legislation on air pollution from ships but do not limit Administrations in comprising additional measures in order to enhance them.

2 A brief history of MARPOL Annex VI main provisions

Air pollution has currently become one of the most significant worldwide environmental issues. Given the fact that maritime transport has a distinctly international character, knowledge acquired via series of research into air pollution from ships confirmed the urgent necessity to develop international regulations to limit their harmful emissions.

In addition to several Conventions[4] of general character which are related to air pollution, framework provisions of the United Nations Convention on the Law of the Sea (UNCLOS), generated in the Montego Bay conference in 1982, specifically deal with pollution of the sea and air in Article 212[7]. The Parties to the UNCLOS have been authorised to adopt national regulations for the prevention, reduction and control of marine pollution from the air, or indirectly through the air, while taking into account internationally recognized norms, standards and recommended practices [7]. This has pledged them to apply national legislation and international rules on the prevention of air pollution in the territory under their sovereignty and to vessels flying their flag [7]. Implementations started in countries and regions, primarily in the European Union and the United States, which are close to busy sea routes and who were therefore more exposed to ships' emissions.

As a result of cumulative scientific evidence, the 1980's saw an increased awareness of the impact of ship emissions on the environment and public health. This prompted the IMO to develop and establish new international policy on the prevention of air pollution from ships. At the International Conference of Parties to the MARPOL Convention, held in London from 15 to 26 September 1997, MARPOL Convention was amended with the Protocol of 1997[5] and a new Annex VI, Regulations for the Prevention of Air Pollution from Ships was introduced [10, 11, 12]. The adoption of Annex VI followed years of negotiation between IMO members and stakeholders of the shipping industry. These negotiations which were driven by political and economic interests impeded and delayed the adoption of the new regulations which came into force on 19 May 2005, 12 months after Samoa fulfilled the ratification requirement[6] of 15 countries which represented at least 50 per cent of the world's gross tonnage (GT) [13, 14]. The provisions of Annex VI, which are of an international character and which all contracting States adhere to, initially set limits on SOx, NOx and PM emissions from ship

---

[2] The International Maritime Organization (IMO) is a specialized organization of the United Nations based in London, which was initially founded with the purpose of promoting maritime safety. It currently has 173 Member States and three Associate Members representing 99.31% of the world's merchant shipping tonnage. It was adopted by the United Nations conference in Geneva in 1948 and came into force in 1958 when it was ratified by 21 states. The first name of the organisation was Intergovernmental Maritime Consultative Organization (IMCO) which was changed in 1982 to IMO. In general, the purpose of the organization is to establish and enhance exchange of information and cooperation between governments in maritime issues, to provide assistance and resources in establishing maritime regulations and standards related to the safety of navigation, ships and people, the protection of the marine environment as well as the elimination of discrimination and unnecessary restrictions in achieving agreement on maritime issues.

[3] MARPOL - the International Convention for the Prevention of Pollution from Ships is a fundamental international agreement covering all types of marine environmental pollution by ships due to operational or accidental causes. It was adopted on 2 November 1973 in London and amended by the Protocol of 1978. The combined instrument finally entered into force on 2 October 1983 and is referred to as MARPOL 73/78 Convention. It was significantly amended again by the Protocol of 1997 when a new Annex VI was added and since then it is referred to as the MARPOL Convention. It presently includes six Annex regulations as follows: Annex I - Prevention of Pollution by Oil, Annex II - Control of Pollution by Noxious Liquid Substances in Bulk, Annex III - Prevention of Pollution by Harmful Substances Carried by Sea in Packaged Form, Annex IV - Prevention of Pollution by Sewage from Ships, Annex V - Prevention of Pollution by Garbage from Ships, Annex VI - Prevention of Air Pollution from Ships. Presently, 156 countries representing 99.42% of the world's merchant shipping tonnage are parties to the MARPOL Convention.

[4] The United Nations Conference on the Human Environment (Stockholm, 1972); The Convention on Long-Range Transboundary Air Pollution (Geneva, 1979) and its seven Protocols; The Vienna Convention for the Protection of the Ozone Layer (Vienna, 1985) and its Montreal Protocol (Montreal 1987); The United Nations Framework Convention on Climate Change (Rio de Janeiro, 1992) and its Kyoto Protocol (Kyoto, 1997).

[5] The requirements of the IMO Protocol of 1997 are in accordance with the Montreal Protocol of 1987, as amended in London in 1990 which is an international environmental agreement, held under the auspices of the United Nations under which the participating nations agreed to reduce production and consumption of ozone depleting substances for the ozone layer protection purposes.

[6] A State to become a Party to MARPOL must ratify MARPOL Annexes I and II. The rest of Annexes VI is voluntary as far as membership to MARPOL Convention is concerned, only MARPOL Annexes I and II must be ratified for the State to become a Party to MARPOL Convention.

exhausts, established Emission Control Areas (ECA) with more stringent limits for SOx and NOx, prohibited deliberate emissions of ODS and regulated VOC emissions as well as shipboard incineration [14].

Since Annex VI represented the starting point of the regulation of air pollution from ships, it was to be expected that due to new scientific evidence, implementation experience and technological improvements, that the provisions laid down in the Protocol of 1997 would need to be revised and amended. Therefore, in the following years, a number of resolutions and amendments including new provisions were introduced by the Marine Environment Protection Committee[7] (MEPC). Revision of Annex VI and Technical Code adopted by MEPC 58 in 2008, came into force on 1 July 2010, was significant progressions in establishing strengthen emission limits [1, 14].

Another crucial regulatory step which was taken by the IMO was the inclusion of binding mandatory energy efficiency measures aimed at reducing greenhouse gas emissions from international shipping, primarily CO_2, the most significant contributor. Resolution A.963(23)[8], adopted by the IMO assembly on the 23 December 2003, initiated the development of these measures by MEPC which finally resulted in the adoption of the related amendments at MEPC 62 in July 2011 when new chapter 4 regulations on the energy efficiency of ships was introduced. This came into force on 1 January 2013[12, 15]. Up to date, Annex VI is ratified by 89 countries, which represents 96.18 % of the world's merchant shipping tonnage [8].

2.1 Current MARPOL Annex VI general provisions

The current version of Annex VI comprises five chapters within which there are 26 regulations arranged. Chapter 1 refers to the definitions and the final application of the rules. Chapter 2 provides guidance for the necessary surveys, certification and means of control. Chapter 3 is devoted to requirements on limits and control mechanisms for all emissions from ships except CO2, which is covered by chapter 4 and its operational and technical energy efficiency measures. Chapter 5 outlines verification of compliance with the provisions of Annex VI during audits [16]. There are currently ten appendices[9] to Annex VI and 30 various continually developing guidelines provided for the smooth and uniform implementation of the regulations by flag State Administrations and other relevant shipping industry stakeholders [12].

Annex VI applies to all ships, fixed and floating drilling rigs and other platforms, except where specified within specific regulations[10]. Starting on 19 May 2005, internationally trading ships of 400 GT or more, which operate in the waters under the jurisdiction of the States that have ratified Annex VI, or fly their flag, must have an International Air Pollution Prevention certificate[11] (IAPPC). For ships of less than 400 GT, the flag Administration may establish appropriate measures to comply with the applicable provisions of Annex VI [16, 17].

Every ship, as defined by Annex VI, is subjected to an initial survey either before the ship is put into service or before the IAPP Certificate is issued for the first time. Periodic surveys are conducted at intervals specified by the flag Administration, but cannot exceed five years. A minimum of one intermediate survey has to be carried out during the period that the certificate is valid [16, 18].

The current Annex VI regulations control and limit the following main areas [12]:

- Emissions of ODS from refrigeration plants and fire fighting equipment (regulation 12)
- NOx emissions from marine diesel engines (regulation 13)
- SOx emissions from ships (regulation 14)
- VOC emissions from oil tankers cargo tanks (regulation 15)
- Shipboard incinerators emissions (regulation 16)
- Marine fuels quality (regulation 18)
- Energy efficiency of ships (chapter 4)

2.2 Regulation 13 - Nitrogen oxides

---

[7] The Marine Environment Protection Committee (MEPC) is the IMO's senior technical body on marine pollution related matters. It is responsible for coordinating all IMO's activities related to prevention and control of the environmental pollutions from ships. In the work and decision-making it is assisted by a number of Sub-Committees sometimes shared with the Maritime Safety Committee. Also, the MEPC sets up working groups that deal with various items of its agenda, *e.g.* ballast water, air pollution, GHG emissions, etc. The Committees and its working groups are supported by the IMO Secretariat that deals with all related administrative aspects.

[8] IMO Policies and Practices Related to the Reduction of Greenhouse Gas Emissions from Ships.

[9] Appendices to Annex VI: Appendix I - Form of IAPP Certificate, Appendix II - Test cycles and weighting factor, Appendix III - Criteria and procedures for designation of Emission Control Areas, Appendix IV - Type approval and operational limits for shipboard incinerators, Appendix V - Information to be included in the Bunker Delivery Note (BDN), Appendix VI - Fuel verification procedure for MARPOL Annex VI, Appendix VII - Emission Control Areas (ECAs), Appendix VIII - Form of International Energy Efficiency (IEE) Certificate Appendix IX - Information to be submitted to the IMO Ship Fuel Oil Consumption database, Appendix X - Form of Statement of Compliance - Fuel Oil Consumption reporting.

[10] Regulations 3, 5, 6, 13, 15, 16, 18, 19, 20, 21, 22 and 22A of Annex VI. Regulation 3 generally stipulates that Annex VI regulations shall not apply to any emission necessary for the purpose of securing the safety of a ship and saving a life at sea, or any emission resulting from damage to a ship or its equipment provided that all reasonable precautions have been taken after a damage occurred.

[11] The International Air Pollution Prevention certificate (IAPPC) is a document certifying ship's compliance with technical and operational requirements of Annex VI after an initial or renewal survey. It is valid for five years and is issued by the flag State Administration or by recognised organization, *i.e.* classification society. The certificate shall be issued in the form and content corresponding to the form of appendix 1 to Annex VI.

Regulation 13 limits NOx[12] emissions from all MDEs with a power output of more than 130 kW which are installed on ships constructed, *i.e.* keel laid, on or after the 1 January 2000[7, 19]. It also applies to engines of the same power which are subject to "major conversion[13]" on or after 1 January 2000 [12, 20]. Boilers and gas turbines are not covered by this provision. However, this regulation does not apply to emergency MDEs, lifeboat engines, any device or equipment intended to be used solely for emergencies or to engines installed on ships exclusively engaged in voyages in waters under the sovereignty or the authority of the country whose flag the ship is flying [16, 18, 19].

The Protocol of 1997 introduced initial NOx emission control levels in 2005 which were applicable to engines installed on the aforementioned ships built before 1 January 2000. The revision of Annex VI in 2008 significantly tightened the NOx emissions allowed by introducing two additional limits of control which apply to MDEs installed on newer ships. These levels of control, known as Tiers, are based on a ship's construction date while the NOx emissions limits, expressed as total weight in g/kWh, depend on an engine's rated speed. Tier III applies only when ships operates in NOx ECAs. When navigating outside of the ECAs, they must comply with either Tier I or Tier II requirements, depending on the ships' construction date. Therefore, the operation of each MDE to which regulation 13 applies is prohibited, except when the engine NOx emissions are within the Tier limits as presented in Table 1 and Figure 1 which illustrates NOx emission limits at different engine speeds [17, 18].

Table 1. MARPOL Annex VI NOx emission limits value [17]

Tier	Ship's construction date (on or after)	Total weighted NOx emission limit value (g/kWh) *n = engine's rated speed / crankshaft revolutions per minute (rpm)*		
		n < 130	n = 130 - 1999	n ≥ 2000
I	1 January 2000	**17.0**	$\mathbf{45 \cdot n^{-0.2}}$	**9.8**
II	1 January 2011	**14.4**	$\mathbf{44 \cdot n^{-0.23}}$	**7.7**
III	1 January 2016* 1 January 2021**	**3.4**	$\mathbf{9 \cdot n^{-0.2}}$	**2.0**

*ships operating in the North American or the US Caribbean ECA ** ships operating in the Baltic Sea or the North Sea ECA

Notwithstanding these provisions, regulation 13 also stipulates that the operation of a MDE is permitted when approved by a flag Administration, an exhaust gas cleaning system or when any other equivalent method is implemented to reduce NOx emissions to within the specified limits [21].

Revision of Annex VI in 2008 introduced a retrospective compliance requirement for all ships built between 1 January 1990 and 30 December 1999 that are equipped with MDEs with a power output greater than 5,000 kW and a per cylinder displacement equal or greater than 90 litres, subject to the commercial availability of an Approved Method[14] to upgrade the engine [12, 17].

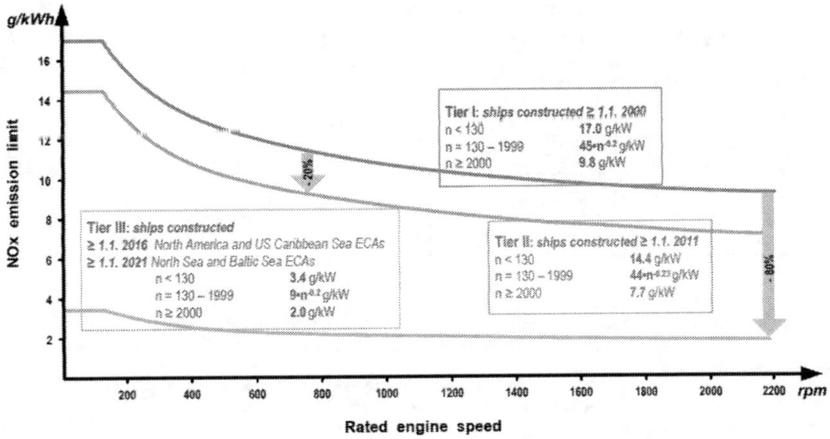

Fig. 1. IMO Annex VI regulation 13 NOx Tier limits

[12] Nitrogen oxides (NOx) as compounds contained in the exhaust gases of marine diesel engines are the result of fuel oil combustion. Their emission into the atmosphere affects the formation of smog and contributes to global warming and acid rains. Together with VOCs and other reactive gases in the atmosphere and with the aid of solar radiation, they live airborne particles participate in the creation of ground-level ozone.

[13] A "major conversion" is the phrase which, for the purpose of regulation 13, implies any engine modification that could cause an increment of emissions above the set limits. It is considered as such if the engine is replaced by a new engine manufactured on 1 January 2000 or later, or if any substantial operational or components modification to the engine as defined in NOx Technical Code 2008, e.g. changing camshaft, fuel injection system, air system, combustion chamber configuration, the engine timing calibration, has been made, or if maximum continuous rating of the engine is increased by more than 10%.

[14] Approved Method is method certified by ship's flag Administration for compliance with the Tier I level and must be applied no later than the first renewal survey which occurs more than 12 months after deposition of the advice to IMO. However, if the ship owner can demonstrate that the Approved Method is not commercially available at that time than it is to be installed no later than the next annual survey after which it has become available.

While regulation 13 of Annex VI sets NOx emission limits, the NOx Technical Code 2008[15] (NTC 2008) provides mandatory procedures for the testing, measurement and certification of applicable MDEs that enable their manufacturers, shipowners and flag Administrations to determine engines' compliance with NOx emission standards set forth in this regulation. Current NTC 2008, mandatory provision udder Annex VI, comprises seven chapters and eight appendices [12].

Each, as per the regulation 13 defined MDE must obtain an Engine International Air Pollution Prevention certificate[16] (EIAPPC) confirming its compliance with the specified NOx emission limits and that it has been tested and inspected in accordance with the requirements of the NTC 2008 [12, 18]. In addition, a certification process includes the approval of a Technical File[17] which is specific to each individual MDE installed on board and which is a record of the engine during its lifetime. Under the NTC 2008 there are three different on board engine verification procedures: engine parameter check method, simplified measurement method and direct monitoring and measurement method. The Engine Technical File contains the one, initially chosen by the engine manufacturer which can in time be changed at the shipowner's discretion. However, as of October 2010, the performance of almost all MDEs is verified using the Parameter check method. All replacements and changes to NOx critical components, settings and operating values are recorded in the Record Book of Engine Parameters which presents the crucial documentation of each MDE [12, 17, 18, 20].

The NTC 2008 testing procedures are different in the case of a new MDE that is to be installed on ship or one that has undergone major conversion. Depending on the NCT 2008 defined conditions, the test for new engines may be carried out on the manufacturer's test bed or upon on board installation. In the second case, the engine manufacturer, shipowner or shipbuilder must submit an application to the flag Administration to perform the test [12]. In addition to the awarding of the individual MDE certification, which is dependent on the approval of the flag Administration, the concepts of Engine Family[18] or Engine Group[19] for a series of engines can be implemented in accordance with one or more of the four duty test cycles as specified in appendix II of the Annex VI [12, 17].

Following the IAPP certification requirements, the MDEs are also subject to: an initial survey before IAPPC is issued for the first time or before the ship is put in service, periodical surveys at intervals specified by the flag Administration which cannot exceed five years, an intermediate survey during the period over which the certificate is valid and an annual survey or an unscheduled survey which is enforced by the flag Administration as an alternative to an annual survey [6, 12 20].

2.3 Regulation 14 - Sulphur oxides and Particulate matter

The SOx[20] and PM[21] emissions from ships are controlled by regulation 14 of Annex VI and apply to all MFOs used on board ships. While regulation 13 only covers NOx emissions from MDEs, SOx and PM regulation applies to all MDEs and all other on board equipment, such as boilers, gas turbines, etc., that use MFOs to operate. The sulphur content limits allowed in MFOs, expressed as a percentage of the mass fraction (% m/m), have been progressively reduced over recent years. The current regulation 14 defines the maximum sulphur content allowed in MFOs used on board ships globally and further tightens these limits in designated ECAs, as shown in Table 2 which also includes the dates when the sulphur limit provisions came or are coming into force [7, 12, 22, 23].

Table 2. Current IMO's MFO sulphur limits with enforcement dates [12]

Global sulphur limits	ECAs sulphur limits
4.50% m/m *prior to* 1 January 2012	**1.50**% m/m *prior to* 1 July 2010
3.50% m/m *on and after* 1 January 2012	**1.00**% m/m *on and after* 1 July 2010

[15] NOx Technical code stands for Technical Code on Control of Emissions of Nitrogen Oxides from Marine Diesel Engines. It was adopted by Resolution 2 on MARPOL Conference in 1997 by as part of Annex VI. It was significantly revised in 2008 at MEPC 58 conference and since then is referred as NOx Technical code 2008.

[16] Engine International Air Pollution Prevention certificate is issued by flag Administration or by recognised organisation/classification society on its behalf after demonstrating compliance with NOx emission limits and it is a prerequisite for the issuance of IAPPC.

[17] The Technical File is a record containing all details of parameters, including components and settings of an MDE which may influence its NOx emission, all in accordance with NTC 2008 requirements.

[18] The Engine Family concept applies to any series produced MDEs which, through their design are proven to have similar NOx emission characteristics and are used as produced. Also, while during installation on board a ship, they do not require any adjustments or modifications which could adversely affect the NOx emissions.

[19] The Engine Group concept may be applied to a smaller series of MDEs produced for similar engine application and which require minor adjustments and modifications during installation or in service. These engines are normally large power engines for main propulsion.

[20] Sulphur oxides (SOx) are a collective name for sulphur dioxides (SO_2) and sulphur trioxides (SO_3). These compounds are known as "acidic" gases because its transformation results in the formation of acidic components that separate from the acid rains causing land acidification. They potentially have a detrimental effect on human's health, vegetation and building constructions. They are usually regarded to have a negligible effect on sea pH far from coastlines because of the seawater alkalinity and their relatively short transportation distances (10-100 km).The SOx emission directly depends on the quality of the MFO, i.e. the sulphur content therein.

[21] Particulate Matter (PM) is a complex mixture of very small particles of soot and ash formed from fuel combustion products and liquid droplets. Additionally, when SOx reaches the atmosphere and forms small aerosol compounds. Most particles form in the atmosphere as a result of complex reactions of chemicals such as sulphur dioxide and nitrogen oxides. The higher the fuel sulphur content, the greater is the particle formation. As they are extremely small, they can be inhaled and cause serious health problems. Particles less than 10 micrometres in diameter pose the greatest problems, because they can get deep into your lungs, and some may even get into your bloodstream.

0.50% m/m *on and after* 1 January 2020	**0.10%** m/m *on and after* 1 January 2015

The IMO commissioned an assessment[22] to determine whether or not the global oil refinery sector would be able to produce MFOs with 0.50% m/m sulphur content in sufficient quantities, before deciding whether to introduce this lower limit in 2020 or 2025. They accounted for trends in global fuel oil markets, the supply and demand for such MFO, demands from other sectors and whether or not the production of these fuels is economically viable in their analyses [15, 24]. On the basis of the audit results, the implementation day for the global 0.50% m/m sulphur limit will be 1 January 2020, as per the MEPC.280(70) resolution which was adopted on 28 October 2016 [25]. Figure 2 illustrates the IMO's progressive reduction of the sulphur content in MFOs over time. The enforcement dates and estimated percentage reduction in overall SOx emissions from ships are also included.

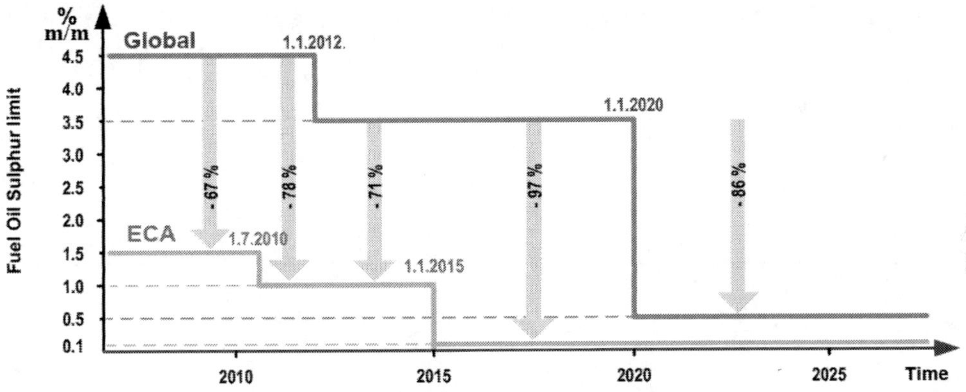

Fig. 2. MARPOL Annex VI regulation 14 - sulphur limits for MFOs [26]

Regulation 14 states that for ships to comply, they must have written on board procedures which detail how to perform the MFO change to enable them to use MFO of differing sulphur content for navigation and operation within and outside the ECAs. This procedure must allow sufficient time for the fuel delivery system to completely flush out all the MFO with sulphur content higher than that permitted for operation within the ECA before the ship enters into it. The date, time and position of the ship after the completion of any such fuel oil change over procedure must be recorded in the logbook prescribed by the flag Administration along with the volume of low sulphur fuel in each tank, either prior to entering the ECA, or to the commencement of such an operation on leaving the ECA area [17, 20].

In addition, one of the controlling instruments of bunkered MFO sulphur content and quality is regulation 18 of Annex VI that contains provisions primarily related to MFOs suppliers and which can be considered to support regulations 13 and 14 regarding aspects beyond the control of the shipowner. This regulation stipulates that the details of each MFO delivered for combustion on board ships must be specified in the Bunker Delivery Note[23] accessible for inspection at all times [12].

Another potential means to comply with emission control requirements, subject to approval by the flag Administration, may be accepted as per Annex VI regulation 4. Any relevant IMO guidelines developed for approving such methods should be taken into account. An exhaust gas cleaning system known as "scrubber" is such a method for which the MEPC.259(68) guidelines have been adopted. This system is called a secondary control method, since the SOx pollutants have already been formed but are subsequently, partially removed by exhaust gas washing water prior to discharge into the atmosphere. This allows high sulphur MFOs to be used if a system performs within acceptable limits as approved by the flag Administration [12, 17].

2.4 Emission Control Areas

An ECA is a specific sea area, including port areas, designated by the IMO in accordance with criteria and procedures set forth in appendix III to Annex VI, in which more stringent emission regulations have been established. For the purpose

[22] An assessment of the availability of MFOs was conducted by CE Delft, independent research and consultancy organisation which specialises in developing innovative solutions to environmental problems. The study considered the estimated demand for MFOs in 2020 based on the fuel consumption of ships in 2012, the projected increase in energy demand, the use of alternative compliance options such as Exhaust Gas Cleaning Systems and LNG. The main result of the study was that for all scenarios the refinery sector has the capability to supply sufficient quantities of MFOs with a sulphur content limited to 0.50% m/m or less and 0.10% m/m or less to meet the predicted demand for these products, while also meeting demand for non-MFOs.

[23] Bunker Delivery Note is a document certifying that the MFO delivered to the ship is in accordance with the requirements of Annex VI Resolution MEPC.176 (58) which has to be kept on board for a minimum of three years. It must include: name and IMO number of receiving ship, MFO loading port, date of delivery, name, address and telephone number of MFO oil supplier, delivered MFO product name, quantity in metric tons, density at 15 °C, kg/m^3, sulphur content in % m/m and a signed and certified declaration by MFO supplier's representative that the supplied product is in accordance with the applicable requirements of Annex VI regulations 14 and 18.

of regulation 13 Tier III emission standards and regulation 14 requirements on SOx emissions limits, there are currently four designated ECAs, as presented in Table 3 which includes adoption, enforcement and effective from dates. The Baltic Sea ECA is defined in MARPOL Annex I, while the North Sea ECA is defined in Annex V. The North American and US Caribbean Sea ECA are detailed in appendix VII to Annex VI [16, 20, 27]. As is apparent from Table 3, the North American and US Caribbean Sea ECAs cover SOx and PM while European ECAs relate only to SOx emissions.

While SOx restrictions are already applicable in all ECAs for all ships, a NOx ECA Tier III standard applies only on new ships built on or after Annex VI defined dates as indicated in Table 3. The latest resolution related to the ECAs was the MEPC.286(71) which was adopted on 7 July 2017 and which designated the Baltic Sea and the North Sea ECAs for NOx Tier III control [16, 20].

Table 3. MARPOL Annex VI ECAs adoption, enforcement and effective from dates of controlled pollutants [27]

ECAs	Controlled compounds	Adopted by IMO	Entry into force	Effective from
Baltic Sea area	SOx	26 Sept 1997	19 May 2005	19 May 2006
	NOx	7 Jul 2017	1 Jan 2019	1 Jan 2021*
North Sea area	SOx	22 Jul 2005	22 Nov 2006	22 Nov 2007
	NOx	7 Jul 2017	1 Jan 2019	1 Jan 2021*
North American Sea area	SOx; PM	26 Mar 2010	1 Aug 2011	1 Aug 2012
	NOx	26 Mar 2010	1 Aug 2011	1 Jan 2016*
US Caribbean Sea area	SOx; PM	26 Jul 2011	1 Jan 2013	1 Jan 2014
	NOx	26 Jul 2011	1 Jan 2013	1 Jan 2016*

* In effect for NOx Tier III compliant ships built on or after this date

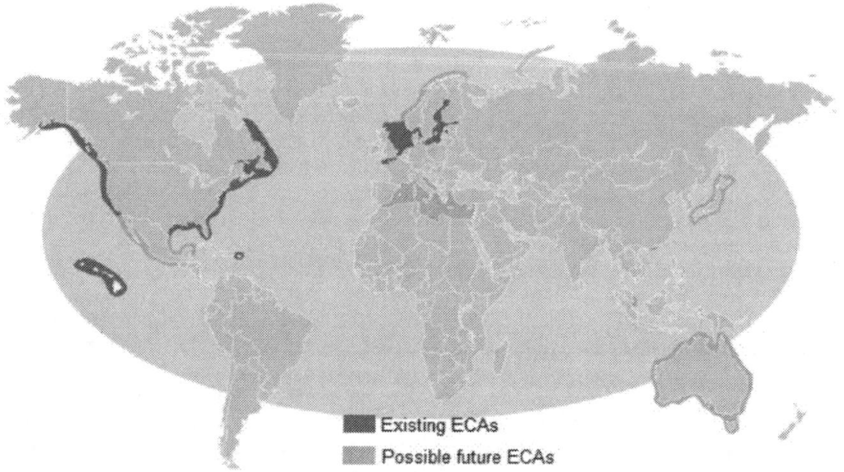

Fig. 3. Present and possible future ECAs [28]

2.5 Chapter 4 - Regulations on energy efficiency for ships

The Kyoto Protocol recognised and acknowledged that greenhouse gas[24] (GHG) emissions from ships cannot be attributed to any particular country or economy due to the complex international nature of the shipping industry. In response , the MEPC 62 session held in July 2011, amended Annex VI with Resolution MEPC.203(62) and a new chapter 4 was introduced which includes regulations to control the energy efficiency of ships. These regulations imply a number of mandatory, cost effective, operational and technical measures for various ship types which are responsible for approximately 85% of GHG emissions, primarily CO_2 which is the most significant GHG effect contributor. They also represent the first mandatory, legally binding regime to reduce global CO_2 emissions from an entire industry sector [5, 29, 30].

[24] Greenhouse gases (GHG) form an air thermal wrap around the Earth that prevents the loss of heat energy into space and also provides favourable climatic conditions for life. Earth reflects 70% of solar radiation due to its surface, and GHG in the atmosphere (CO_2, H_2O, CH_4, N_2O) absorbs part of that radiation, resulting in atmospheric warming, called the greenhouse effect which maintains a favourable climate for all living beings. Human activity, in particular the emission of CO_2, a combustion product of fossil fuels, is the biggest contributor to the increase in the unnatural concentration of greenhouse gases in the atmosphere. Some chlorofluorocarbons from refrigeration devices also fall into this category.

Chapter 4 came into force on 1 January 2013. It has been amended over the years, to contain regulatory provisions, including guidelines which were further developed to assist in the implementation of the regulations. It currently comprises six regulations and 13 related guidelines which apply to all internationally trading ships of 400 GT and more [12, 21].

The main technical measure of compliance with the regulations is the Energy Efficiency Design Index[25] (EEDI) which is a performance based mechanism for a new ship or one which has undergone a major conversion and represents its minimum required energy efficiency calculated using an IMO developed formula [29]. It is left to the ship designers and builders' discretion to choose adequate and cost effective technologies the ship to comply with the Attained EEDI[26] which has to be equal to or less than the Required EEDI[27] [12]. This mechanism includes a schedule which sets out stringent CO_2 emission levels for every five years period, as outlined in Figure 4. This progressive measure continuously stimulates technical development and component innovation and provides an incentive s to improve ship's energy efficiency [29, 31, 32]. However, IMO is in progress of reviewing the status of technological developments and, if proven necessary, will amend the time periods and the EEDI reference line parameters for relevant ship types as well as the reduction rates set out in this regulation [33].

The Ship Specific Energy Efficiency Plan (SEMP) is a second crucial measure which is required by regulation 22 for existing and new ships of 400 GT and above, with the purpose of establishing operational mechanisms in terms of fuel efficiency to improve ship's energy efficiency performance. It also provides an approach for shipping companies to manage the entire fleet efficiency over the time by incorporating a number of operational management tools such as: improved voyage planning, speed and power optimization, optimized ship handling, improved fleet management, improved cargo handling, energy management and monitoring tools like Energy Efficiency Operational Indicator[28] (EEOI) [15, 29, 33]. The SEEMP should be developed as a ship-specific plan and adjusted to the needs of individual companies by taking into account relevant guidelines[29] adopted by IMO. It is not subject to approval by the flag Administration. However, it must be on board before the issuance of the first International Energy Efficiency certificate (IEEC) which is required as per Annex VI chapter 2 requirements. Its form if provided as appendix VIII to this Annex 12].

The latest significant milestone in the consistent reduction of GHG emissions from ships by IMO is a new mandatory requirement, adopted by MEPC 70 on 28 October 2016. Amendments to chapter 4 of Annex VI, which came into force on 1 March 2018, introduced new Regulation 22A on the Collection and Reporting of Ship Fuel Oil Consumption Data. These monitoring requirements obligate ships of 5,000 GT and more to collect consumption data for each type of fuel oil that they use. They also include provisions for flag States in relation to data collection, surveys and certification. The amendments also encompass two new related appendices to Annex VI. These newly enforced regulations will enable a global centralised database to be established which will enable more accurate analyses to be performed and inform the decision making process [14, 34, 35].

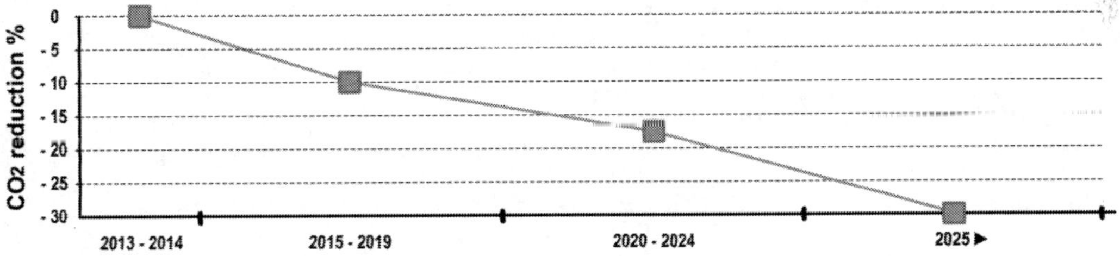

Fig. 4. CO_2 reduction based on EEDI implementation phases [31]

3 Conclusion

Since air pollution has been scientifically proven to negatively impact climate change, and has become a serious global issue, emissions of harmful pollutant gases from land based sources gradually started to be regulated and reduced. As a

[25] The Energy Efficiency Design Index (EEDI) provides a specific figure for an individual ship design expressed in grams of CO_2 per tonne mile (g/t·nm). The smaller the EEDI, the more energy efficient is the ship design. It is calculated using a formula based on the technical design parameters for a given ship.

[26] The Attained EEDI calculation is specific to each ship type as defined in Annex VI regulation 2 and it is performed by the method specified in the latest amended MEPC.281(70) guidelines. The Attained EEDI indicates the estimated energy efficiency performance of the ship and has to be accompanied by the EEDI technical file that contains all the information necessary to be able to calculate the Attained EEDI and which also shows the process of calculation. It must be verified, based on the EEDI technical file, by the flag Administration.

[27] The Required EEDI must be greater than, or equal to the Attained EEDI. Its calculation depends on X - correction factor (%) specified for each ship type and size and which represents the EEDI relative to the EEDI reference line given in Table 2 of the same regulation.

[28] The Energy Efficiency Operational Indicator (EEOI) is a monitoring tool which enables operators to measure the fuel efficiency of a ship in operation and to gauge the effect of any operational changes, e.g. improved voyage planning or more frequent propeller cleaning, or the introduction of technical measures such as waste heat recovery systems or a new propeller.

[29] New 2016 Guidelines for the Development of a Ship Energy Efficiency Management Plan were adopted on 28 October 2016 and include the Ship Fuel Oil Consumption Data Collection Plan as the new provision of the SEEMP.

result, focus has turned to air pollution from international shipping which accounts for 90% of total world trade, and is undoubtedly a significant contributor to global atmospheric pollution.

Hence, under the IMO initiative, 26 of September 1997, the MARPOL Convention was revised and amended with the 1997 Protocol. A new Annex VI was introduced which included rules regulating the emission of air pollutants from ships, with a particular emphasis on emissions of NOx, SOx and ODS. It is important to emphasize that, after the regulations came into force, objective flexibility was applied to parameters such as the inertness of fleet renewal, the average age of a ship as well as the adaptation of MDEs, systems and equipment in existing ships.

Following on from the adoption of Annex VI, further research into air pollutants informed additional amendments, and resulted in the adoption of new regulations and certificates. The most significant example was the introduction of chapter IV and its role in reducing CO_2 emissions from ships, the last remaining significant atmospheric pollutant to be controlled. The IAPP certificate is a crucial document that verifies a ship's compliance with Annex VI regulations. Annex VI is still a work in progress, a regulatory provision which continues to be enhanced with new regulations and more stringent revisions to existing ones.

Shipowners and other relevant stakeholders in the international maritime industry have to accept the fact that stringent rules regulating emissions of atmosphere pollutants from ships are enforceable and are not optional. They represent important preconditions and obligations to the international maritime industry and have to be complied with.

The introduction of these regulations has had a significant effect on both, MDEs manufacturers and marine fuel oil suppliers. Furthermore, they require shipowners to be more involved in ensuring that the necessary operative procedures are in place, are correctly implemented and effectively applied, both on ships and in companies' offices ashore.

90% of merchant ships on international voyages use MDEs as their prime mover whose resultant exhaust gases are a major contributor to air pollution. Therefore, to maintain MDEs as the prime and the most efficient propulsion engine in use, it is imperative for manufacturers to work intensively to seek out new technological solutions and alternatives to further reduce harmful emissions and to comply with increasingly stringent air pollution regulations. New knowledge gained through future research will enable existing propulsion systems to be more efficiently altered to meet ecological demands.

References

1. *Marine Environment*, International Maritime Organization, Available: http://www.imo.org/en/OurWork/Environment/Pages/Default.aspx (March 2018)
2. *Air pollution from ships*, Clean Shipping Coalition (2011), Available: http://www.cleanshipping.org/download/111128_Air%20pollution%20from%20ships_New_Nov-11(3).pdf (March 2018)
3. *Shipping sector cuts nitrogen loads to the Baltic Sea*, HELCOM (2016), Available: http://www.helcom.fi/news/Pages/Shipping-sector-cuts-Nitrogen-loads-to-the-Baltic-Sea.aspx (March 2018)
4. *Development of a Methodology to Measure and Asses Shi Emissions*, Australian Maritime College, IAMU 2015 Research Project (No. 20150202) (2016), Available: https://gmn.imo.org/wp-content/uploads/2017/10/20150202_AMC_measure-and-assess-ship-emissions.compressed.pdf (March 2018)
5. *Third IMO GHG Study 2014, Executive Summary and Final Report* (2015), International Maritime Organization, Available: https://gmn.imo.org/wp-content/uploads/2017/05/GHG3-Executive-Summary-and-Report_web.pdf, (March 2018)
6. M. Anantharaman, V. Garaniya, F. Khan, B. Lewarn: *Marine Engines and their Impact on the Economy, Technical Efficiency and Environment*, JIME Vol. **50**, No. **3** (2015), Available: https://www.researchgate.net/publication/310666162_Marine_Engines_and_their_Impact_on_the_Economy_Technical_Efficiency_and_Environment (March 2018)
7. B. Milošević Pujo, N. Jurjević: *Onečišćenje mora iz zraka emisijom ispušnih plinova*, NM **51 (5-6)**, p.p. 178-184, Međunarodni znanstveno-stručni časopis za more i pomorstvo, Dubrovnik, (2004), Available: https://hrcak.srce.hr/index.php?show=clanak&id_clanak_jezik=12802&lang=en (March 2018)
8. *Status of Treaties*, International Maritime Organization, Available: http://www.imo.org/en/About/Conventionsb/StatusOfConventions/Documents/StatusOfTreaties.pdf (March 2018
9. *Introduction to IMO*, International Maritime Organization, Available: http://www.imo.org/en/About/Pages/Default.aspx (March 2018)
10. *Focus on IMO, MARPOL - 25 years*, International Maritime Organization (1998), Available: http://www.imo.org/en/KnowledgeCentre/ReferencesAndArchives/FocusOnIMO(Archives)/Documents/Focus%20on%20IMO%20-%20MARPOL%20-%2025%20years%20(October%201998).pdf (March 2018)
11. *MARPOL 73-78 Briefhistory - list of amendments to date and where to find them*, International Maritime Organization, Available: http://www.imo.org/en/KnowledgeCentre/ReferencesAndArchives/HistoryofMARPOL/Documents/MARPOL73-78Brief History-ListOfAmendmentsAndHowToFindThem.htm (March 2018)
12. *MARPOL Annex VI and NTC 2008 with guidelines for implementation 2017 edition*, International Maritime Organization (2017)
13. *Annex VI of MARPOL 73/78 - Regulations for the Prevention of Air Pollution from Ships*, Insight **176** Gard AS, Arendal, Norway (2004), Available: http://www.gard.no/web/updates/content/53059/annex-vi-of-marpol-7378-regulations-for-the-prevention-of-air-pollution-from-ships (March 2018)

14. *Module 1 – Climate Change and the Shipping Response, IMO Train the Trainer (TTT) Course on Energy Efficient Ship Operation*, International Maritime Organization (2016), Available: http://www.imo.org/en/OurWork/Environment/PollutionPrevention/AirPollution/Documents/Air%20pollution/M1%20Climate%20change%20and%20shipping%20response%20final.pdf (March 2018)
15. Hughes, E.: *MARPOL Annex VI – an update of international regulatory developments for prevention of air pollution and the energy efficiency of ships*, Global Shippers' Forum AGM, Colombo, Sri Lanka (2016), Available: https://globalshippersforum.com/media/1252/edmund-hughes-imo-29-july-2016_final.p (March 2018)
16. *MARPOL Consolidated Edition 2017*, International Maritime Organization, London, UK, (2017)
17. *MARPOL How to do it 2013 Edition*, International Maritime Organization, London, UK, (2011)
18. *MARPOL 73/78 Annex VI - Regulations for the prevention of Air Pollution from ships – Technical and Operational implications*, Det Norske Veritas, Available: http://hulpinnood.nl/wp-content/uploads/2015/03/BIJLAGE3_Marpol-annex-VI.pdf (March 2018)
19. M. Yamasaki: *MARPOL Annex VI Regulations – Prevention of Air Pollution from Ships*, TRACECA Regional Seminar on MARPOL Awareness and Implementation, Moldova (2015), Available: http://www.traceca-org.org/fileadmin/fmdam/TAREP/68ta/1/Session_8_MY_MARPOL_Annex_VI___chapters_1_to_3__and_Black_Carbon_Final.pdf (March 2018)
20. *MARPOL 73/78, Regulations for the Prevention of Air Pollution from Ships Annex VI*, Autoridad Marítima de Panamá, Available: http://amp.gob.pa/amp/asi/online%20course%20module/MODULE%208A.pdf (March 2018)
21. I. Djadjev: *How to comply with, MARPOL 73/78*, DOI: 10.13140/RG.2.1.4715.7922 (2015), Available: https://www.researchgate.net/publication/277708775_Complying_with_MARPOL_7378, (March 2018)
22. *IMO and the Environment*, International Maritime Organization, Available: http://www.imo.org/en/OurWork/Environment/Documents/IMO%20and%20the%20Environment%202011.pdf (March 2018)
23. *Particulate Matter (PM) Pollution*, United States Environmental Protection Agency, Available: https://www.epa.gov/pm-pollution (March 2018)
24. J. Faber, S. Ahdour, M.'t Hoenet, et. al.: *Assessment of Fuel Oil Availability Final report*, Publication code: **16.7G68.71**, CE Delft (2016) Available: http://www.imo.org/en/OurWork/Documents/MEPC%2070-INF.6%20%20Assessment%20of%20fuel%20oil%20availability.pdf (March 2018)
25. *RESOLUTION MEPC.280(70), Effective Date of Implementation of the Fuel Oil Standard in Regulation 14.1.3 of MARPOL Annex VI*, MEPC **70/18/Add.1**, International Maritime Organization (2016), Available: http://www.imo.org/en/OurWork/Environment/PollutionPrevention/AirPollution/Documents/280%2870%29.pdf (March 2018)
26. Almeida R.: *Have the IMO and the International Chamber of Shipping Overlooked the Obvious?* (2012), Available: http://gcaptain.com/international-chamber-shipping/#.VfaXak-9S71 (March 2018)
27. *Special Areas under MARPOL*, International Maritime Organization, Available: www.imo.org/en/OurWork/Environment/SpecialAreasUnderMARPOL/Pages/Default.aspx (March 2018)
28. Gard: *Annex VI of Marpol 73/78 - Regulations for the Prevention of Air Pollution from Ships*, Insight **176**, Gard (2004), Available: http://www.gard.no/web/updates/content/53059/annex-vi-of-marpol-7378-regulations-for-the-prevention-of-air-pollution-from-ships (March 2018)
29. *Energy Efficiency Measures*, International Maritime Organization, Available: http://www.imo.org/en/OurWork/Environment/PollutionPrevention/AirPollution/Pages/Technical-and-Operational-Measures.aspx (March 2018)
30. *Učinak staklenika*, Croatian Ministry of Environment and Nature protection, Available: http://klima.mzoip.hr/default.aspx?id=4 (March 2018)
31. J-F Helfre, P. A. Couto Boot: *Emission Reduction in the Shipping Industry: Regulations, Exposure and Solutions*, Sustainalytics (2013), Available: http://www.sustainalytics.com/sites/default/files/shippingemissions_july2013.pdf (March 2018)
32. E. Hughes: *A new chapter for MARPOL Annex VI – requirements for technical and operational measures to improve the energy efficiency of international shipping*, International Maritime Organization, London, UK, February (2013), Available: http://www.imo.org/en/KnowledgeCentre/PapersAndArticlesByIMOStaff/Documents/A%20new%20chapter%20for%20MARPOL%20Annex%20VI%20-%20E%20Hughes.pdf (March 2018)
33. *Module 2: Ship Energy Efficiency Regulations and Related Guidelines, IMO Train the Trainer Course*, International Maritime Organization, Available: http://www.imo.org/en/OurWork/Environment/PollutionPrevention/AirPollution/Documents/Air%20pollution/M2%20Energy%20Efficiency%20Regulations%20%20IMO%20TTT %20course%20presentation%20final1.ppt (March 2018)
34. *List of amendments expected to enter into force this year and in the coming years*, International Maritime Organization, Available: http://www.imo.org/en/About/Conventions/Pages/Action-Dates.aspx (March 2018)
35. New requirements for international shipping as UN body continues to address greenhouse gas emissions, International Maritime Organization (2016), Available: http://www.imo.org/en/MediaCentre/PressBriefings/Pages/28-MEPC-data-collection--.aspx (March 2018)
36. H. Wang: *Cutting carbon from Ships*, The International Council On Clean Transportation (2012), Available: https://www.theicct.org/blogs/staff/cutting-carbon-ships (March 2018)

Analysis of the market concentration of the container shipping markets – selected issues

Jędrzej Charłampowicz[1]

[1]Department of Logistics and Transports Systems, Gdynia Maritime University, 81-222 Gdynia, Poland

Abstract. The value of maritime freight transport grows yearly both in terms of value and quantity. The dynamic development of container shipping is responsible for this state of affairs. The maritime container transport market is strongly integrated on the supply side. The aim of this paper is to determine, based on empirical data, whether the market is still competitive for its participants on the supply side. For this purpose, the 2 indicators of concentration were calculated, namely CRx-concentration for 4 operators with the largest market share, and Herfindahl-Hirshmann Index for TOP 10 and TOP 20 operators, based on AXS Alphaliner TOP100 and various issues of UNCTAD Review of Maritime Transport. The results of the calculations indicate that the market is still competitive, although it is heading towards a direction in which several large entities will dictate the conditions to other participants.

Key words: Herfindahl-Hirshmann Index, maritime container shipping market, competitiveness, concentration

1 Introduction

Maritime container shipping is one of the fastest developing transport branch. Introducing containers to the transportation process implied the need to change the existing transportation model in the way to enable integration of every branch [1, 2]. This modification was conducted for every member of the transportation process. Firstly, it was made through readjustment of vessels for shipping containers and as a consequence also through readjustment of ports for their handling. This type of adaptive activities significantly reduced transportation costs [3]. Reducing transportation costs and time, thanks to markets integration, also accelerated further markets liberalization of global trade as well as opening up and developing new markets. Due to its characteristics containerization is considered to be the main factor and catalyst for globalization in 20th century [4]. The dynamic growth of the new transportation branch caused an increase in a number of markets members on the supply side as well as on the demand side. The supply side of the market took actions to strengthen their market shares and risk sharing [5]. In 21th century those activities were carried out by establishing strategic alliances that led to the reduction costs and increase market shares battening on the smaller entities. For needs of this paper the supply side of maritime container shipping markets (MCSM), taken into account during conducted research, is defined as follows: all vessel operating common carriers transporting containers (filled or empty), with various but standardized sizes/dimensions regardless of the content, put into service on a regular basis, loading and discharging at specified ports.

The aim of this paper is to verify, based on available empirical data, such as market share of the maritime container carriers, whether the supply side of the maritime container shipping markets is still globally competitive for its participants and detect the trend of market concentration.

This paper is divided as follows:

- Section 2 provides brief characteristics of the supply side of the MCSM;
- Section 3 describes methodology and data;
- Section 4 presents results of calculations;
- Section 5 outlines research limitations and further research directions;
- Section 6 shows final conclusions.

2 The supply side of the maritime container shipping markets

The average increase in sea freight during period 2005-2016, in million loaded tones, was 5,2 per cent. In 2016 global containerized trade expanded 3,2 per cent year-to-year, with volumes attaining an estimated 140 million TEU's [6]. [7] notes that MCSM are currently a strongly integrated markets, the main reasons of this state of affairs are increasing transportation capacity and increasing share of high-value goods in transported cargoes. Table 1 presents the change of quantity and capacity of container vessels in last decade. Despite of decrease the number of vessels scheduled to be deployed in 4Q2017 by around 4,3 per cent compared to 4Q 2007, during period 2007-2017 there was significant increase in total deployed capacity from 123,2 mil TEU to 183,8 mil TEU, which is almost 50 per cent growth [8].

Table 1. Number of vessels based on their capacity

	Ship size (TEU)	4Q 2007	4Q 2017
Number of vessels	<5000	4589	3523
	5000-7499	328	469
	7500-9999	172	527
	10000-12499	0	96
	12500-14999	8	172
	15000+	0	92
Total no of vessels		5097	4879

Source: [8]

Placing higher orders for container ships in recent years has caused an oversupply. The trend of ordering gearless ships has been sustained. Although, in 2016, TEU capacity per delivered ship was 12 per cent lower than in 2015, the new vessels are larger than existing fleet [6]. This state of affairs continues to exert pressure on ports to accommodate to larger vessels and, what is important, due to cascading effect, this trend applies also on smaller ports. The main reason of introducing mega container ships was the need of achieving the economies of scale. Oversupplied market characterized by the mega container ships (over 18000 TEU) and weak demand growth effected in the need for consolidation and rationalization to optimize capacity and reduce costs [6]. Ways to achieve those goals were mergers and acquisitions (M&A) and shake-ups in the liner shipping alliances. Implementation of mega container ships changed overall dynamics of MCSM.

The MCSM has to face very volatile and rapidly changing market realities [9], the way of dealing with these conditions is strong concentration on the supply side. It can be expressed in twofold: subjective capital integration and organizational and functional form expressed through establishing strategic alliances. Main reasons for those agreements are risk-sharing, knowledge exchange, economies of scale and strengthening the market position [5]. Concluding, the above settlements have high impact on MCSM. First, it is reducing cost and risk for alliance participants, that are generally top operators. This also affects on smaller enterprises, which cannot draw benefits from economies of the scale and risk-sharing.

The MCSMs have witnessed unique transformations unheard of in any other transportation sector. Through M&A, as well as formation of alliances, the three major alliances, namely 2M Alliance (Maersk, MSC), THE Alliance (NYK, MOL, K Line, Yang Ming, Hapag-Lloyd) and Ocean Alliance (CMA CGM, Evergreen, OOCL, COSCO), represents almost 80% of global container capacity. Some M&A requires the alliance to be reorganized [3].

3 Methodology and data

3.1 Market concentration

Market concentration refers to the extent of which small number of enterprises account for large proportion of provision of goods or/and services [10]. The theory of industrial organization economics suggest that higher level of market concentration will lead entities to monopolistic practices, that will result in misallocation of resources and poor economic performance. The ownership concentration, which is related to the supply side of market, measures the degree of market control and can be evaluated using concentration indexes [10]. To determine the degree of concentration two concentration indicators are calculated.

3.2 Concentration ratio

Concentration ratio (CR_x) is one of the most popular and simplest methods for judging the degree of concentration. Generally the CR_x is used for computing the relative share of industry accounted for the x largest firms. One of the most prominent indicators is the four-firms concentration ratio (CR_4) which involves totaling up the market shares of four largest entities in the industry. Mathematical formula for calculating CR_4 can be written as follows:

$$CR4 = \sum_{n=1}^{4} s_i \tag{1}$$

where s_i represents the percentage market share of the ith carrier. The main disadvantage of this indicator is lack of ability to show the total number of competitors on the market, since it is focused only on the top entities [10].

3.3 Herfindahl-Hirshman Index

Herfindahl-Hirshman Index (HHI) is based on the total number and market share of enterprises in the industry. This indicator is typically used by the regulators to provide an assessment of market concentration [10, 11]. It is computed as a sum of the squares of percentage market shares of entities in the studied business. Mathematical formula for HHI is as follows:

$$HHI = \sum_{i=1}^{n} s_i^2 \quad (2)$$

where n stands for total number of examined carriers and s_i represents the percentage market share of the ith carrier. The maximum value of HHI, according to equation 2, is 10000, which stands for pure monopoly [10]. HHI below 1000 represents unconcentrated competitive market, value between 1000 and 1800 indicates moderately concentrated market and value above 1800 shows highly concentrated market [11].

3.4 Data description

Data used for calculating CR4 and HHI comes from publicly available sources such as: AXS Alphaliner TOP100, and various issues of Review of Maritime Transport published yearly by UNCTAD. The data used for calculations includes market share of examined operators. The concentration indexes are computed over the period of 2010-2018. For computing HHI only operators having at least 1% of market share were taken into account. The reason for this procedure is the fact, that squaring market shares of less than 1% has no significant impact on total HHI. As the result, where the MCSM is concerned, only TOP 20 operators market share can be calculated without affecting the whole image of market concentration.

4 Results

Computation has been made using data concerning market share[6, 12-19]. Table 2 presents results of calculations of HHI TOP10 and TOP 20 operators and CR4. Conclusion made on basis of this calculation shows clearly that average increase of concentration, year-to-year, is around 13 per cent for TOP10 operators and 12 per cent for TOP20 carriers. Figure 1 presents the evolution of the Herfindahl-Hirshman Index.

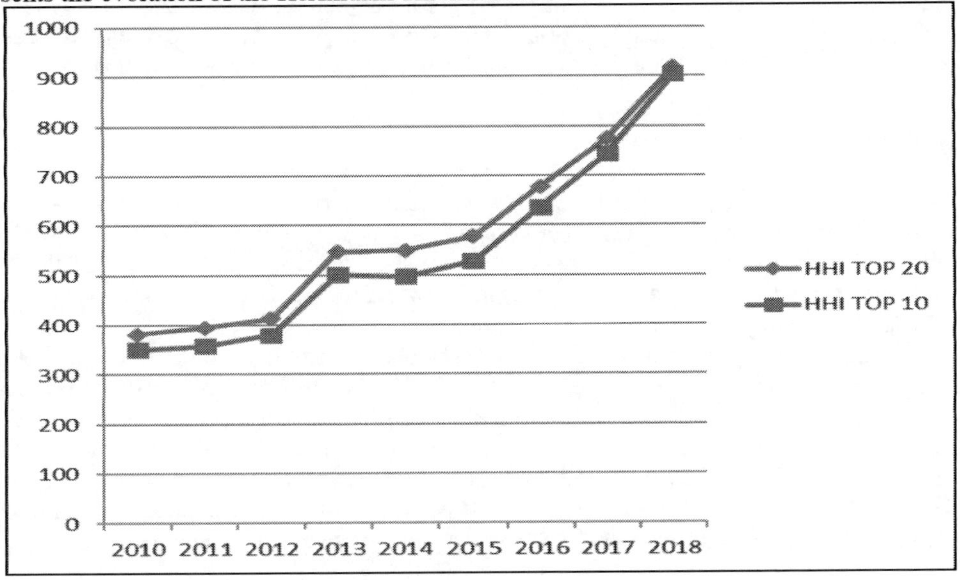

Fig. 1. Evolution of Herfindahl-Hirshman Index

Table 2. Measurement of concentration

	2010	2011	2012	2013	2014	2015	2016	2017	2018
HHI TOP10	350	357,54	379,89	501,12	497,12	527,66	635,52	744	904,34
HHI TOP20	381,18	394,92	413,2	546,57	549,93	577,08	678,52	774,74	918,49
CR4	32,1%	32,3%	33%	38%	38,8%	39,75%	45,5%	49,7%	54,2%

Current value of HHI for TOP 10 and TOP 20 is still below value 1000, which means that market is still competitive, although CR4 in March 2018 is 54,2 per cent, that refers to situation where more than half of the market is controlled by four largest entities. In 2010 50,2 per cent of the market was controlled by top 10 operators [19], in 2013 7 largest liner companies had 49,8 per cent [16].

5. Research limitations and further research directions

5.1 Research limitations

The ability to utilize and have a full access to *Science Direct* journals made it possible to review the subject in interesting areas. None the less, the fact of relying on only one database greatly limited the possibility of conducting a more extensive literature research.

Another research limitation is the lack of access to full database of AXS Alphaliner and possibility to conduct research based on publicly available reports like UNCTAD Review of Maritime Transport and briefs reports such as MDS Transmodal Container Shipping Bulletin and, mentioned above, AXS Alphaliner.

5.2 Further research directions

It is necessary to conduct two-directional research about the impact of strategic alliances on global market concentration as well as on the main trade lanes. Additionally, it is crucial to determine the impact of strategic alliances on efficiency of the supply chains passing through MCSM [2], which are the main catalyst for the global trade development.

6. Conclusions

MCSM have to face the volatile, rapidly changing global trade environment, which is propelled by increasing customers requirements, technology advances, political disruptions and intensifying competition. For markets participants reducing cost has become the most important factor. The adaptation of the supply side of the MCSM was made by deploying ever-larger container vessels and consolidation.

The aim of this paper was to determine the global market concentration on the MCSM using 2 concentration indexes, namely Herfindahl-Hirshman Index and Concentration Ratio for 4 largest entities. The results of the calculations indicate that market is still competitive for its participants. Current value of the TOP20 HHI is below 1000, although it is clear that this rate is growing rapidly in period 2010-2018 having average pace of about 12 per cent. The TOP20 HHI value of 918,49 (919,9 for TOP50) represents fragmentized, unconcentrated, competitive market. Albeit, the largest liner companies capture dominant market shares. It can be stated that this trend of increasing market concentration will continue.

The main conclusions of this paper are:

- MCSM heading towards to more concentrated market, where few operators will dictate the conditions for other participants;
- It is important to determine the impact of strategic alliances on market concentration and market competitiveness with respect to global trade and main trade lanes.

7 References

1. T. E. Notteboom, *Container Shipping* in W. K. Talley (Ed.) *The Blackwell Companion to Maritime Economics*, 32 (2012)
2. J. Charłampowicz, *Proceedings of the 17th International Scientific Conference Business Logistics in Modern Management*, 13 (2017)
3. C.-Y. Lee, P.-D. Song, Trans. Res. P. B, **95**, 33 (2017)
4. D. M. Bernhoffen, Z. El-Sahli, R. Kneller, J. of Int. Ec., **98**, 14 (2016)
5. P. Rau, S. Spinler, Trans. Res. P. E, **101**, 21 (2017)
6. UNCTAD, Rev. of Mar. Trans., (2017)
7. A. S. Grzelakowski, Int. Bus. and Gl. Ec., **32**, 20 (2013)
8. MDS, Transmod. Cont. Ship. Bul., November (2017)
9. T. E. Notteboom, Rev. of Net. Ec., **3**, 21 (2004)
10. http://www.oecd.org/regreform/sectors/2376087.pdf - access March, 4, 2018
11. U. Qazi, M. Jahanzaib, W. Ahmad, S. Hussain, Ren. and Sust. En. Rev., **70**, 13 (2017)
12. https://alphaliner.axsmarine.com/PublicTop100/ - access February, 14, 2018
13. UNCTAD, Rev. of Mar. Trans., (2016)
14. UNCTAD, Rev. of Mar. Trans., (2015)
15. UNCTAD, Rev. of Mar. Trans., (2014)
16. UNCTAD, Rev. of Mar. Trans., (2013)
17. UNCTAD, Rev. of Mar. Trans., (2012)
18. UNCTAD, Rev. of Mar. Trans., (2011)
19. UNCTAD, Rev. of Mar. Trans., (2010)

Legal approach to marine accidents: a comparative research between Turkish and international perspectives

Nil Kula Degirmenci [1,*]

[1] Izmir Dokuz Eylul University, Maritime Faculty, Department of Maritime Business Administration, Izmir, Turkey

Abstract. The importance of prevention of marine accidents increases parallel to the technological improvements in maritime transportation and the rise of sensitiveness to environmental protection. Legal approach to marine accidents can be gathered under three important topics as follows: legal approach to general average, legal approach to collision and legal approach to salvage. The international perspective on the issue has been shaped through international conventions and advisory rules, as applied in the field of general average, and some of their modernized amendments. Turkey which is one of the most important countries in maritime transportation especially as a result of its geographical position, on the one hand aims to protect its existing legal regime to prevent the occurrence of marine accidents and on the other hand follows the updated legal developments on the issue. In the study, it is aimed to point the important facts of general average, salvage and collision and analyze the Turkish and international perspectives comparatively. Thus the success in ensuring uniformity which is targeted through the international studies on these issues is tried to be measured for the Turkey example.

1. Introduction

With the increased frequency of shipping activities, navigation safety has become a major concern especially when economic losses, human casualties and environmental issues are considered [1]. Investigations into the causes of shipping accidents show that over 30% of the accidents are caused by poor weather, and an additional 25% remain completely unexplained [2]. On the other hand the need for qualified seamen is one of the major problems in improving safety in the shipping industry [3].

About 63% of the world's shipping accidents are recurrent—they occur to ships that have already experienced at least one prior accident [4]. Therefore, reducing recurrent accidents or adopting a legal framework to determine the applications upon the occurrence of marine accident can contribute significantly to maritime safety. Thus, in the study, legal approach to marine accidents has been analyzed in terms of general average, collision and

*Corresponding author: nilkuladegirmenci@gmail.com

salvage through the framework of international conventions and domestic legislation of Turkey.

2. Legal Approach to Marine Accidents

In this part of the study, legal approach to marine accidents has been analyzed in terms of general average, collision and salvage.

2.1. Legal Approach to General Average

Because the marine adventure is viewed as a joint project risk of the parties including the ship owner, master, crew, and cargo owners, involved, all are deemed as sharing the risk [5]. Due to the complexity of activities which can generate cargo surveys in respect of transport, maritime ships can be in such situations that will not allow carrying out activities onboard under normal conditions, leading to damages to the ships or carried cargo, thus entailing litigations between participants in such transport [6]. A general average act usually comes into play when a part of the cargo is jettisoned, by order of the ship's master, in order to save the ship, the remaining cargo, and all who are in the ship. General average is one of the most historic issues governed under maritime law. There are no international conventions that govern general average through a mandatory approach, however Comité Maritime International (CMI) has adopted several groups of rules, namely York Antwerp Rules (YAR), which are frequently adopted and applied by most countries all over the world, even though they have advisory characteristics. All the standard forms of contracts of affreightment and bills of lading refer to the application of YAR. These rules have been renewed at certain periods of time, such as 1924, 1950, 1974, 1990, 1994, 2004 and the latest, 2016.

The substance of the YAR—in some cases the actual text—has been incorporated into the law of many countries. Unless otherwise agreed by the parties, Turkish Commercial Code (TCC) directly refers to the application of the updated version of YAR for the general average adjustment (TCC art.1273, para.1), for instance. As for the consideration of YAR by the Turkish courts, an updated version of the rules shall be translated into Turkish by an expert committee and the translation shall be published in the Official Gazette along with the original text of the rules (TCC art.1273, para.2).[2] Additionally, TCC has placed some supplementary provisions (from art.1272 to art.1285), generally related to the procedure for general average adjustment, to complete the application of YAR in Turkey which will be examined in the following parts of the study.

Components of general average can be counted as follows: common maritime adventure (in which the ship is wholly or partially loaded with cargo), common peril (any danger threatening the ship, cargo, other property and the freight),[3] extraordinary sacrifice or expenditure (made or incurred for the common safety), intention (even though the 'master' frequently decides whether there is a general average or not, it is not a requirement in

---

[2] As the time of writing, YAR 2016 has not been translated in Turkish by the expert committee and published in the Official Gazette, thus; unless their application is contractually preferred by the parties to contract of affreighthment or bill of lading, the rules shall have no effect as for the Turkish law. However YAR 2016 will soon be translated by the expert committee and for this reason, they have been considered for the explanations on the general average in this study.

[3] The principle of "imminent peril" operates in a wider ambit in circumstances where a danger exists and doing nothing will inevitably require the insurer to respond under the terms of the policy. See *Garden View Restaurant Ltd. v. Portage La Prairie Mutual Insurance Co.*, 2014 CarswellNS 994, 43 C.C.L.I. (5th) 110, 92 C.E.L.R. (3d) 100 Nova Scotia.

accordance with the referred legal text), reasonability (which is the pre-condition for the existence of general average) and beneficial result (in which the ship or the cargo is partially or wholly saved).

General average contribution is defined to be a contribution by all parties in a sea adventure to make good the loss sustained by one of their number due to voluntary sacrifices of part of the ship or cargo to save the remainder and the lives of those on board from an impending peril, or for extraordinary expenses necessarily incurred by one or more of the parties for the general benefit of all the interests embarked in the enterprise [7]. Rights to contribution in general average shall not be affected, though the event which gave rise to the sacrifice or expenditure may have been due to the 'fault' of one of the parties to the common maritime adventure, but this shall not prejudice any remedies or defenses which may be open against or to that party in respect of such fault.[4]

There are three groups of articles in YAR 2016: numbered rules, lettered rules and Rule of Paramount. The Rule of Interpretation which is placed as the first provision of the rules states the line between these rules. According to the rule, except as provided by the Rule Paramount and the numbered Rules, general average shall be adjusted according to the lettered Rules. Thus, the line between the groups of articles placed in YAR 2016 shall be as follows: (1) Rule Paramount[5] (2) numbered Rules and (3) lettered Rules.

The numbered rules give the examples of the sacrifices that can be included in the general average such as; jettison of cargo (under YAR 2016 I), loss or damage by sacrifice for the common safety (under YAR 2016 II), extinguishing fire on shipboard (under YAR 2016 III), cutting away wreck (under YAR 2016 IV) and voluntary stranding (under YAR 2016 V).

In addition to the above pointed sacrifices, one party may affect payments for the common interest of saving the ship and the cargo such as; salvage remuneration (under YAR 2016 VI), damage to machinery and boilers (under YAR 2016 VII), expenses of lightening a ship when ashore and consequent damage (under YAR 2016 VIII), cargo, ship's materials and stores used for fuel (under YAR 2016 IX), expenses at port of refuge (under YAR 2016 X), wages and maintenance of crew and other expenses putting into and at a port of refuge (under YAR 2016 XI), damage to cargo in discharging (under YAR 2016 XII), deductions from cost of repairs (under YAR 2016 XIII), temporary repairs (under YAR 2016 XIV), loss of freight (under YAR 2016 XV), amount to be allowed for cargo lost or damaged by sacrifice (under YAR 2016 XVI), contributory values (under YAR 2016 XVII), damage to ship (under YAR 2016 XVIII) and undeclared or wrongfully declared cargo (under YAR 2016 XIX). As per YAR, the above noted expenditures also can be included in the general average.

2.2. Legal Approach to Collision

Since the weight and speed of ships have increased, and iron and steel have been used for their construction instead of wood in time, the damage which is suffered as a result of collisions has increased accordingly [8]. Thus, the regulation of collisions, especially under international conventions, has gained more importance.

---

[4] See YAR 2016, Rule D.

[5] In no case shall there be any allowance for sacrifice or expenditure unless reasonably made or incurred (YAR 2016, Rule Paramount).

A crash of two or more ships is called collision.[6] Components of the collision can be listed as the ship (being a commercial ship is not required),[7] the crash, and the non-contractual relationship between the crashed ships.

The issue of collision is governed under international conventions, which can be classified as conventions that aim to prevent the occurrence of collision and conventions that govern the liabilities of the parties after collision. The International Convention for the Safety of Life at Sea (London, 01.11.1974; SOLAS)[8] specifies minimum standards for the construction, equipment and operation of ships, compatible with their safety, and is deemed to be one of the conventions placed under the first classification. In addition, Convention on the International Regulations for Preventing Collisions at Sea (London, 20.10.1972; COLREG)[9] which recognizes traffic separation schemes and determines safe speeds, the risk of collision and the conduct of ships operating in or near traffic separation schemes, is also deemed to be one of the conventions placed under that classification.

The Convention for the Unification of Certain Rules of Law with respect to Collisions between Vessels (Brussels, 23.09.1910; Collision Convention),[10] which aims to set certain uniform rules of law with respect to collisions, is deemed to be one of the conventions placed under the second classification. Related articles of this convention have been considered while governing the related articles of TCC on collision (TCC art.1286 to art.1297).

2.3. Legal Approach to Salvage

A "salvage service" is a service rendered to a vessel or other marine property in need of assistance. The need for salvage has existed since the time that ships commenced performing voyages at sea [9]. Salvage has been governed under domestic legislations and international conventions. The first attempt to establish a legally mandatory regime governing salvage was the adoption of the Convention for the Unification of Certain Rules of Law respecting Assistance and Salvage at Sea (Brussels, 23.09.1910; Salvage Convention 1910). However as years passed, the convention was not adequate to cover the needs of modern salvage practices. Thus, the International Convention on Salvage (London, 28.04.1989; Salvage Convention 1989)[11] has been adopted.

Turkey has become a contracting state to both of the above mentioned conventions, and while governing the related articles of TCC on salvage (TCC art.1298 to art.1319), provisions of the latter convention has been considered. Hence Independenta (1979), Nassia (1994), TPAO (1997) and Volgoneft 247 (1999) marine accidents have highlighted the importance of the qualified salvage services, the requirement of prevention of marine pollution, and thus, the codification of salvage in Turkey.

---

[6] The English cases apparently take the view that the term "collision" applies only to the striking together of two navigable ships, and that the term is not applicable to a case where a ship runs into a sunken object. See, *Cline v. Western Assur. Co.* (1903) 101 Va. 496, 44 S.E. 700, *Union Marine Ins. Co. v. Borwick* (1895) 64 L. J. Q. B. N. S., *Baum v. Girard Fire & Marine Ins. Co.*, 228 N.C. 525, 46 S.E.2d 324 (1948).

[7] Unlike the Collision Convention art. 11, warships are also subject to the provisions of TCC governing collision in accordance with TCC art.935, para.2 (b).

[8] Date of Official Gazette: 25.05.1980, Number of Official Gazette: 16985. Turkey is also contracting state to Protocol of 1988 relating to the International Convention for the Safety of life at Sea, 1974. Date of Official Gazette: 29.05.2013, Number of Official Gazette: 28661.

[9] Date of Official Gazette: 18.11.1984, Number of Official Gazette: 18579.

[10] Date of Official Gazette: 22.02.1955, Number of Official Gazette: 8937.

[11] Date of Official Gazette: 24.05.2014, Number of Official Gazette: 29009.

Components of salvage can be listed as: the activity which is directed to assist a vessel or any other property,[12] the danger which has the risk of material and economical loss,[13] the advantageous consequence and the performance of the activity on a volunteering basis.

3. Comparative Research Between Turkish and International Perspectives

In this part of the study, Turkish domestic legislation and international conventions that govern marine accidents have been analyzed, comparatively.

3.1. In terms of General Average

There is a general average act when, and only when, any extraordinary sacrifice or expenditure is intentionally and reasonably made or incurred for the common safety for the purpose of preserving from peril the property (ship, cargo, other property and freight) involved in a common maritime adventure and only such losses, damages or expenses which are the direct consequence of the general average act shall be allowed as general average (TCC art.1272, para.1).[14] Any additional expense incurred in place of another expense which would have been allowable as general average shall be deemed to be general average and so allowed without regard to the saving, if any, to other interests, but only up to the amount of the general average expense avoided (TCC art.1272, para.2).[15]
Personal debtors of general average are: the owner of the ship at the time of general average, collector of the freight and owner of the other property at the time of discharge (TCC art.1274, para.1). The consignee of the cargo which is subject to adjustment, shall be personally liable up to the value of that cargo at the time of delivery, and with the rate of adjustment share that shall be found with the liquidation of that cargo in case of non-delivery, provided that he is aware that the cargo was part of the contributing fund during delivery (TCC art.1274, para.2).[16]

Unlike the previous TCC (art.1209) which obliged the 'master' to provide the execution of the adjustment, TCC states that the 'owner' shall provide the execution of the adjustment, without delay (TCC art.1278, para.1). Hence, none of the owners in today's commercial life leave providing the execution of the adjustment to the master. In the breach of this duty, the owner shall be liable to each of the interested persons (TCC art.1278, para.1). When the adjustment is not executed in time, all the interested persons, including the insurer, has the right to demand and provide the execution of it (TCC art.1278, para.2). If the adjustment demand is rejected by the adjuster due to its non-compliance with the features of general average, the court shall decide, upon the demand of each of the interested persons including the insurer, whether the adjustment is required or not (TCC art.1278, para.3). The adjustment shall be executed in the place of arrival. When the ship has not arrived at the place of arrival, the adjustment shall be executed in the place where the voyage has been completed (TCC art. 1279).

---

[12] "Property damage", generally defined as injury to or destruction of tangible property, including loss of use. See *Garden View Restaurant Ltd. v. Portage La Prairie Mutual Insurance Co.*, 2014 CarswellNS 994, 43 C.C.L.I. (5th) 110, 92 C.E.L.R. (3d) 100 Nova Scotia.
[13] Therefore, in order to establish a salvage claim, the salvor must establish that the vessel was in marine peril. See American Law Reports, 70 A.L.R. Fed. 2d 251 (Originally published in 2012).
[14] See YAR 2016, Rule A and Rule C
[15] See YAR 2016, Rule F.
[16] YAR Rules 2016, Rule G, para.4.

Adjustment shall be executed by one or more adjusters who have been appointed by consensus of the interested persons. If consensus cannot be reached, the court which will execute the adjustment transactions shall appoint the adjuster(s) (TCC art.1280, para.1). Each of the interested persons shall submit the documents, especially charter parties, bills of lading and invoices that are required for the execution of adjustment (TCC art.1280, para.2). Upon the demand of the adjuster, the court shall order the delivery of these documents which shall be submitted legally (TCC art.1280, para.3). The adjuster shall permit the examination of adjustment by the interested persons and submit one copy of it, provided that these persons cover the costs (TCC art.1280, para.4).

The adjustment report shall consist of four parts as follows: (1) explanation of the general average case, (2) determination of the collectors (owners of the extraordinary sacrifices or expenditures), (3) determination of the debtors (owners of the saved values) and (4) calculation of the adjustment share (total of the extraordinary sacrifice and expenditures – total values ratio) [10].

All the interested persons, including the insurers, have the right to demand the approval of the adjustment report or object to the type of average or average calculations through a petition which shall be submitted to the authorized court noted above (TCC art.1281, para.1). The petition shall include the names and surnames of the interested persons who will be called to the hearings (TCC art.1281, para.2). Upon the submission of the petition, the court shall demand the documents that have force to prove the adjustment and claims from the adjuster or other interested persons (TCC art.1281, para.3). All the interested persons shall be called to the hearings (TCC art.1281, para.4). These persons shall be notified at least 15 days before the hearing (TCC art.1281, para.4). Objections to the adjustment report shall be made explicitly and thoroughly in the first hearing, at the latest (TCC art.1281, para.5). In case of the existence of fair reasons that prevent the submission of the objection on time, the judge shall give sufficient time to the interested person to submit his objection just once (TCC art.1281, para.5). In case of non-submission of the objection, the adjustment report shall be approved (TCC art.1282, para.1). In the case of submission of the objection, it shall be heard and if its fairness is approved by the court, the adjustment report shall be approved through the amendment of the report in accordance with that objection (TCC art.1282, para.1). When the court cannot decide on the objection, all the remaining parts of the adjustment report which are not subject to objection shall be approved by the court, and hearings on the objection shall proceed (TCC art.1282, para.2)

The adjustment report shall have the effect of a court decree with the finalization of the decision on the approval of that report (TCC art.1284, para.1). The same applies to the decision of the court on the adjustment report which has not been subjected to objection (TCC art.1284, para.1). A court decision on the approval of the adjustment report shall have no effect on interested persons who have not been called to the hearings (TCC art.1284, para.2).

Claims arising from a general average adjustment shall be time barred in 1 year (TCC art.1285, para.1). The time bar commences with the arrival of the ship at the place of arrival, or, if she cannot, at the place where the voyage is completed (TCC art.1285, para.2). The provision is the only provision which is in contradiction with the YAR. Hence according to Rule XXIII of YAR, the period of 1 year will commence when the general average adjustment is issued. However, in no case shall such an action be brought after six years from the date of termination of the common maritime adventure.

Table 1 is figured according to the above placed explanations and shows the similar, supplementary and conflicting provisions upon the comparison of provisions governing general average in Turkish legislation and international legislation.

Table 1: Comparison of Provisions Governing General Average (TCC vs. YAR)

Comparison of Provisions Governing General Average (TCC vs. YAR)	Similar Provisions	Supplementary Provisions	Conflicting Provisions
Definition of general average	TCC art.1272		
General average adjustment	TCC art. 1273		
Determination of personal debtors of general average	TCC art.1274, para.2	TCC art. 1274, para.1	
Owner's obligation to execute the adjustment		TCC art.1278	
Appointment of adjusters		TCC art.1280	
Approval of the adjustment report by the authorized court and objection to the report		TCC art.1281 and art.1282	
Time bar			TCC art.1285

Source: Author

3.2. In terms of Collision

Where a collision occurs between two or more ships, the compensation due for damages caused to the ships or to any things or persons on board thereof shall be settled in accordance with the provisions which are placed under TCC art.1286 to art.1297 (TCC art.1286, para.1). In addition, provisions governing collision shall also be applied when damage which a ship has caused to another ship, or to cargo or persons on board either ship, either by the execution or non-execution of a maneuver or by the non-observance of the regulations, even if no collision had actually taken place (TCC art.1286, para.2).[17] However if the damage is caused to the coastal facilities, provisions of TCC governing collision shall not be applied. In such cases, provisions governing the liability of the owner for his seamen's and pilots' faults (under TCC art.1062, para.1) shall be applied.

Collision can be classified in terms of the fault of the parties as; faulty collision and faultless collision. If the collision is accidental, if it is caused by *force majeure*, or if the cause of the collision is left in doubt, the damage shall be borne by those who have suffered it (TCC art. 1287, para.1).[18] This provision is applicable notwithstanding the fact that the ships, or any one of them, may be at anchor at the time of the accident (TCC art.1287, para.2). Parallel to the Collision Convention art.2, para.2, the Turkish lawmaker admits that this provision shall also be applied when the ships 'otherwise made fast'; however the issue is not explicitly stated under TCC due to the absence of that wording in the original French text. Unless the existence of fault is proved, faultless collision is deemed to exist. Thus, the person who asserts that he has suffered damage due to the fault of the counterparty in collision shall prove that fault.

If the collision is caused by the fault of one of the ships' owners or seamen,[19] liability to compensate for the damage incurred as a result of collision shall attach to the one which has

---

[17] See Collision Convention art. 13.

[18] See Collision Convention art.2, para.1.

[19] For instance, master's failure to use ship's radar to obtain early warning of risk of collision, after observing ship on near reciprocal course, has been accepted to violate international regulations for preventing collisions at sea. Additionally ship equipped with radar was under duty to use it intelligently and fully, and master or pilot who failed to do so had heavy burden of proving that fault

committed the fault (TCC art.1288). Unlike the related article placed in the Collision Convention which is seeking the fault of the ships; [20] TCC art.1288 seeks the fault of ship owners and seamen.

If the collision is caused by the fault of two or more ships' owners or seamen, liability to compensate for the damage incurred as a result of collision shall attach to each owner, in proportion to the degree of their fault (TCC art.1289, para.1). If, having regard to the circumstances, it is not possible to establish the degree of the respective fault, or if it appears that the fault is equal, the liability shall be apportioned equally (TCC art.1289, para.1).[21] Owners are not severally liable in terms of these compensation demands (TCC art.1289, para.1). However, in respect of damage caused by death or personal injuries, the ships' owners shall be jointly as well as severally liable to third parties (TCC art.1290, para.1). Still, if, having regard to the circumstances, it is not possible to establish the degree of the respective fault, or if it appears that the fault is equal, the liability shall be apportioned equally (TCC art.1290, para.1). Each owner shall be liable to the degree of his fault in the recourse process (TCC art.1290, para.2).

The right of recovery for damage resulting from a collision is not subject to the issuance of any protest or the fulfillment of any other special formality (TCC art.1293).[22] Any presumption shall not be considered in regard to determination of fault in collision (TCC art.1294). Actions for the recovery of damage shall be time barred after an interval of 2 years from the date of the casualty (TCC art.1297, para.1).[23] The period within which a recourse case must be instituted for enforcing the right to obtain contribution under TCC art.1289 and 1290, shall be 1 year from the date of payment (TCC art.1297, para.2).

Table 2 is figured according to the above placed explanations and shows the similar, supplementary and conflicting provisions upon the comparison of provisions governing collision in Turkish legislation and international legislation.

Table 2: Comparison of Provisions Governing Collision (TCC vs. Collision Convention)

Comparison of Provisions Governing Collision (TCC vs. Collision Convention)	**Similar Provisions**	**Supplementary Provisions**	**Conflicting Provisions**
Applicable law	TCC art.1286, para.2		TCC art.1287, para.2
Application to "warships"			TCC art.935, para.2 (b)
Classification of collision	TCC art. 1287, para.1		
Seeking the fault of ships' owners in faulty collision			TCC art.1288
Apportionment of faults	TCC art.1289, para.1		
No requirement of any formal transactions to direct compensation demand	TCC art.1293		

did not contribute to collision. See *Complaint of G & G Shipping Co., Ltd. of Anguilla*, 767 F.Supp. 398 (1991).

[20] If the collision is caused by the fault of one of the vessels, liability to make good the damages attaches to the one which has committed the fault (Collision Convention, art.3).

[21] See Collision Convention art.4, para.1.

[22] See Collision Convention, art. 6, para.1.

[23] See Collision Convention, art.7, para.1.

Time bar	TCC art.1297, para.1		

Source: Author

3.3. In terms of Salvage

TCC has placed several provisions with regard to the definition and components of salvage. Most of the provisions are parallel to the Salvage Convention 1989. First of all, 'salvage operation' is defined as any act or activity undertaken to assist a vessel or any other property in danger in navigable waters and is subjected to TCC art.1298 to art.1319 (TCC art.1298, para.1).[24] 'Vessel' means any ship or craft or any structure capable of navigation (TCC art.1298, para.2).[25] 'Property' means any property not permanently and intentionally attached to the shoreline and includes freight at risk, other than: (a) fixed or floating platforms or mobile offshore drilling units when such platforms or units are on location engaged in the exploration, exploitation or production of sea-bed mineral resources and (b) prehistorical, archeological or cultural monuments that has historical value on sea-beds (TCC art.1298, para.2 and para.3). Additionally, provisions of TCC governing salvage shall be applied notwithstanding that the salvor has obligation to save under legislation or the salved vessel and the vessel undertaking the salvage operations belong to the same owner (TCC art.1299). These provisions will also be applied to warships (TCC art.935, para.2 (b)).[26]

On the other hand, activities shall not be regarded under the context of salvage activity if they are performed: (a) in contravention of the clear and reasonable deprecation of the owner of the vessel or master or owner of the property or (b) by the persons who are employed on the vessel that is in danger or (c) to fulfill the duties arising from a contract which is signed before the occurrence of the danger (TCC art.1298, para.4).

The master is entitled to conclude contracts for salvage operations on behalf of the owner of the vessel (TCC art.1300, para.1).[27] In addition, the master or the owner of the vessel is entitled to conclude such contracts on behalf of the owner of the property on board the vessel (TCC art.1300, para.2).[28] These authorities include determination of the authorized court and arbitration (TCC art.1300, para.2).

Since there are no forms required between the articles of TCC governing salvage contracts for validity; salvage contracts can even be concluded orally. However in international practice, parties of salvage operations frequently prefer to use standard forms of salvage contract such as Lloyd's Standard Form of Salvage Agreement (universally known as Lloyd's Open Form, or LOF) [11].

A contract or any terms thereof may be annulled or modified by the court upon demand if: (a) the contract has been entered into under undue influence or the influence of danger and its terms are inequitable or (b) the reward under the contract is to an excessive degree too large or too small for the services actually rendered (TCC art.1301).[29] This article is one of the mandatory provisions of TCC governing salvage. Hence parties of a salvage contract cannot act against this article by removing the rights of the parties with regard to annulling

[24] See Salvage Convention 1989, art.1, para.1 (a).
[25] See Salvage Convention 1989 art.1, para.1 (b).
[26] Even though the preference of Turkey is against the Salvage Convention 1989 art.4, para.1, the convention has recognized the right of a contracting state to apply the provisions of the convention to warships, provided that Secretary-General has been notified by that state (Salvage Convention 1989 art.4, para.2).
[27] See Salvage Convention 1989, art.6, para.2.
[28] See Salvage Convention 1989, art.6, para.2.
[29] See Salvage Convention 1989 art.7.

or modifying the contract (TCC art.1302, para.2). Another mandatory provision, which cannot be changed by the parties of a salvage contract, governs the obligation of the parties to exercise due care to prevent or minimize damage to the environment set under TCC art. 1303 (TCC art.1302, para.2).

Salvage operations which have had a useful result give right to payment (TCC art.1304, para.1).[30] Except as otherwise stated under the related articles of TCC governing salvage, no payment is due if the salvage operations have had no useful result (TCC art.1304, para.2). The payments, exclusive of any interest and recoverable legal costs that may be payable thereon, shall not exceed the salved value of the property during the salvage activity (TCC art.1304, para.3).

When the salvage reward is not determined by the parties or the amount which has been determined is claimed to be modified by the court, the reward shall be fixed with a view to encouraging salvage operations, taking into account the following criteria without regard to the order in which they are presented: (a) the salved value of the vessel and other property after salvage activity, (b) the skill and efforts of the salvors in preventing or minimizing damage to the environment, (c) the measure of success obtained by the salvor, (d) the nature and degree of the danger, (e) the skill and efforts of the salvors in salving the vessel, other property and life, (f) the time used and expenses and losses incurred by the salvors, (g) the risk of liability and other risks run by the salvors or their equipment, (h) the promptness of the services rendered, (i) the availability and use of vessels or other equipment intended for salvage operations and (j) the state of readiness and efficiency of the salvor's equipment and the value thereof (TCC art.1305, para.1). Unlike the Salvage Convention 1989 art. 13, the above noted criteria shall only be considered by the Turkish courts when the salvage reward is not determined by the parties, or modification of the amount which has been determined by the parties has been claimed from the court [12].

Expenditure and charges of official institutions, customs taxes and other dues that shall be paid for the salved value and costs incurred for the storage, protection, valuation and sale of these values are not included under the salvage reward (TCC art.1305, para.2). The salvage reward shall be calculated over money. Unless otherwise agreed, the salvage reward shall not be determined as a percentage of the salved value (TCC art.1305, para.3).

The debtors of the salvage reward are the owners of the vessel and other property at the time of salvage activity (TCC art.1306, para.1). The salvage reward shall be paid by all of the vessel and other property interests in proportion to their salved values (TCC art.1306, para.2). Liability of the debtors of the salvage reward shall not be jointly and severally allocated (TCC art.1306, para.2). The consignee, on the other hand, shall be personally liable for the payment of the salvage reward in proportion to the salved value that will be calculated upon the liquidation of that cargo in case of non-delivery, provided that he has knowledge of the payment of salvage reward during delivery of the cargo (TCC art.1307, para.1). When other properties are saved with the delivered cargo, the liability of the consignee for the payment of the salvage reward shall not exceed the amount which will be charged up to the delivered cargo as a proportion of all costs among all property (TCC art.1307, para.2).

A single salvage reward shall be determined for all activities performed during the period, beginning with the danger that leads to the salvage activity until the demand for re-delivery of the salved property (TCC art.1308, para.1). All the salvors who demand a proportion of the salvage reward shall take their proportions from that single salvage reward (TCC art.1308, para.1). The apportionment of the salvage reward between salvors shall be made in accordance with the criteria stated under TCC art. 1305 with regard to their attendance to the salvage activity (TCC art.1309).

---

[30] See Salvage Convention 1989 art. 12, para.1.

The apportionment of the salvage reward between the owner, master and other persons in the service of each salving vessel shall be determined by the law of the flag of that vessel (Salvage Convention 1989 art.15, para.2). The Turkish lawmaker has preferred the execution of the apportionment between master and other seamen in accordance with the criteria stated under TCC art.1305, by reserving the amount of damage to the salvor vessel and the costs incurred after the salvage activity (TCC art.1310, para.1). However, when the salvage operation is performed by a ship which is assigned for salvage purposes or by a trailer, seamen who are in the service of these ships cannot demand a proportion of the salvage reward (TCC art.1310, para.5).

Upon the finalization of the salvage activity, the owner shall prepare a 'share table' that indicates the shares of the master and other seamen, and notifies it to these persons (TCC art.1310, para.2). The master and other seamen are entitled to object to this share table in front of the authorized court located in the place of arrival in Turkey within 15 days beginning from the arrival at that port (TCC art.1310, para.3). Upon listening to the interested persons, the court shall approve the share table directly or through amendment (TCC art.1310, para.4). The decision of the court is final (TCC art.1310, para.4). When the salvage activity is performed by a vessel other than a ship, the salvage reward shall be allocated between the salvor and his servants in accordance with the contract between them, and in the absence of this contract, in accordance with the criteria set under TCC art.1305 (TCC art.1310, para.6). Finally, a salvor may be deprived of the whole or part of the salvage reward due to the extent that the salvage operations became necessary or more difficult due to the fault or neglect on his part or if the salvor has been guilty of fraud or performed other dishonest conduct (TCC art.1311).[31]

The right of the salvor to interest on any payment due under the Salvage Convention 1989 shall be determined in accordance with the law of the state in which the tribunal dealing with the case is situated (Salvage Convention 1989 art.24). The Turkish lawmaker has determined to apply interest to all the claims of the salvor beginning from the date of delivery of the salved property, and when the property cannot be delivered, from the date of finalization of the salvage activity in terms of special compensation (TCC art.1313). Upon the request of the salvor, a person liable for salvage reward shall provide satisfactory security for the claim, including interest and judicial costs incurred by the salvor (TCC art.1314).[32]

The salvor has the right of retention on the salved property in accordance with TCivC art.950 to art.953 and the right of maritime lien on the salved vessel (TCC art.1315, para.1). The salvor may not enforce the right of maritime lien when satisfactory security for his claim, has been duly tendered or provided. Hence, the owner of the salved vessel shall use his best endeavors to ensure that the owners of the cargo provide satisfactory security for the claims against them including interest and costs (TCC art.1315, para.2).[33] The salved vessel and other property shall not, without the consent of the salvor, be removed from the port or place at which they first arrive after the completion of the salvage operations until satisfactory security has been put up for the salvor's claim against the relevant vessel or property (TCC art.1315, para.3).[34] The tribunal or court having jurisdiction over the claim of the salvor may, by interim decision, order that the salvor shall be paid on account such amount as seems fair and just, and on such terms, including terms as to security where appropriate, as may be fair and just according to the circumstances of the case (TCC

[31] See Salvage Convention 1989 art.18.
[32] See Salvage Convention 1989 art.21, para.1.
[33] See Salvage Convention 1989 art.21, para.2.
[34] See Salvage Convention 1989 art.21, para.3.

art.1316).[35] In the event of an interim payment, the security provided shall be reduced accordingly (TCC art.1316).[36]

The rule is that salvage operations which have had an advantageous consequence shall reveal the right of salvage reward.[37] However there is an exception to this rule. Hence, as noted above, the Turkish lawmaker explicitly highlighted that there are exceptions to the rule which states that no payment shall be due if the salvage operations have had no useful result (TCC art.1304, para.2).

The Turkish lawmaker has stated that when the salvor has carried out salvage operations in respect of a ship which by itself or its cargo threatened damage to the environment and has failed to earn a payment under the related provisions at least equivalent to the special compensation assessable in accordance with TCC art.1305, he shall be given special compensation from the owner of that vessel equivalent to his expenses as defined hereunder (TCC art.1312, para.1).[38] The court or tribunal is under no duty to fix a salvage reward up to the maximum salved value of the vessel and other property before assessing the special compensation to be paid (TCC art.1312, para.1).[39]

If, in the circumstances set out above for special compensation, the salvor by his salvage operations has prevented or minimized damage to the environment, the special compensation payable by the owner to the salvor may be increased up to a maximum of 30 per cent of the expenses incurred by the salvor (TCC art.1312, para.2).[40] However, the court or the tribunal, if it deems it fair and just to do so and bearing in mind the relevant criteria to determine salvage reward, may increase such special compensation further, but in no event shall the total increase be more than 100 per cent of the expenses incurred by the salvor (TCC art.1312, para.2).[41]

The salvor's expenses for the purpose of the above noted paragraphs means the out-of-pocket expenses reasonably incurred by the salvor in the salvage operation and a fair rate for equipment and personnel actually and reasonably used in the salvage operation, taking into consideration the criteria set out in TCC art. 1305, para. 1 (h), (i) and (j) (TCC art.1312, para.3). The total special compensation under this article shall be paid only if and to the extent that such compensation is greater than any reward recoverable by the salvor under TCC art. 1305 (TCC art.1312, para.4). If the salvor has been negligent and has thereby failed to prevent or minimize damage to the environment, he may be deprived of the whole or part of any special compensation due under this article (TCC art.1312, para.5). Nothing in this article shall affect any right of recourse on the part of the owner of the ship (TCC art.1312, para.6). No payment made under this article is subject to general average adjustment (TCC art.1312, para.7).

Provided that the amount of his reward shall not be prejudiced should it be found that such a request was unreasonable, the salvor shall owe a duty to the owner of the vessel or

---

35 See Salvage Convention 1989 art.22, para.1.

36 See Salvage Convention 1989 art.22, para.2.

37 An award of salvage is generally appropriate when property is successfully and voluntarily rescued from marine peril. See *Sunglory Maritime, Ltd. v. PHI, Inc.*, 212 F.Supp.3d 618 (2016). In other words, successful salvage claim requires three proofs: (1) marine peril; (2) voluntary service rendered when not required as existing duty or from special contract; and (3) success in whole or in part, or contribution to success of operation. See *U.S. v. EX-USS CABOT/DEDALO*, 297 F.3d 378 (2002). Finally, to establish a salvage claim, it is not necessary for the claimant to actually complete the work of saving the property; it is sufficient if he endeavors to do so, and his efforts have a causal relation to the property's eventual preservation. See *In re Iowa Fleeting Service, Inc.*, 211 F.Supp.2d 794 (2002).

38 See Salvage Convention 1989 art. 14, para.1.

39 See Salvage Convention 1989 Attachment 1.

40 See Salvage Convention 1989 art.14, para.2.

41 See Salvage Convention 1989 art.14, para.2.

other property in danger: (a) to carry out the salvage operations with due care, (b) to exercise due care to prevent or minimize damage to the environment, while performing the duty specified in (a), (c) to seek assistance from other salvors, whenever circumstances reasonably require, and (d) to accept the intervention of other salvors, when reasonably requested to do so by the owner or master of the vessel or other property in danger (TCC art.1303, para.1).[42]

The owner and master of the vessel or the owner of other property in danger shall owe a duty to the salvor: (a) to co-operate fully with him during the course of the salvage operations, (b) in so doing, to exercise due care to prevent or minimize damage to the environment, and (c) when the vessel or other property has been brought to a place of safety, to accept delivery when reasonably requested by the salvor to do so (TCC art.1303, para.2).[43] In the application of the above noted provisions 'damage to the environment' refers to substantial physical damage to human health or to marine life or resources in coastal or inland waters or areas adjacent thereto, caused by pollution, contamination, fire, explosion or similar major incidents (TCC art.1303, para.3).[44]

Every master is bound, so far as he can do so without serious danger to his vessel and persons thereon, to render assistance to any person in danger of being lost at sea (TCC art.1317, para.1).[45] The owner of the vessel shall incur no liability for a breach of that duty by the master (TCC art.1317, para.2).[46]

According to Salvage Convention 1989, no remuneration is due from persons whose lives are saved (Salvage Convention 1989 art.16). The convention has also recognized the discretion of the contracting states by stating that provisions of domestic law governing salvage reward upon salvage of persons shall not affect the provisions of domestic law on this subject. The Turkish lawmaker has preferred to act in accordance with the related provision of the convention. Thus, no remuneration is due from persons whose lives are saved in terms of Turkish law (TCC art.1318, para.1). On the other hand, a salvor of human life, who has taken part in the services rendered on the occasion of the accident giving rise to salvage, is entitled to a fair share of the payment awarded to the salvor for salving the vessel or other property or preventing or minimizing damage to the environment (TCC art.1318, para.2).[47]

Any case relating to salvage reward payment shall be time-barred if judicial or arbitral proceedings have not been instituted within a period of 2 years (TCC art.1319, para.1).[48] The limitation period commences on the day on which the salvage operations are terminated (TCC art.1319, para.2).[49] The person against whom a claim is made may at any time during the running of the limitation period extend that period by a declaration to the claimant for several times (TCC art.1319, para.3).

Table 3 is figured according to the above placed explanations and shows the similar, supplementary and conflicting provisions upon the comparison of provisions governing salvage in Turkish legislation and international legislation.

---

[42] See Salvage Convention 1989 art.8, para.1.
[43] See Salvage Convention 1989 art.8, para.2.
[44] See Salvage Convention 1989 art.1, para.1 (d).
[45] See Salvage Convention 1989 art.10, para.1.
[46] See Salvage Convention 1989 art.10, para.3.
[47] See Salvage Convention 1989 art.16, para.2.
[48] See Salvage Convention 1989 art.23, para.1.
[49] See Salvage Convention 1989 art.23, para.1.

Table 3: Comparison of Provisions Governing Salvage (TCC vs. Salvage Convention 1989)

Comparison of Provisions Governing Salvage (TCC vs. Salvage Convention 1989)	Similar Provisions	Supplementary Provisions	Conflicting Provisions
Definition of "salvage operation" and "vessel"	TCC art.1298		
Application to "warships"		TCC art.935, para.2 (b)	
Master's authority to conclude contracts of salvage	TCC art.1300, para.1		
Annulment or modification of salvage contract	TCC art.1301	TCC art.1302, para.2	
Salvage reward	TCC art.1304, para.1		
Consideration of the criteria in determining the amount of salvage reward			TCC art.1305, para.1
Debtors of the salvage reward		TCC art.1306, para.1	
Execution of the apportionment between master and other seamen		TCC art.1310, para.1	
Interest on any payment		TCC art.1313	
Providing security to collect salvage reward	TCC art.1315		
Special compensation	TCC art.1312, para.1		
Duties of the salvor and owner of the vessel or other property in danger	TCC art.1303, para.1		
Salvage of human life		TCC art.1318, para.1	
Time bar	TCC art.1319		

Source: Author

4. Conclusions

The occurrence and effects of marine accidents carry intensive international characteristics. Accordingly, ensuring international uniformity on the legal approach to marine accidents is important. Hence when it is achieved, the courts of the different states are expected to give nearly the same decision on the same conflict. If the effect of uniformity is assumed to be a 'chain' consisting of three rings, the first ring of the chain will be 'predictability' that originates from the parties' knowledge of the consequences of their acts; the second ring of the chain will be 'reliability' that is a consequence of a predictable liability regime; and the last ring of the chain will be 'continuance' that is supported by reliable and secure relationships.

In the study the main topics of marine accidents such as; general average, collision and salvage have been analyzed in terms of their general aspects and their regulation under domestic and international approaches. Thus the achievement of ensuring uniformity in the legal approach to marine accidents for Turkish application is tried to be figured out.

As a consequence, it has been found out that Turkey is following the updated legal developments on the issues by considering the international regulations governing marine accidents and referring to their applications explicitly or adapting the related articles of them in its domestic regulations. However, upon the examination of the above placed three tables, it is clear that Turkish lawmaker has placed some provisions other than the ones placed in these international regulations, especially while governing the procedural issues and applicability of the articles. There would be no concern if these provisions only had supplied the gaps of the international conventions (such as determination of personal debtors of general average; owner's obligation to execute the adjustment; appointment of adjusters; approval of the adjustment report or objection to it; annulment or modification of the salvage contract; debtors of the salvage reward; execution of the apportionment between master and other seamen; interest on any payment or salvage of human right) or if such provisions are only governing procedural issues (such as consideration of the criteria in determining the amount of salvage reward). However especially in terms of collision, number of provisions governing the essentials of the issue and placed under Collision Convention and TCC are conflicting in terms of applicability and liability. This may arise as a result of the outdated feature of the convention. Whatever the reason is, it is obvious that these conflicts will damage the aim of ensuring international uniformity in the area. Additionally, the difference in between the provisions governing time-bar in YAK and TCC may cause the same problem even though YAK is not an international convention but has been referred to in all standard forms of contracts of affreightment.

References

1. Zhiwei, Z., Xiao-Ming, L., Nat. Hazards Earth Syst. Sci. **17**, 2041 (2017).
2. Faulkner, D., Proceedings- Institute of Marine Engineering Science and Technology Part B Journal of Marine Design and Operations **5,** 37 (2004).
3. Ugurlu, O., International Journal of Industrial Ergonomics **55**, 11 (2016).
4. Meifeng, L., Sung-Ho, S., Young-Tae, C., Maritime Policy & Management **44**, 603 (2017).
5. Kendall, W. R., Boykin, N. H., Nathan A., Transportation Journal **50,** 219 (2011).
6. Felicia, S., Constanta Maritime University's Annals **18**, 349 (2013).
7. Kendall, Boykin, Heller, 219.
8. Kender, R., Çetingil, E ., Yazıcıoğlu, E, *Deniz Ticareti Hukuku (Temel Bilgiler)* (14th Edition, Istanbul 2014, p. 281).
9. Kender, Çetingil, Yazıcıoğlu, p. 293.
10. Kender, Çetingil, Yazıcıoğlu, p. 278.
11. Gaskell, N.J.J., Tul. Mar. L.J. **16,** 7 (1991).
12. Kula Degirmenci, N. *Maritime Law in Turkey* (Istanbul 2017).

Court Decisions

Baum v. Girard Fire & Marine Ins. Co., 228 N.C. 525, 46 S.E.2d 324 (1948).
Cline v. Western Assur. Co. (1903) 101 Va. 496, 44 S.E. 700.
Complaint of G & G Shipping Co., Ltd. of Anguilla, 767 F.Supp. 398 (1991).
Garden View Restaurant Ltd. v. Portage La Prairie Mutual Insurance Co., 2014 CarswellNS 994, 43 C.C.L.I. (5th) 110, 92 C.E.L.R. (3d) 100 Nova Scotia.
In re Iowa Fleeting Service, Inc., 211 F.Supp.2d 794 (2002).
Mitsui & Co Ltd v Beteiligungsgesellschaft LPG Tankerflotte, 2017 WL 04792011.
Sunglory Maritime, Ltd. v. PHI, Inc., 212 F.Supp.3d 618 (2016)
Union Marine Ins. Co. v. Borwick (1895) 64 L. J. Q. B. N. S..
U.S. v. EX-USS CABOT/DEDALO, 297 F.3d 378 (2002).

SHS Web of Conferences **58**, 01007 (2018) https://doi.org/10.1051/shsconf/20185801007
GLOBMAR 2018

A research on the factors of the new globalization model influencing the maritime transport

Soner Esmer[1,*]

[1]Dokuz Eylul University, Maritime Faculty, Tinaztepe Buca/Izmir, Turkey

Abstract. After the global financial crisis in 2008, world economy and trade have taken place in a radical change. Especially after 2012, global economic development is far from the rapid growth rates recorded in the 90s and 2000s. This term is called "New Globalization Era" and this era emphasizes the real-time economy. The aim of this study is to identify the variables that define the new globalization model from the point of view of the container shipping and port industry and to rank these variables by their importance. According to the results, it is expected that the most influential variables of the new globalization model that will affect the maritime transport in the future will be cybersecurity threat, Industry 4.0 technologies and climate change.

1 Introduction

In the world, the manufacturing industry has slowed down in response to the global financial crisis in 2008, especially after 2012, global commodity trade has almost fallen to GDP growth rates. In 2016, for the first time since 2001, commodity trading has remained below the world GDP growth rate of 1.3%. Between 2001 and 2008 container transport increased by an average of 10.8% per year, while growth between 2011 and 2016 was only 3.9% [1]. These statistics are far from the usual statistics in the 1990s and in the first half of the 2000s. It is understood that from these data, the world classical globalization model is changing.

It is inevitable that logistics and maritime transport sector like every sector will be influenced by this change. While maritime transport being always in an important position in global logistics activities, it is not clear which variables will affect the maritime future in the new global economic model, which is gradually felt at this point. This uncertainty constitutes the basic motivation of the research.

In this study, the basic variables of the new globalization model and the effects of these variables on the maritime transportation are questioned. The literature has revealed fundamental variables, followed by interviews with experts from the maritime industry and variables that will affect the maritime industry in the future. With the Fuzzy AHP study

* Corresponding author: soneresmer@gmail.com

conducted with the same experts, the rank and importance of the variables have been determined.

2 Literature Review

Globalization has a very fundamental link with transport and logistics [2]. From a mega container ship to a small truck, the entire distribution system is integrated with production activities in global markets. Thus, globalization is supported and spread by the development of modern transport systems [3]. However, the role of transport in the global economy since the beginning of the 21st Century is in a significant change. The motivators of this change in the global transport system can be classically grouped into six main groups: politics, society and demography, environment and energy, economics, finance and technology [4]. These variables play important roles together and separately. The process referred to as the "New Globalization Model" after the 2008 global financial crisis differs with the fact that it is more knowledge-based rather than labour or capital intensive [5].

The main reasons of the new globalization model are the slowing down of economic growth in developing countries including China (1), the tendency to decrease in demand for products as a result of the increase in average age ratios in developed countries and the importance of the service sector (2), the productivity gains achieved in production technologies, (3) and increasing countries' protectionism policies in response to the classic model of globalization [1-6].

It is expected that this change will cause significant changes in global commerce and maritime transport that is used 90% in the global supply chain. The International Transport Forum (ITF), for example, emphasized that economic geography in developing countries will be affected by changes in consumption and local demand, and the new economic model will be pioneering in the coming period where the low-cost labour-based production model will no longer apply. According to the ITF, the quantities of goods carried in the world between the years 2015 and 2030 will increase by 4.2% per annum, and after 2030 it will fall to 3.3%. While the flow rate of goods among developed countries is low, the flow of goods among the developing countries themselves will be high [7].

So which elements of this new globalization model will most impact the maritime industry? There is no clear answer to this question yet. Aggelakakis et al. [8] prepared a report for the EU Commission entitled "Future of the Transport Industry" which is set out the variables that will affect the transport industry. These variables are predominantly derived from the Bernardino et al. [9] study. The variables include climate change, globalization, urbanization, ageing, increasing knowledge of the community, individuality, migrations, connectivity, urgent needs (here&now), slow movement, women's strengthening in society, awareness, consciousness, the trends of being young, seeking for experience and do it yourself. Bernardino et al. [9] called these variables "key drivers of transport demand evolution".

3 Methodology

In this study, it is aimed to determine the key future factors affecting maritime transport. A research process involving both qualitative and quantitative methods has used to collect and analyse data [10-11]. A semi-structured interview technique is an ideal method for in-depth understanding of the topic in qualitative research [12]. In this study, semi-structured interview form has applied to experts for finding basic variables, then the variables have ranked by Fuzzy Analytic Hierarchy Process (Fuzzy AHP).

3.1 Sampling

Boddy [13] says that 12 interviews are sufficient to reach the theoretical satisfaction in a qualitative research. On the other hand, when theoretically new variables are not produced, the sample size is assumed to be at a satisfactory level [14].

According to Rodrigue [4] "*Maritime transport is composed of the maritime shipping and the ports dimensions*". 12 of the 24 interviews conducted in this study were carried out by experts from maritime shipping sector and the remaining 12 were conducted by experts in the port sector. While 12 of these experts living in Turkey, the rest 4 in Dubai, another 4 in Singapore, 3 in China and 1 is living in Hamburg. Half of the interviewees work in global container line operators and the others work in global container terminal operators as a senior manager. The main reason for selection of an equal number of administrators by shipping line and port is a balanced determination of the distinction between both sides. Interviewees have selected by judgemental sampling method so that the choice of the people who could answer the questions has secured [15].

3.2 Qualitative Research Process: Semi-structured Interview

The experts were asked about the variables related to the new globalization model that is expected to affect maritime transport in the future. Twelve interviews were conducted face-to-face, while the remaining twelve were conducted by video call. Interviews lasted approximately 50 minutes and records were kept. As a result of the interviews, 9 variables were determined and the quantitative step of the research was passed.

3.3 Quantitative Research Process: Fuzzy AHP

In this study, the Fuzzy AHP method was applied to find the weights of the 9 variables obtained from the interviews. Saaty was used the expression "the method of determining priority orders with pairwise comparisons of judges obtained from experts" when defining AHP method [16]. Although the method was developed by Saaty in 1977 [17], it has been widely used in many academic disciplines [18].

There are 3 basic levels in a typical AHP structure. At the top level, the objectives of the research are shown, while the second level is the criteria. Although there are alternatives to choose from at the bottom, there is no third level in this work due to the purpose of the study.

The questionnaire prepared for this research has two sections. The first part consists of the profile questions about the participants and the second part contains the pairwise comparison of the variables obtained from the interviews before.

In the questionnaires, a 9-point comparative scale was used as suggested by Saaty [16]. Interviews were conducted in January and February 2018 and provided 24 replies to the questionnaire from the experts.

Then the Fuzzy AHP method was applied. The reason for not applying the classical AHP method at this point is that it is criticized because it is inadequate to handle the uncertainty and indecision situations [19]. The focal point of these criticisms is that the acceptance of "something right or wrong" in classical logic does not overlap much with real life, and does not reflect the way human thinking works. In real life, however, there could be many situations between right and wrong.

The article entitled "Fuzzy Sets" published by Zadeh in 1965 [20] has been adapted to the AHP method as well as many other scientific methods. The Fuzzy AHP method, which is able to make more consistent decisions because it is very close to the logic of human thinking, has a more advanced technique than Saaty's classic AHP [21-22].

Fuzzy numbers do not form a natural order like real numbers. For this reason, there are many studies in the literature for ranking the fuzzy numbers [23, 24, 25, 26, 27, 28, 29, 30]. In addition to these studies, Chang [31] developed a new approach and used triangular fuzzy numbers for pairwise comparison scales in the Fuzzy AHP. In this study, the widely used Fuzzy AHP method proposed by Chang [31] was used. The choice of this method is due to the low computational requirement and the use of classic AHP steps.

4 Research Findings

4.1 Interview results

According to the results obtained from the interviews, there are 9 basic variables that will affect maritime transport and the port in the future. These variables and the basic opinions expressed by interviewees regarding these variables are as follows:

- *Climate change:* Increasing precipitation affects the smooth operation of radar and radio equipment on ships, causing drainage systems in ports to be inadequate and storage areas to be damaged. Increased average temperatures cause damage to railways and roads, increasing the energy consumption of refrigerated containers. Storms cause the container to overturn in the port, increase the wave power and coastal erosion. On the other hand, storms cause the wave breakers and breakers in the ports to be inadequate. Rapid escalating winds adversely affect navigation and communication equipment, delay vessel port operations, damage old buildings and warehouse.
- *Change in business model:* Factors such as changes in customer needs and expectations, threats created by emerging and emerging competitors, and increased costs of risk reduction put pressure on managers to create a much more dynamic business model. This new business model has to be resilient to the growing fragility of global markets. It is thought that these business models will also affect the business models of maritime transport.
- *Political imbalances and protectionism policies:* It is estimated that in 2015, there will be an additional cost of $ 56 billion in the global supply chain due to unexpected weather changes, terrorism, forced migrations and criminal activities in the world [32]. In addition, the increasing complexity of regulations, the excessive and unpredictable political decisions of governments, the increase in safety and security rules, slowdown of operations due to safety concerns, protectionism policies that restrict competition, foreign trade customs procedures and taxes, changes in macro environmental factors, it will continue to create constraints in future commercial activities and in the transport sector.
- *Digital vulnerability and cybersecurity threat:* With the impact of mobile devices, digitalization has begun in every area of the world, and digital markets have become the most important markets. Technically it is not possible to switch to Industry 4.0 technology without digitalization. However, digitalization is bringing about security issues. Increasing dependence on digital markets requires measures against risks. Cyber-attacks for digital data and data breach violations have become a major threat. With cyber-attacks, IT systems are collapsing, which can lead to significant loss and disruption in business processes. As a matter of fact, it is stated that Maersk suffered a loss of 300 million dollars in the third quarter due to the cyber-attack in 2017 [33]. While it is difficult to keep up with this fast digital change process and technological progress, it is very much reliant on these technologies. Much more sophisticated models of operation emerge in the business world where interconnectivity is constantly increasing.

- *Industry 4.0 technologies*: Industry 4.0 is the industry's digital transformation. Intelligent robots that can respond quickly to customer requests with flexibility and productivity in production refer to 8 basic technologies such as the Internet of Things (IoT) and big data analysis. It is thought that these technologies will deeply affect not only the production but also the service sector. For example, with IoT, the transparency, efficiency and reliability of the supply chain and environmental awareness will increase with efficient resource planning. 3D printers will reduce transport demands, and transport will be predominantly directed at raw material shipments. With autonomous devices, the demand for labour force will be reduced and the performance of delivery processes will increase. With industrial robots, production efficiency will increase, production cost and dependence on labour force will decrease.
- *Return of production (Localization)*: With the reduced dependence on labour, it is expected that the return of production to the nearest point in developed countries will have a shortening effect on the global supply chain.
- *Ageing*: Especially in the developed countries, the increase in the average age causes the service sector to gain importance and increase the usage habit (Consumption 2.0) instead of buying the products, which is considered as an element that can negatively affect the global commodity trade.
- *Urbanization*: With controlled and uncontrolled migration, the population is concentrated in large cities. This is another important reason for the production facilities to be deployed around these cities. Especially urban logistics is expected to gain much more importance.
- *Disruptive innovation*: Many destructive innovations are expected in the future (Hyperloop, unmanned ships, drones, etc.) and new technologies are a threat to existing transportation vehicles and infrastructures.

4.2 The Results of Fuzzy AHP

In order to be able to determine the significance ratings of the criteria, the same decision makers have asked to answer the 9-point scale questionnaires based on pairwise comparisons. Then the data has transformed into a triangular fuzzy number by reducing the evaluations of these 24 decision makers to a single value to indicate the outcome of each decision maker's criterion (Table 1).

Table 1. Fuzzy comparison matrix.

	C1	C2	C3	C4	C5	C6	C7	C8	C9
C1	1 1 1	1.02 1.69 2.38	1.66 2.29 3.00	1.52 2.21 2.99	1.16 1.67 2.26	1.85 2.49 3.12	1.28 1.61 2.11	3.78 4.82 5.85	6.24 7.26 8.27
C2	0.42 0.59 0.98	1 1 1	1.32 2.23 3.09	2.05 2.96 3.84	1.16 1.61 2.19	2.69 3.76 4.80	2.85 3.94 4.99	5.36 6.38 7.40	5.65 6.57 7.47
C3	0.33 0.44 0.60	0.32 0.45 0.76	1 1 1	0.85 1.26 1.85	0.34 0.47 0.70	1.39 2.00 2.60	0.65 0.87 1.24	1.72 2.43 3.20	2.87 3.72 4.63
C4	0.33 0.45 0.66	0.26 0.34 0.49	0.54 0.79 1.18	1 1 1	0.70 1.01 1.47	1.91 2.69 3.45	1.75 2.31 2.89	4.03 5.05 6.06	3.60 4.61 5.56
C5	0.44 0.60 0.87	0.46 0.62 0.86	1.42 2.13 2.91	0.68 0.99 1.43	1 1 1	3.12 4.02 4.96	1.59 2.12 2.75	3.14 4.96 7.39	3.94 5.57 7.52
C6	0.32 0.40 0.54	0.21 0.27 0.37	0.38 0.50 0.72	0.29 0.37 0.52	0.20 0.25 0.32	1 1 1	0.48 0.67 0.91	0.96 1.24 1.54	1.16 1.59 2.18
C7	0.47 0.62 0.78	0.20 0.25 0.35	0.81 1.16 1.54	0.35 0.43 0.57	0.36 0.47 0.63	1.10 1.50 2.07	1 1 1	1.73 2.52 3.30	2.23 3.00 3.78
C8	0.17 0.21 0.26	0.14 0.16 0.19	0.31 0.41 0.58	0.16 0.20 0.25	0.14 0.20 0.32	0.65 0.80 1.04	0.30 0.40 0.58	1 1 1	1.27 1.69 2.20
C9	0.12 0.14 0.16	0.13 0.15 0.18	0.22 0.27 0.35	0.18 0.22 0.28	0.13 0.18 0.25	0.46 0.63 0.86	0.26 0.33 0.45	0.46 0.59 0.79	1 1 1

Finally, the weights of each criterion have calculated by the Fuzzy AHP method (Table 2).

Table 2. The overall importance of all responses.

Criteria	Weight (%)
C2- Digital vulnerability and cybersecurity threat	24,1
C1- Industry 4.0 technologies	22,3
C5- Climate change	17,5
C4- Disruptive innovation	13,4
C3- Political imbalances and protectionism policies	10,6
C7- Change in business model	8,8
C6- Return of production (Localization)	5,6
C8-Urbanisation	4,0
C9-Ageing	3,0
Number of respondents	24

Given the general conclusions of the 24 decision makers, it has been determined that the three most important variables that are expected to affect the maritime industry in the future are the cybersecurity threat, Industry 4.0 technologies and climate changes. These three variables are followed by disruptive innovations, political imbalances and protectionism.

On the other hand, the importance ranks of the two groups have calculated to see the disagreement of the participant groups is shown in Table 3.

Table 3. Respondent groups' comparisons.

Criteria	Ship Operators		Terminal Operators	
	Weight (%)	Rank	Weight (%)	Rank
C1- Industry 4.0 technologies	28,0%	1	17,4%	3
C2- Digital vulnerability and cyber security threat	23,9%	2	23,9%	1
C5- Climate change	15,2%	3	19,9%	2
C4- Disruptive innovation	12,8%	4	13,8%	4
C3- Political imbalances and protectionism policies	10,0%	5	11,1%	5
C7- Change in business model	7,4%	6	10,3%	6
C6- Return of production (Localization)	5,5%	7	5,7%	7
C8- Urbanisation	4,0%	8	3,9%	8
C9- Ageing	2,4%	9	3,8%	9
Number of respondents	12		12	

Although the order of the first three variables is different in the two groups, the order of the other variables have remained the same. While ship operators think that the maritime industry will be the most influential variable in the future for Industry 4.0 technologies, port operators think that the most important variables are "digital vulnerability" and "cybersecurity threats".

5 Conclusion

The logistics sector is directly affected by economic activity around the world. In recent years, fundamental changes in the basic statistics on world economy and trade have revealed some doubts about the future of the logistics industry. It is difficult to determine how much the economic, technological, social, demographic and legal factors that are often mentioned today will affect the logistics industry in the future. However, it is relatively possible to identify these variables and to find their significance. In this study, interviews with experts of the maritime industry revealed potential variables that will affect the maritime industry in the future. These variables have then ranked according to their significance from the point of view of both port and shipping sector representatives.

Taking into account all the responses of both ship and port operators, it is clear that the threat of cybersecurity is considered to be the most likely future element in the maritime industry. The second important variable, Industry 4.0, is in fact directly related to the first variable. Because Industry 4.0 is at the heart of digitalization and the digitalization of all business processes brings together digital vulnerability and cybersecurity threats. These two variables also at the top rank, when the responses of the ship operators are taken into consideration. When the only responses of port operators are taken into consideration, it is seen that the first two rows are accompanied by the cybersecurity threat and the climate change. Considering the possible negative effects of the climate change on the ports, this can be regarded as expected.

The most important limitation of this study is that it is made only for the container transport sector. It may be advisable to carry out studies that take into account other types of cargo and other modes of transport.

References

1. S. Saxon, S. Stone, *Container Shipping, The next 50 years*. McKinsey&Company. Travel, Transport & Logistics, (2017)

2. L.A. Tavasszy, C.J. Ruijgrok, M.J.P.M. Thissen, *Emerging global logistics networks: Implications for transport systems and policies*. Growth and Change **34(4)**, 456–472, (2003)

3. J.P. Rodrigue, *Transportation and Globalization*. Encyclopaedia of Globalization, Edited by Roland Robertson and Jan Aart Scholte. Routledge (2006)

4. J.P. Rodrigue, *Maritime Transportation: Drivers for the Shipping and Port Industries*. International Transport Forum, Paris. (2010)

5. L.D. Tyson, S. Lund, *Globalization isn't in retreat. It's just gone digital*. https://www.weforum.org/agenda/2017/02/why-globalization-isnt-it-in-retreat-its-gone-digital, (2017).

6. A.Bhattacharya, D. Khanna, C. Schweizer, A. Bijapurkar, *The New Globalization: Going Beyond The Rhetoric*. Boston Consulting, (2017)

7. OECD/ITF, *ITF Transport Outlook 2017*, OECD Publishing, Paris. http://dx.doi.org/10.1787/9789282108000-en, (2017)

8. J. Aggelakakis, M. Bernandino, P. Boile, A. Christidis, M. Condeco, A. Krail, M. Papanikolaou, J. Reichenbach. *The Future of the Transport Industry*. European Commission Joint Research Centre Institute for Prospective Technological Studies, (2015)

9. J. Bernardino, J. Vieira, H. Garcia, *Factors of evolution of demand and methodological approach to identify pathways*. FUTURE Deliverable D3.1. Project financed by the 7th Framework Programme, (2013)

10. K.M. Eisenhardt, *Building theories from case study research*. Academy of management review, **14(4)**, 532–550, (1989)

11. R.B. Johnson, A.J. Onwuegbuzie, *Mixed methods research: A research paradigm whose time has come*. Educational Researcher, **33(7)**, 14–26, (2004)

12. A. Wilson, *Marketing research: an integrated approach*, FT Prentice Hall. (2011)
13. C.R. Boddy, *Sample size for qualitative research*. Qualitative Market Research: An International Journal, **19(4)**, 426–432, (2016)
14. A. Bryman, *Social research methods*, Oxford university press, (2008)
15. N.K. Malhotra, *Marketing research: An applied orientation*, 5/e, Pearson Education India. (2008).
16. T.L. Saaty, *Decision making with the analytic hierarchy process*. International journal of services sciences, **1(1)**, 83–98, (2008)
17. T.L. Saaty, *A scaling method for priorities in hierarchical structures*. Journal of mathematical psychology, **15(3)**, 234–281, (1977)
18. O.S. Vaidya, S. Kumar, *Analytic hierarchy process: An overview of applications*. European Journal of operational research, **169(1)**, 1–29, (2006)
19. H. Deng, *Multicriteria Analysis with Fuzzy Pairwise Comparison*, International Journal of Approximate Reasoning, **21**, 215-231(1999)
20. L.A. Zadeh, *Fuzzy Sets*. Information and Control, **8**, 338- 353, (1965)
21. C.S. Yu, *A GP-AHP Method for Solving Group Decision-Making Fuzzy AHP Problems*, Computers & Operations Research, **29(14)**, 1969-2001, (2002),
22. J.B. Sheu, *A Hybrid Fuzzy-Based Approach for Identifying Global Logistics Strategies*, Transportation Research, **40(1)**, 39-61, (2004)
23. R. Jain, *Decision Making in the Presence of Fuzzy Variable*, IEEE Trans Systems Man and Cybernet SMC, **6**, pp. 698-703. (1976)
24. G. Bortolan, R. Degani, *A Review of Some Methods for Ranking Fuzzy Numbers*, Fuzzy Sets and Systems, **15**, 1-19, (1985)
25. S.M. Baas, H. Kwakernaak, *Rating and Ranking of Multiple Aspect Alternatives Using Fuzzy Sets*, Automatica, **13**, 47-58, (1977)
26. SJ. Chen, C.L. Hwang, *Fuzzy Multiple Attribute Decision Making Methods. In: Fuzzy Multiple Attribute Decision Making*. Lecture Notes in Economics and Mathematical Systems, **375**. Springer, Berlin, Heidelberg, (1992)
27. M. Abdel-Kader, D. Dugdale, *Evaluating Investments in Advanced Manufacturing Technology: A Fuzzy set Theory Approach*. British Accounting Review, **33**, 455-489 (2001)
28. R. Ezzati, T. Allahvıranloo, S. Khezerloo, M. Khezerloo, *An Approach for Ranking of Fuzzy Numbers*, Expert Systems with Applications, **39**, 690-695, (2012)
29. T.S. Liou, M.J. Wang, *Ranking Fuzzy Numbers with Integral Value. Fuzzy Sets and Systems*, **50(3)**, 247-255, (1992)
30. C.K. Kwong, H. Bai, *Determining the Importance Weights for the Customer Requirements in QFD Using a Fuzzy AHP with an Extent Analysis Approach*. IEE Transactions, **35(7)**, 619-626, (2003)
31. DY. Chang, *Applications of the extent analysis method on fuzzy AHP*. European Journal of Operational Research, **95(3)**, 649-655, (1996)

32. BSI, *the British Standards Institution report*, https://www.bsigroup.com/Documents/about-bsi/financial-performance/2016/BSI-Annual-report-and-financial-statements-2016.pdf (2016)
33. C. Wienberg, *Maersk Says June Cyberattack Will Cost It up to $300 Million*. https://www.bloomberg.com/news/articles/2017-08-16/maersk-misses-estimates-as-cyberattack-set-to-hurt-third-quarter (2017)

Towards navigational safety of ecosystem based Maritime Spatial Planning solutions

Mihhail Fetissov[1, 2], *Robert* Aps[1*], *Kaidi* Katus[3], *Anni* Konsap[4], *Liisi* Lees[1], *Nele Kristin* Meyer[5], *Anneliis* Peterson[1], *Olev* Tõnismaa[2] and *Riku* Varjopuro[6]

[1]University of Tartu, Estonian Marine Institute, Tallinn, Estonia
[2]Tallinn University of Technology, Estonian Maritime Academy, Tallinn, Estonia
[3]Estonian Maritime Administration, Tallinn, Estonia
[4]Ministry of Finance, Spatial Planning Department, Tallinn, Estonia
[5]Federal Maritime and Hydrographic Agency, Hamburg, Germany
[6]Finnish Environment Institute, Helsinki, Finland

Abstract. According to EU Directive establishing a framework for Maritime Spatial Planning (MSP) the main purpose of MSP is to promote sustainable development and to identify the utilization of maritime space for different sea uses as well as to manage spatial uses and conflicts in marine areas. The offshore wind energy production is considered to be one of the main drivers of MSP in the Baltic Sea Region. However, the impact on navigational safety from offshore renewable energy installations (OREI) should be assessed and taken into account in the MSP processes. In a course of implementation of the INTERREG BSR Baltic LINes project, this planning issue is addressed by analysis of the potential impact on navigational safety from planned OREI off the Hiiumaa Island (Estonia) in the Baltic Sea. As a result, the safe distance of turbines' boundary from shipping route is evaluated referring as an example to UK relevant regulations determining the manoeuvring space that vessels need in order to keep a safe distance from multiple structures such as OREI. Accompanying environmental safety issues related to marine space allocation for OREI developments off the Hiiumaa Island are addressed.

1 Introduction

Human activities in marine areas are increasing in numbers and intensity, with the patterns of sea use changing as a result of political, economic, and societal developments. The Integrated Maritime Policy for the EU [1] identifies Maritime Spatial Planning (MSP) as a cross-cutting policy tool enabling public authorities and stakeholders to apply a coordinated, integrated and trans-boundary approach. According to EU Directive establishing a framework for Maritime Spatial Planning (MSP) [2], the main purpose of MSP is to promote sustainable development and to identify the utilization of maritime space for different sea uses as well as to manage spatial uses and conflicts in marine areas. The offshore wind energy production is considered to be one of the main drivers of MSP in the Baltic Sea Region.

It is stated [3] that there are some issues of critical importance to be considered by shipping industry when engaging in an MSP process: 1) when considering the rerouting of shipping lanes or the placement of MSP limitations on sea space (e.g., aquaculture and offshore energy installations) the manoeuvring characteristics of vessels must be considered both for normal and abnormal conditions, 2) the constraint should be observed so that four

* Corresponding author: robert.aps@ut.ee

ships should safely be able to pass each other in a shipping lane and a distance between overtaking and meeting vessels of two ship's lengths should be normally maintained as a minimum passing distance, 3) anything that might obscure visibility or radar conspicuity (i.e. a physical object, electronic interference or even light pollution, either at sea or on the shoreline) must be taken into account when assessing the impact on shipping by other marine users under an MSP plan, 4) enforcement of isolation zones for different ocean users such as commercial shipping, fishing and leisure craft in cases of increased traffic density in increasingly constricted water space, 5) in addition to navigational safety risks, assessment of the impact rerouting may have on the environment and commercial operations.

A System Theoretic Accident Modelling and Processes (STAMP) approach was originally developed for software and space engineering applications [4]. STAMP considers accident occurrence as the result of a lack of, or inadequate enforcement of, constraints imposed on the system design and operations at various system levels. The STAMP methodology has been used to map the hierarchical regulatory levels - from global to local - of the maritime transportation and environment safety management system [5]. The STAMP-Mar approach is recommended as a basis for safety management of a sustainable eco-socio-technical maritime transportation system [6]. The STAMP based Systems Theoretic Process Analysis (STPA) [4, 7] as a powerful new hazard analysis method has been successfully applied e.g. to space engineering applications [8] as well as to analysis of maritime traffic safety in the Gulf of Finland [9].

In a course of implementation of the INTERREG BSR Baltic LINes project the impact on navigational safety from offshore renewable energy installations (OREI) is identified as one of the critical transnational and cross sectoral planning issues. This planning issue is addressed by analysis of the potential impact on navigational safety from planned OREI off the Hiiumaa Island in the Baltic Sea.

The UK OREIs related safety of navigation guidance requirements [10] based STPA hazard analysis is applied to identify the system level hazards related to planned OREI areas, corresponding safety constraints and the potentially unsafe control actions that may lead to hazardous allocation of OREIs areas.

2 Assessing safe distances between OREI boundaries and shipping routes

Marine navigation safety issues within and close to OREI exist in many situations, and particularly where there is a combination of high traffic levels, different vessel operations and constrained water spaces. Referring to [10] the following factors should be evaluated and taken into account: 1) concentrations of fishing vessels, or leisure traffic, will create requirements for manoeuvre and course alteration by other through traffic and also restrict sea room in the shipping lane consequently increasing the risk of further vessel to vessel conflict, 2) displacing a group of traffic into space utilised by other users where available sea room is already confined and especially where leisure traffic is forced to use the same sea space as much larger and faster commercial vessels, 3) the existence of submarine cables or other seabed obstructions may affect the ability of a vessel to anchor safely away from other traffic and this may be another consideration when assessing sea room requirements, and 4) dependant on the proximity to wind turbine towers, and the location of radar scanners aboard the vessel, some vessels may experience degradation of the radar display by false echoes reducing the ability of the bridge team to identify other vessels, including crossing vessels at the extremities of the lanes, which may require avoiding action.

2.1 The planning issue

The planned OREI sea areas off the Hiiumaa Island (Estonia) in the Baltic Sea are crossed with intensive maritime traffic according to the Automatic Identification System (AIS) based maritime traffic visualization (Figure 1). This spatial overlap is creating the critical transnational and cross sectoral planning issue that needs to be resolved before the actual implementation of OREI in this sea area.

Fig. 1. Planned OREI sea areas off the Hiiumaa Island (Estonia) in the Baltic Sea are crossed by intensive maritime traffic (source - HELCOM Secretariat, based on HELCOM AIS data).

As an example, the possible practical solution to the planning issue presented in Figure 1 can be based on the UK OREIs related safety of navigation guidance [10] requirements. In Hiiumaa case the OREIs safety buffer areas design is not meeting these guidance requirements and is identified as the system-level hazard. Further, the guidance related OREIs buffer areas design safety criteria are considered as the system safety constraints to be imposed on the OREIs buffer areas design.

According to UK OREIs related safety of navigation guidance [10] requirements if the distance of turbine boundary from shipping route 1) is less than <0.5nm (<926m) – intolerable (red on Figure 1), 2) is between 0.5nm – 3.5nm (926m – 6482m) - tolerable (yellow on Figure 1) if the risk is being reduced to as low as reasonably practicable (ALARP) - additional risk assessment and proposed mitigation measures required, and 3) is more than >3.5nm (>6482m) – broadly acceptable.

2.2 STPA hazard analysis

The OREIs safety buffer areas design related system-level hazards and safety constraints to be enforced are presented in Table 1.

Table 1. The OREIs safety buffer areas design related system-level hazards and safety constraints.

System level hazards related to OREIs safety buffer areas design	OREIs safety buffer areas design safety constraints according to UK OREIs related safety of navigation guidance requirements
OREIs safety buffer areas design is not meeting the UK OREIs related safety of navigation guidance requirements	The distance of turbine boundary from shipping route 1) is less than <0.5nm (<926m) - intolerable, 2) is between 0.5nm – 3.5nm (926m – 6482m) - tolerable if the risk is being reduced to as low as reasonably practicable (ALARP) - additional risk assessment and proposed mitigation measures required, and 3) is more than >3.5nm (>6482m) – broadly acceptable.

As the first step of STPA, the potentially unsafe design control actions that may lead to hazardous design of OREIs safety buffer areas are identified and presented in Table 2.

Table 2. Potentially unsafe OREIs safety buffer areas design control actions that may lead to the OREIs safety buffer areas hazardous design.

Control action required	Action required but not provided	Action provided unsafe	Action provided too late
Impose OREIs safety buffer areas design constraints according to UK OREIs related safety of navigation guidance requirements	Hazardous design – OREIs safety buffer areas design constraints according to UK OREIs related safety of navigation guidance requirements are not imposed	Hazardous design – OREIs safety buffer areas design constraints according to UK OREIs related safety of navigation guidance requirements are not properly imposed	Hazardous design – OREIs safety buffer areas design constraints according to UK OREIs related safety of navigation guidance requirements are imposed too late

The second step of STPA hazard analysis is performed based on expert opinions to identify the causal factors for potentially unsafe control actions. With the aim to avoid potentially unsafe OREIs safety buffer areas design flaws the implementation of safety-guided design is suggested. It is suggested also to use STPA in a proactive way by integrating the design and hazard analysis into the safety-guided design processes from the very beginning.

In a case that the Hiiumaa Island OREIs safety buffer areas will be designed according to UK OREIs related safety of navigation guidance requirements the special attention should be paid to the critical consequences for the maritime traffic management - the unavoidable displacement of a group of traffic into space utilised by other users where available sea room is already confined and especially where leisure traffic is forced to use the same sea space as much larger and faster commercial vessels. Further, concentrations of fishing vessels and/or leisure traffic, will create requirements for manoeuvre and course alteration by other through traffic and also restrict sea room in the shipping lane consequently increasing the risk of further vessel to vessel conflict.

2.3 Environmental safety

The environmental background of the planned OREIs sea areas off the Hiiumaa Island in the Baltic Sea is characterized by high and sensitive benthic biodiversity (Figure 2).

Biodiversity enhances the efficiency by which ecological communities capture resources, produce biomass, maintain water properties [11] and recycle nutrients [12]. In addition, it increases the stability of ecosystem processes through time [13, 14] and is vitally important for maintaining ecosystem integrity and sustainability [15, 16].

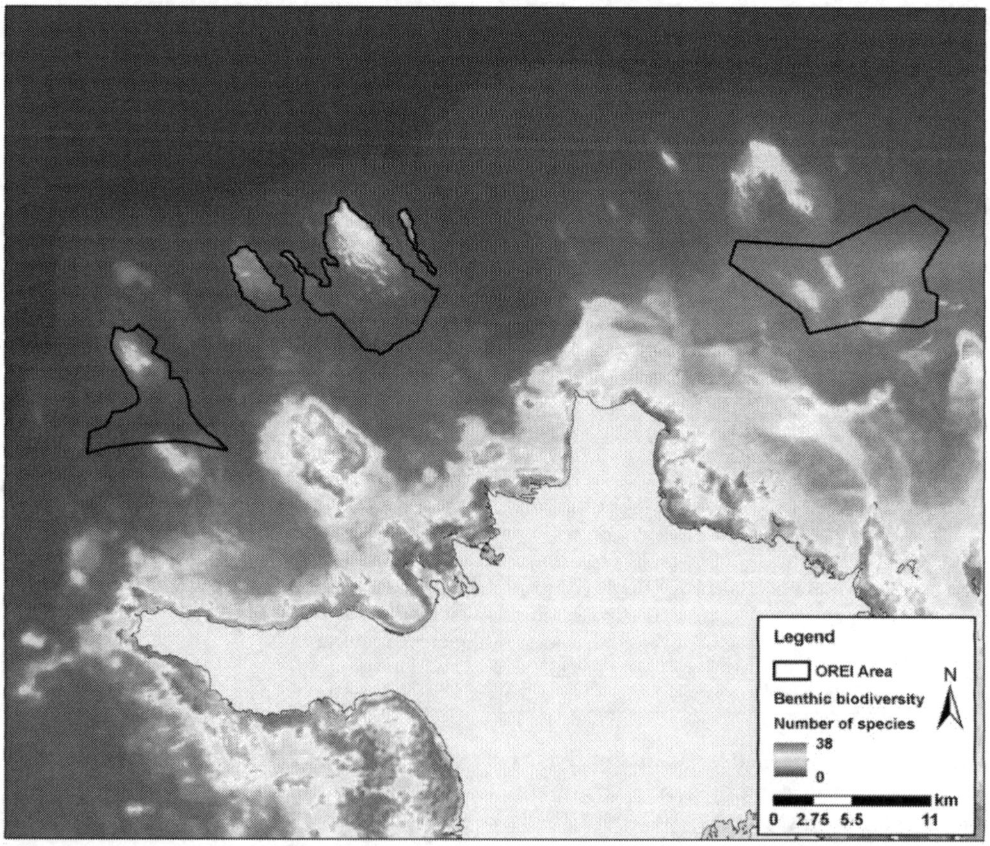

Fig. 2. Benthic biodiversity spatial distribution related to planned OREIs sea area off the Hiiumaa Island (Estonia) in the Baltic Sea.

As an example, referring to diverse and sensitive environmental background of the Hiiumaa Island planned OREIs sea area the main environmental safety requirement is that the design should satisfy, the UK OREIs related safety of navigation guidance requirements. However, spatial attention should be paid to environmental risk management caused by unavoidable displacement of a group of traffic into space utilised by other users where available sea room is already confined. It is also stated [10] that where larger developments have to provide corridors between sites to allow safe passage of shipping a detailed assessment will be required to establish the minimum width of the corridor.

According to IMO [17] in planning to establish multiple structures at sea, including but not limited to wind turbines "Governments should take into account, as far as practicable, the impact these could have on the safety of navigation, including any radar interference. Traffic density and prognoses, the presence or establishment of routeing measures in the area, and the manoeuvrability of ships and their obligations under the 1972 Collision Regulations should be considered when planning to establish multiple structures at sea. Sufficient manoeuvring space extending beyond the side borders of traffic separation schemes should be provided to allow evasive manoeuvres and contingency planning by ships making use of routeing measures in the vicinity of multiple structure areas".

The MSP is becoming an increasingly important tool for adaptive safety management of maritime transportation being ecosystem-based, integrated, place-based or area-based, adaptive, strategic, anticipatory and participatory iterative process [3]. Thereby, the objective is to integrate the issues that are most vital to maritime transportation - safety, continued operation, business success and efficiency of sustainable maritime transportation system into the MSP options.

3 Conclusions

To avoid potentially unsafe OREIs safety buffer areas design flaws the implementation of safety-guided design is suggested with the aim to embed the cost-effective safety effort into the OREIs safety buffer areas design processes from the very beginning and to design safety into the system as the design decisions are made.

It is suggested to use STPA in a proactive way guiding OREIs safety buffer areas design by integrating the design and hazard analysis into the safety guided design processes. As a result, it is expected that the OREIs safety buffer areas design safety constraints will be satisfied in a course of ecosystem based development and implementation of the integrated transboundary MSP solutions.

Spatial attention should be paid to efficiency of environmental risk management related to unavoidable displacement of a group of traffic into space utilised by other users where available sea room is already confined and risk of further vessel to vessel conflict is increased.

Acknowledgements

Authors would like to offer special thanks to Manuel Frias Vega and Florent Nicolas from HELCOM Secretariat for providing the HELCOM AIS data used by this study.

This study is supported by European Regional Development Fund, INTERREG Baltic Sea Region project Baltic LINes "Coherent Linear Infrastructures in Baltic Maritime Spatial Plans".

References

1. EC, Directive 2008/56/EC of the European Parliament and of the Council of 17 June 2008 establishing a framework for community action in the field of marine environmental policy - Marine Strategy Framework Directive (2008)
2. EC, Establishing a Framework for Maritime Spatial Planning, Directive 2014/89/EU, European Council, Brussels (2014)
3. D. Patraiko and P. Holthus, The shipping industry and marine spatial planning - a professional approach. The Nautical Institute and the World Ocean Council (2013)
4. N. Leveson, Engineering a safer world - systems theory applied to safety, Cambridge, MA, MIT Press (2011)
5. R. Aps, M. Fetissov, F. Goerlandt, J. Helferich, M. Kopti and P. Kujala, Towards STAMP based dynamic safety management of eco-socio-technical maritime transport system, Procedia Engineering, **128** (2015)
6. R. Aps, M. Fetissov, F. Goerlandt, M. Kopti and P. Kujala, STAMP-Mar based safety management of maritime navigation in the Gulf of Finland (Baltic Sea), European Navigation Conference 2016, Helsinki, 30 May-2 June 2016, IEEE Xplore (2016
7. J.P. Thomas, Extending and Automating a Systems-Theoretic Hazard Analysis for Requirements Generation and Analysis, Sandia Report SAND2012-4080, Sandia National Laboratories (2012)
8. T. Ishimatsu, N. Leveson, J. Thomas, C. Fleming, M. Katahira, Y. Miyamoto, R. Ujiie, H. Nakao, N. Hoshino, Hazard Analysis of Complex Spacecraft Using Systems-Theoretic Process Analysis, Journal of Spacecraft and Rockets, **51 (2)** (2014)
9. R. Aps, M. Fetissov, F. Goerlandt, P. Kujala, A. Piel, Systems-Theoretic Process Analysis of maritime traffic safety management in the Gulf of Finland (Baltic Sea), Procedia Engineering, **179**, Elsevier (2017)
10. MCA, Offshore Renewable Energy Installations (OREIs) - Guidance on UK Navigational Practice, Safety and Emergency Response Issues, MGN 371 (M+F) (2016)
11. B. Worm, EB. Barbier, N. Beaumont, JE. Duffy, C. Folke and BS. Halpern, Impacts of Biodiversity Loss on Ocean Ecosystem Services. Science **314** (2006)
12. BJ. Cardinale, JE. Duffy, A. Gonzalez, DU. Hooper, C. Perrings, P. Venail et al Biodiversity loss and its impact on humanity. Nature **486** (2012)
13. D. Tilman, PB. Reich and JMH. Knops, Biodiversity and ecosystem stability in a decade-long grassland experiment. Nature **441**(2006)
14. V. Campbell, G. Murphy and TN. Romanuk, Experimental design and the outcome and interpretation of diversity–stability relations. Oikos **120** (2011)
15. DU. Hooper, FS. Chapin, JJ. Ewel, A. Hector, P. Inchausti et al, Effects of biodiversity on ecosystem functioning: a consensus of current knowledge. Ecol. Monograph **75** (2005)
16. F. Boero and E. Bonsdorff, A conceptual framework from marine biodiversity and ecosystem functioning. Mar. Ecol. **28** (2007)
17. IMO, Guidance Note on the preparation of proposals on ships' routing systems and ship reporting systems, MSC/Circ.1060 (2003)

A method to estimate the demand for freight transport in absence of historical data. A case study of the Oder Waterway

Ludmiła Filina-Dawidowicz[1,*] , *Izabela* Kotowska[2] , *Marta* Mańkowska[3] , and *Michał* Pluciński[3]

[1]West Pomeranian University of Technology, Faculty of Maritime Technology and Transport, 41 Piastów Ave., 71065 Szczecin, Poland
[2]Maritime University of Szczecin, Faculty of Engineering and Economics of Transport, 11 Henryka Pobożnego St., 70507 Szczecin, Poland
[3]University of Szczecin, Faculty of Management and Economics of Services, 8 Cukrowa St., 71004 Szczecin, Poland

Abstract. The aim of the research described in this article is to work out a method to estimate the demand for freight transport in a situation when no historical data are available, thus rendering it impossible to apply methods based on time series data. The method presented in this article was developed and verified on the basis of an analysis of potential inland shipping operations on the Oder Waterway to/from the seaports in Szczecin and Świnoujście, assuming that the waterway has been upgraded to navigability class III. The analysis was predicated on a demand survey performed among cargo shippers. The obtained research results made it possible to specify the advantages and drawbacks of forecasting based on qualitative methods, and to identify the factors which significantly reduce the reliability of freight transport forecasts.

1 Introduction

Studies of demand for transport services are of key importance in the theory of transport economics. They are used to initiate the process of forecasting the development of a transport system, and they constitute the basis for decision-making with regard to planning any linear and point infrastructure. Reliable determination of transport needs is necessary to rationalise transport processes in compliance with the idea of sustained development [e.g. 1, 2, 3]. The answer to these challenges is promoting the modal shift from road transport to inland shipping [e.g. 4, 5, 6, 7, 8, 9].

Recently, in European countries more and more emphasis is put on inland shipping development. This situation is also observed in Poland. To that end, the major waterways in Poland are to be upgraded to a navigation class of at least IV [10]. In terms of investment measures, the highest priority has been assigned to the Oder Waterway (OW). Due the long-standing negligence resulting in a very bad technical condition of the waterway, the estimated

* Corresponding author: ludmila.filina@zut.edu.pl

cost of investment in the OW amounts to: PLN 2.9 billion until 2020, and PLN 27.8 billion from 2021 to 2030 [11].

The considerable capital intensity of the planned investment measures should be justified by anticipated freight transport volumes, the cargo structure, and the directional structure of freight transport on the OW after it's been made fully navigable. In a situation when input data are available from official sources (e.g. EUROSTAT), cargo flow modelling and transport volume estimation is usually done by means of quantitative methods. However, application of a methodology predicated on time series for the purpose of estimating the demand for inland shipping in Poland is constrained and unreasonable mainly because of the lack of historical data. Apart from the sporadic local use of inland waterways, currently inland shipping is not used in serving cargo flows in Poland.

The aim of the studies described in this article is to work out an optimal method to estimate the demand for inland shipping, taking into account the constraints related to its current condition (the lack of historical data) and the plans of upgrading the major inland waterways in the future. The method described in this article was developed and verified on the basis of an analysis of potential inland shipping operations on the Oder Waterway to/from the seaports, assuming that the waterway has been upgraded to navigability class III. The obtained research results also made it possible to determine the advantages and drawbacks of forecasting based on qualitative methods and to identify the factors which significantly reduce the reliability of freight transport forecasts. Although the article focuses specifically on the case study of the OW and Szczecin and Świnoujście sea ports, the proposed method could be used in other countries and in relation to other modes of transport in the absence of historical data. Thus, the research results presented in this article provide generalized conclusions.

2 Literature overview

According to Rolbiecki [12], the fundamental issue in estimating the demand for transport services is to identify any transformations taking place in the transport system (endogenous factors) and in its environment (exogenous factors). The intra-transport factors such as e.g. changes in the transport infrastructure availability are particularly important in estimating the demand for transport services with regard to the individual transport modes, also in the case of assumed modal shifts of the demand to alternative modes of transport.

Reliability of transport demand estimates is also affected by the choice of research methods and by the source data selection. Source data may be obtained from secondary and primary sources. Primary data provide more precise information on the demand structure, however, it may be difficult and time-consuming to obtain them. Also, as opposed to secondary data sourced from official statistics databases, primary data faster become out-of-date.

A large portion of research studies in the area of forecasting and modelling the demand for transport pertains to passenger traffic, including the one in conurbations [e.g. 13, 14, 15, 16, 17, 18, 19]. Freight transport issues are addressed to a lesser extent [e.g. 20, 21, 22, 23, 24, 25], which is connected with the lesser availability of comparable source data. The basic groups of methods applied to estimate the demand for transport include the methods predicated on quantitative and qualitative techniques [21, 26, 27, 28, 29, 30]. The quantitative methods include the linear trend extrapolation method (examining the direction and rate of changes), the comparative (ratio) analysis method, and econometric modelling [12]. The two former methods are traditional methods based on the time function and an assumption that in the future the demand for transport will be showing a trend that is similar to the one observed so far, and the factors determining this trend will not change. The econometric modelling methods make it possible to identify a correlation between one or more analysed phenomena and other factors, and enable forecasting their future behaviours.

The qualitative methods, in turn, consist in observation of facts and obtaining the knowledge which is the basis for formulation of theorems and generalised regularities [12, 31]. Within this group, we distinguish between the demoscopic and expert methods. The demoscopic methods consist in obtaining, directly from cargo shippers, data regarding the volume and structure of their transport needs, and also information on the subjective motifs of transport users' preferences, e.g. with respect to transport time and costs. When researchers apply this method, they select a group of major cargo shippers who generate the biggest transport volumes and who are based within a specified area. The expert methods, in turn, consist in a logical analysis of a problem by a team of experts having appropriate expertise, skills and experience.

Estimation of transport demand by means of the specified groups of methods has its advantages and drawbacks [e.g. 20, 32, 33, 31, 34, 22]. The main advantage of the quantitative methods is their objectivity, whereas the drawbacks include insufficient identification of changes of the qualitative nature in the demand structure, and of changes taking place in the external environment. Forecasts based on the quantitative methods are short-term by nature [27, 35]. The methods are useless if there are no historical data on transport. The weaknesses of the quantitative methods are at the same time the strengths of the qualitative methods. The qualitative methods enable identification of changes in cargo shippers' decision-making processes, and consequently more precise estimation of the demand.

Estimation of the demand for inland shipping in Poland requires an innovative methodological approach based mainly on the qualitative methods. This is due to the fact that this mode of transport has so far been used in cargo operations only to a negligible extent, and there are no input data that could be applied to make forecasts by means of the quantitative methods.

3 Methodology

The methodology applied in developing the forecast of inland shipping volumes (on the Oder Waterway) was predicated on the assumption of moving some of the cargoes from the Oder corridor (road and rail haulage) to inland shipping to/from the seaports of Szczecin and Świnoujście. Four stages were involved in working out the methodology.

The first stage pertained to evaluation of the hinterland potential and the intermodal shift possibilities. To that end, we analysed the trading volume and structure of the businesses located in the areas that gravitate to inland shipping. The area covered by the study was limited to the areas located no further than 50 km from the Oder Waterway. The study included selected districts (poviats) in five provinces (voivodeships): Lubuskie, Wielkopolskie, Dolnośląskie, Opolskie, and Śląskie. The presented analysis of the hinterland potential was performed on the basis of the input data obtained from the Analytical Centre of the Customs Administration (ACCA).

The second stage covered the demand research studies which were performed twice, i.e. in 2015 and 2017. The studies run in 2015 applied the ACCA data for 2014, while the studies of 2017 used the ACCA data for 2016. At this stage, the techniques of demoscopic and expert methods were applied. The ACCA data helped identify the major exporters and importers who made use of maritime transport and were based within the area of the districts covered by the study. The study involved businesses which generated annual import/export volumes of more than 10 thou tonnes (identification as per the data for 2014 and 2016). Both industrial plants and trading companies were found among the major exporters/ importers located within the studied area. Expert interviews were held with a selected group of entities: 48

entities in 2015, and 42 entities in 2017. The interviews were conducted in the course of personal meetings, video conferences and conversations on the telephone. In those interviews, the interviewees representing the entities were usually directors/ managers of the units responsible for delivery/ distribution logistics. In the course of the interviews, the shippers were asked: *In case the whole inland waterway has been upgraded to navigability class III, and the navigable season is guaranteed, would that incline you to shift some of the transported cargoes to inland shipping*? When a cargo shipper showed interest in such a modal shift, subsequent questions regarded *the expected level of the price advantage of inland shipping (water-land chain) in relation to the rail and road haulage, and the scale (volume) of the possible modal shift to inland shipping (water-land chain).*

In the course of the interviews, the entities declared that they were willing to shift some cargoes (which presently were carried by other transport modes) to inland shipping or to a land-water chain.

Due to the fact that the declarations obtained in 2017 involved considerable changes in the transport directions declared by the entities' representatives in relation to 2015, the transport directions were verified, mainly in terms of their durability.

The third stage was devoted to the comparative analysis of costs of transport over selected routes to/from the seaports of Szczecin and Świnoujście, involving various transport modes: road and rail haulage and inland shipping, or – in the case any given entity did not have a direct access to a waterway – a water-land chain. The analysis made it possible to evaluate the competitive position of inland shipping in relation to other transport modes and to estimate transport forecasts.

Finally (4th stage), the volume and structure of intersectoral shifts and inter-corridor shifts in expected demand were estimated.

The research material obtained as a result of the previous stages provided us with a detailed knowledge about:

- current level of transport volume, type of cargo and route of transport of each of the shippers
- the threshold value (%) of transport costs reduction which induces the shippers to make a decision on the modal shift
- the cost of transport of different types of cargo on individual routes
- percentage of cargo shifted to inland shipping in case the OW operational parameters are improved (ensuring safety and reliability of transport)
- investment plans resulting in changes in the volume of transport and in the organisation of supply chains.

Bearing in mind the above, the potential annual demand for transport on the OW was estimated. The forecast only includes those cargo groups for which the cost simulation has shown that after the waterway modernization the following condition will be met:

$$TC_{is} < (1 - t_i) \cdot TC_{r/rt} \quad (1)$$

TC_{is} – transport costs of cargo in inland shipping after the OW modernisation [PLN]
$TC_{r/rt}$ – costs of road/rail transport [PLN]
t_i –value of transport costs reduction for shippers to make a decision on modal shift [%].

The potential volume of transport $[VOL_{is}]$was estimated based on the formula:

$$VOL_{is} = \sum_{i=1}^{m}(VOL_i \cdot p_i + FVOL_i) \quad (2)$$

VOL_i – annual transport of cargo to/from the seaports by shipper i (by road and rail transport) [t]
p_i – declared percentage of cargo volume shifted to inland shipping after the modernisation [%]
$FVOL_i$ – future potential of cargo individually declared by shipper i to inland shipping, [t]
m - number of interviewees.

4 Results

4.1 The forecast of demand for inland shipping to/from Szczecin and Świnoujście seaports – empirical studies

The selected entities were located within the Oder Waterway belt covered by the study, in the provinces (voivodeships): Śląskie, Opolskie, Dolnośląskie, Wielkopolskie, and Lubuskie, and generated international transport volumes at the level of ca. 70 million tonnes per year, while in 2016 they were ca. 10% higher than in 2014. The majority of the international cargoes were carried by rail and by road. Only a small percentage of the cargoes – ca. 10 million tonnes – was carried by sea (Table 1).

Over the years 2014-2016, the volume of maritime transport increased by mere 6%. The increase was noticeable in all cargo groups apart from chemical products, machinery and devices.

The biggest cargo group was *mineral products*, i.e. coal, coke, iron ore, aggregate, this group also included liquid bulk cargoes, i.e. fuels. The transport volume of these cargoes was at the level of 7-7.5 million tonnes, out of which 6.5 million tonnes were destined to/from shippers from the Śląskie province. This cargo group is characterised by a low unit value, which makes it particularly suitable for inland shipping.

The second biggest cargo group in terms of volume was *base metals and articles thereof*. Nearly 1 million tonnes of these were transported via the seaports, mainly as imports. The cargoes were transported mainly to/ from the voivodeships: Dolnośląskie, Śląskie, and Opolskie.

The third major cargo group was *chemical products*, including chemical fertilisers. Over the years 2014-2016, a drop of over 30% was observed in the transport volume for this group via the seaports. The provinces of Dolnośląskie and Opolskie had the biggest share in chemical products transport volumes.

As for grain and feed, ACCA statistics do not reflect the actual cargo flows, since what they show is the trade of companies registered in the studied area, not the actual production. In 2016, the maritime trade of agricultural products from the studied area amounted to less than 50 thou tonnes. The other cargoes in this group were carried mainly by road to/from European destinations, which is due to the considerable availability of this transport mode and the considerable market fragmentation.

Other cargo constituted general cargo. Some of the cargo, e.g. cellulose, wood, and project cargo, largely lend themselves to inland shipping in both technical and economic terms.

Nearly 0.3 million tonnes of cellulose (classified as *wood pulp, paper and cardboard*) were transported annually, mainly to Kostrzyn-on-Oder where paper mills are located.

Inland shipping lends itself to transporting project cargoes, as these usually involve heavy and oversized items, in particular machines and equipment. Ca. 0.2 million tonnes of such cargoes were transported per year.

The biggest increase in transhipment volumes in the years covered by the study was observed in *other cargo* group. This group comprised general cargo, i.e. textiles, foodstuffs, plastics, various finished goods, which are not suitable for inland shipping, unless it is containerised general cargo (in the case of navigability class III, container transport is constrained).

Table 1. Trade by sea in the analysed OW hinterland (tonnes, 2014 and 2016), (Source: own work based on ACCA data)

Products	Year	Śląskie	Opolskie	Dolnośląskie	Wielkopolskie	Lubuskie	Total
Mineral products	2014	6 547 195	314 835	242 663	1 862	33 013	7 139 568
	2016	6 773 674	416 465	283 058	2 454	43 139	7 518 790
Base metals and articles thereof	2014	239 069	168 599	246 400	146 329	11 108	811 505
	2016	277 415	212 515	315 888	128 455	12 993	947 266
Products of the chemical industry	2014	48 934	115 384	299 027	146	2 412	465 903
	2016	51 079	109 259	155 470	2 364	1 366	319 538
Vegetable products	2014	12 637	539	11 932	949	458	26 515
	2016	11 556	1 483	22 132	8 996	260	44 427
Machinery and mechanical appliances, electrical equipment	2014	106 300	7 733	91 373	7 439	21 850	234 695
	2016	85 591	7 788	85 758	7 001	17 923	204 061
Wood pulp, paper and paperboard and articles thereof	2014	5 025	137	5 992	9 611	236 570	257 335
	2016	8 157	250	8 457	19 879	228 582	265 325
Wood and articles of wood	2014	13 018	34 323	2 029	673	1 614	51 657
	2016	7 896	62 638	1 989	4 718	930	78 171
Other cargo	2014	280 307	52 474	163 586	91 010	81 557	668 934
	2016	306 897	71 346	187 146	155 658	145 741	866 788
Total	2014	7 252 485	694 024	1 063 002	258 019	388 582	9 656 112
	2016	7 522 265	881 745	1 059 896	329 525	450 933	10 244 364

The analysis presented in Table 1 shows that the cargo potential of a given region, which may be shifted to inland shipping, has actually not changed in the two analysed years. This would mean that simple forecasts based on time series and involving data of 2014 and 2016 should not differ significantly. Unfortunately, forecasts predicated on historical data without accounting for additional conditions will be considerably biased.

In fact, there are many economic and political factors which are decisive for the transport volume that may be shifted to inland shipping, and they are not reflected in statistics or forecasts predicated on time series. This is proved by the forecasts regarding inland shipping to/from the seaports of Szczecin and Świnoujście, developed on the basis of direct interviews (Table 2).

Table 2. Comparison of estimated demands for inland shipping to/from the Szczecin – Świnoujście port complex after the OW has been upgraded to at least class III, based on interviews of 2015 and 2017 (Source: own work)

Cargo group	2015	2017
Solid mineral products	8.00	2.85
Base metals and articles	0.40	0.10
Chemical products	0.05	0.30
Cellulose	0.10	0.10
Diesel oil	0.60	0.50
Timber	0.20	0.20
Stone	0.01	0.01
Total	9.36	4.06

The demand forecasts prepared on the basis of interviews conducted in 2015 and 2017 differ considerably. In 2017, it was estimated that the transport volume at the level of 4 million tonnes, whereas in 2015 the amount exceeded 9 million tonnes. The biggest change was observed in the case of the transport volume forecast for *solid mineral products*: it fell from 8.0 million tonnes to 2.85 million tonnes (mainly coal). This is the largest and the most important cargo group in terms of a possible shift to inland shipping. Moreover, a significant increase in forecast volumes was seen in the *chemical products* group: from 50 thou tonnes to 300 thou tonnes, whereas there was a drop in the *base metals and articles thereof* group. The differences in the transport volumes forecasts can be attributed to two significant premises: the changes in the economic and political environment.

4.2 The factors affecting the changes in the estimated volume and structure of the demand

4.2.1 The factors deriving from the changes in the economic environment

In the study conducted in 2015, the most important maritime exporter of Polish coal generated transport volumes of 5 million tonnes via Polish seaports. This entity was also very much interested in the possibility of an intermodal shift to inland shipping on the Oder Waterway, and an inter-corridor shift (e.g. from the Trójmiasto ports corridor to the Oder Waterway), even if the savings on the total transport costs were less than 10% in relation to the level before the time of the study. At that time (end of March 2015), 8.2 million tonnes of coal had been accumulated on coal mine dump sites in Poland, and a similar amount could be found on storage sites of Polish coal-fired power stations. It was the time when the total coal production in Poland was at the level of 70 million tonnes per year, and coal imports amounted to ca. 10 million tonnes per year. This situation made the power coal mining and trading companies search for foreign – also overseas – sales markets.

2017 saw a change in the situation in the Polish coal mining industry. As a result of the lack of new investments in coal mines and the process of phasing out the non-profitable mines or their parts, there was a fall in coal supply. The mines transferred or planned to be transferred to the Mine Restructuring Company accounted for extraction of ca. 10-12 million tonnes of coal per year. At the same time, there was a growing demand for Polish coal on the neighbouring markets, e.g. in the Czech Republic and in Germany. The combined impact of the above listed factors led to the practically total extinction of maritime exports of Polish coal in 2017. What is more, there are no grounds to expect maritime exports to revive (and thus give rise to barge transport of coal to seaports) in the years to come.

The extinction of overseas exports of Polish coal has also led to a significant decrease in the railway transport rates. Compared to 2015, the 2017 rates for rail haulage from the south of Poland to the seaports dropped by 1/4, in some cases even by 1/3. This means decreased cost competitiveness of inland shipping with regard to this cargo group.

A new category among cargoes declared in 2017 for shifting to inland shipping was coal imports. The shipper of this cargo obtained a consent from the Polish government for such imports in the second half of 2017. First consignments of the cargo were used to make mixes with Polish coal and finally they ended up on the markets of the Czech Republic, Slovakia, Hungary, Austria and Germany. In the context of the shrinking supply of Polish coal, subsequent consignments of imported coal will possibly be delivered to domestic recipients. It is hard to estimate the scale of this import in terms of the potential direction of deliveries. The closest coal supply market is Russia, with which Poland has been unsuccessfully negotiating a reduction in the natural gas prices. In view of this fact, majority of the entities covered by the study did not provide official declarations regarding the volume of imports possible in the years to come, which could be shifted to inland shipping.

Analogously to the study of 2015, also in 2017 none of the entities from the leading group in the metallurgical industry declared an intention to shift other mineral cargoes (transported in the Oder corridor) to inland shipping. Similarly, in the case of imported coal and iron ore cargoes as well as coke exports, nobody declared shifting them, even partially, to inland shipping. However, a new argument came up that is unfavourable to the modal shift, i.e. the restructuring that is under way in the production plants in the metallurgical industry. This results in reluctance to incur any additional costs that are necessary at the initial stages of introducing new transport and logistics solutions.

Another factor related to the market changes observed in the years 2015-2017 was connected with the continuing process of the so-called shift in the directional structure of Polish coke trading. Polish coke plants have been increasing the extent of their products sales on the European markets. The decrease in the maritime exports of coke is definitely a factor that discourages anybody from making decisions on changing the mode of transport to seaports.

4.2.2 The factors deriving from the changes in the political environment

In the interviews conducted in 2015 with representatives of power plants located within the Oder Waterway region, the respondents pointed out to the economic account as the only basis for decision-making with regard to any changes in raw material supply sources and transport solutions. Knowing the situation on the international coal markets, the outcomes of the discussions inclined us to outline a realistic scenario for taking over by inland shipping ca. 50% of coal deliveries volume destined to the power plants located in the Oder Waterway belt (i.e. ca. 4 million tonnes via seaports after 2019).

In the second half of 2015, a parliamentary election was held in Poland. The declarations made by the new Polish government strongly emphasised making use of Polish coal, and even creating a permanent bond between mines and power generation companies. Thus specified policy of the new government with regard to the mining industry had a decisive effect on the declarations made by representatives of the same power plants during the survey of 2017. The significant impact of this factor is due to the fact that the State Treasury is the majority shareholder in those entities. The political factor has obviously had an impact on increased interest of the energy sector in the possibilities of diversifying the modes of transport to be used in deliveries of Polish coal to individual power plants. In particular this regards replacement of a part of rail deliveries with rail-barge chains, without deliveries via seaports. The declarations obtained in 2017 from representatives of the power plants in question were finally adjusted to result in a system of barge deliveries in mixed directions, i.e. 1.5 million tonnes – barge deliveries of coal via the seaports of Szczecin/Świnoujście and 1.5 million tonnes of coal deliveries from Polish mines.

Another issue is the discrepancy in the declared volume of raw material needed for the functioning of Opole power plant beyond 2019. Currently, there are four power units operating in Opole, which use ca. 3 million tonnes of coal per year. After 2019, two new power units numbered 5 and 6 will be commissioned, and their demand for coal will amount to 4.1 million tonnes per year, which adds up to more than 7 million tonnes of coal per year. However, the commissioning of the new power units will be connected with limited operation of the old power units. Still, there is no unambiguous declaration regarding the extent of this limitation. The 2015 survey forecast that the total coal consumption of Opole power plant after 2019 would be ca. 6 million tonnes. The 2017 survey indicated the level of 5 million tonnes. Any doubts and discrepancies regarding the coal consumption level are connected mainly with the planned increase in the prices of CO_2 emissions allowances. For power plants, an increase in the prices will have an adverse effect on the profitability of power generation from coal, and a positive effect on the profitability of energy produced from natural gas and RES, which consequently will have an impact on further investment policies.

The political factors also influenced the changes in the volume and structure of rolled steel supplied to Poland. The main direction of maritime imports of this semi-product to the production plants based in the Oder belt and covered by the study was Russia. In the second half of 2016, the European Commission imposed an anti-dumping duty for cold rolled steel imported from Russia. Also, an analogous regulation is being considered with regard to hot rolled steel. As a result of the change in the trade policy of the European Union, importers of rolled steel were forced to search for new supply markets.

The presented economic and political factors have had a significant effect on inland shipping forecasts made for the Oder Waterway in view of its upgrading to navigability class III. Inland shipping in class III waterways is predisposed to full ship-load transport, particularly bulk cargo. In the Order Waterway corridor, the cargo that has the greatest transport potential is coal, while the major producers and consumers of coal are companies in which the State Treasury has a holding.

4.2.3 Factors resulting from changes in the group of entities involved in the study

Estimation of the transport volumes was also affected by changes that took place over the years 2015-2017 within the structure of the group of entities involved in the study. The major changes consiserated in the research included:

- Change in the research sample. The survey involved the entities which generated the biggest cargo volumes in maritime trading. The threshold was determined at the level of 10 thou tonnes of cargo transported to/from seaports. Some of the entities that met the criterion in 2015 no longer met it in 2017, and vice versa. Also, the 2017 survey featured new cargo shippers that were absent from the study of 2015.
- Changes taking place in the ownership structure of the entities involved in the study. The representative of a company from the mineral products sector, which in 2015 operated within a major industrial group, but in 2017 was a self-standing entity, had a different view regarding the plausibility of shifting some of their cargoes to inland shipping, compared to their opinion declared in 2015.
- Changes in the respondent group. Expert interviews are characterised by subjectivity of received responses. Personnel changes in the surveyed entities led to obtaining different responses regarding the scale of any possible intermodal shift in 2017, in relation to the answers received in 2015.
- Changes with regard to reasons for undertaking the research. The declarations made by cargo shippers differed depending on whether the survey was conducted for merely

orientation purposes or in connection with implementation of concrete infrastructural investment projects. In 2015, at the beginning of each interview it was stressed that the survey was not connected with financing any investment projects, and the respondents tended to think that restoring the navigability of inland waterways in Poland was rather unlikely. In 2017, interviews were started with communicating that the data were being gathered in connection with implementation of investments contributing to upgrading the Malczyce – Widuchowa section of the Oder Waterway to class III. In 2017, the respondents were aware of the governmental plans to restore the navigability of major Polish rivers. The fact of referring to the concrete infrastructural projects co-financed by the EU inclined some of the respondents to formulate their views in a more balanced (reserved) manner compared to the situation in 2015. The interviewees paid more attention to the factors that do not favour a modal shift of some cargoes to inland shipping.

5 Conclusions

The developed and applied qualitative method of forecasting the volume of inland shipping to/from seaports after upgrading the technical parameters of the waterways makes it possible to precisely specify the volume and structure of cargo flows in the future. The method takes into account many external factors, including those economic and political, which are absent from forecasts predicated on time series. The obtained results of the research conducted using the developed method may be the basis for making rational decisions on planning an extension of a transport network. Application of the method is reasonable particularly in a situation when there are no historical data on transport volumes, as in the case of developing a transport mode that so far has been practically non-existent, e.g. inland shipping in Poland.

However, the discussed method also has its drawbacks. The conducted research has proved that changes in the economic and political environment as well as changes in the structure of the research group have a significant effect on the volume and structure of the forecast transport. Moreover, the effect is visible in a relatively short time. This was proved by the significant discrepancies between the volume of intermodal shifts of demand declared by cargo shippers in 2015 and 2017, respectively. The greater the impact of those factors on the demand volume, the greater importance of the cargo group in the estimated demand structure. In the analysed case study, this was exemplified by the considerable decrease in the volume of maritime trading of coal via seaports.

The conducted research study has also proved that the purpose of the survey affects the decision-making process of the business entities with regard to shifting their cargoes to alternative modes of transport. Linking the research results to implementation of a concrete investment project leads to a more reserved approach taken by cargo shippers to declarations made with regard to intermodal shifts of the demand.

Regardless of the indicated limitations and drawbacks of the developed method to estimate the demand for freight transport, it constitutes the best solution in case there are no historical data to rely on. The listed factors that affect the estimated results are to a large extent dependent on the "human factor". Therefore, application of this method requires verification of the obtained results in terms of durability of individual cargo flows and will direct our future research. Expert verification of the research results will make it possible to eliminate the risk of being outdated and will have a positive effect on the reliability of estimated volumes and structure of transport.

This research outcome has been achieved under the research projects MUS 8/S/IZT/2017 and US 503-2500-230 675 financed from a subsidy of the Ministry of Science and Higher Education for statutory activities.

References

1. R. A. Daziano, D. Bolduc, Incorporating pro-environmental preferences towards green automobile technologies through a Bayesian hybrid choice model. Transportmetrica A: Transp. Sc., **9**, 74-106 (2013)
2. W. Y. Szeto, Guest Editorial: Special Issue on Quantitative Approaches to Environmental Sustainability in Transportation Networks. Netw. and Spat. Econ., **16**, 1-8 (2016)
3. W.Y. Szeto, Y. Jiang, D.Z.W. Wang, A. Sumalee, A sustainable road network design problem with land use transportation interaction over time. Netw. and Spat. Econ., **15**, 791-822 (2015)
4. S. Cheon, E. Deakin, Supply chain coordination for port sustainability: lessons for new institutional designs. Transp. Res. Rec.: J. of the Transp. Res. Board, **2166**, 10-19 (2010)
5. N. S. Kim, B. Van Wee, The relative importance of factors that influence the break-even distance of intermodal freight transport systems. J. of Transp. Geogr., **19,** 859-875 (2011)
6. C.-C. Chang, C.-M. Wang, Evaluating the effects of green port policy: Case study of Kaohsiung harbor in Taiwan. Transp. Res. Part D: Transport and Environment, **17**, 185-189 (2012)
7. J.-B. Sheu, T.-L. Hu, S.-R. Lin, The key factors of green port in sustainable development. Pakist. J. of Stat., **29**, pp. 755-767 (2013)
8. S. Wang, T. Notteboom, The role of port authorities in the development of LNG bunkering facilities in North European ports. WMU J. of Marit. Aff., **14**, 61-92 (2015)
9. B. Wiegmans, R. Konings, Intermodal Inland Waterway Transport: Modelling Conditions Influencing Its Cost Competitiveness. The Asian J. of Shipp. and Log., **31**, 273-294 (2015)
10. M. Hann, L. Piotrowski, K. Woś, A new concept for utilising the Oder waterway in intermodal container transport. Sc. J. of Marit. Univ. in Szczecin, **47**, 129-135 (2016)
11. Ministry of Maritime Economy and Inland Navigation (2016)
12. R. Rolbiecki, Badania popytu na przewozy ładunków transportem wodnym śródlądowym. Zeszyty Naukowe Uniwersytetu Gdańskiego. Ekonomika Transportu i Logistyka, **43,** 41-50 (2012)
13. G. De Jong, A. Daly, M. Pieters, C. Vellay, M. Bradley, F. Hofman, A model for time of day and mode choice using error components logit. Transp. Res. Part E: Log. and Transp. Rev., **39**, 245-268 (2003)
14. M. Börjesson, Joint RP–SP data in a mixed logit analysis of trip timing decisions. Transp. Res. Part E: Logistics and Transportation Review, **44**, 1025-1038 (2008)
15. J. Willigers, M. de Bok, Updating and extending the disaggregate choice models in the Dutch National Model. *Paper Presented at the 2009 European Transport Conference*, Noordwijkerhout (2009)
16. P.R. Koster, The cost of travel time variability for air and car travellers. Doctoral Theses - Economics and Business Administration (2012)
17. M. Gelhausen, P. Berster, D. Wilken, A new model of long term forecasting air passenger demand and the number of air transport movements of Germany. *2016 Atrs World Conference* (2016)
18. X. Feng, M. Saito, Y. Liu, Improve urban passenger transport management by rationally forecasting traffic congestion probability. Int. J. of Prod. Res., **54**, 3465-3474 (2016)
19. R. Żochowska, G. Karoń, R. Janecki, A. Sobota, Selected Aspects of the Methodology of Traffic Flows Surveys and Measurements on an Urban Agglomeration Scale with Regard to ITS Projects. In: Macioszek E., Sierpiński G. (eds) Recent Advances in Traffic

Engineering for Transport Networks and Systems. TSTP 2017. Lecture Notes in Networks and Systems, Springer, **21**, pp. 37-49 (2017)

20. W. Abdelwahab, M. Sargious, Modelling the demand for freight transport: a new approach. J. of Transp. Econ. and Pol., **26**, 49-70 (1992)
21. G. De Jong, H. Gunn, W. Walker, National and international freight transport models: an overview and ideas for future development. Transp. Rev., **24**, 103-124 (2004)
22. G. De Jong, M. Kouwenhoven, K. Ruijs, P van Houwe, D. Borremans, A time-period choice model for road freight transport in Flanders based on stated preference data. Transp. Res. Part E: Log. and Transp. Rev., **86**, 20-31 (2016)
23. A. Valdas, R. Ruus, R. Põldaru, J. Roots, Forecasting road freight transport alternatives for sustainable regional development in Estonia. *Economic Science for Rural Development Conference Proceedings*, **42,** 171-178 (2016)
24. M. Postan, L. Filina-Dawidowicz, Dynamic Optimization Model for Planning of Supply, Production, and Transportation of Perishable Product. In: Suchanek M. (eds) Sustainable Transport Development, Innovation and Technology. TranSopot 2016. Springer Proceedings in Business and Economics, Springer, 235-244 (2016)
25. C. W. Howe, J. L. Carroll, A. P. Hurter Jr, W.J. Leininger, S. G. Ramsey, N.L. Schwartz, E. Silberberg, R. M. Steinberg, *Inland waterway transportation: studies in public and private management and investment decisions* (Routledge 2016)
26. B. H. Archer, Forecasting demand: quantitative and intuitive techniques. Int. J. of Tour. Manag., **1**, 5-12 (1980)
27. E. I. Vlahogianni, J. C. Golias, M. G. Karlaftis, Short-term traffic forecasting: Overview of objectives and methods. Transp. Rev., **24**, 533-557 (2004)
28. J. Liu, W. Guan, A summary of traffic flow forecasting methods. J. of High. and Transp. Res. and Dev., **3**, 82-85 (2004)
29. B. Atasoy, A. Glerum, R. Hurtubia, M. Bierlaire, Demand for public transport services: Integrating qualitative and quantitative methods. *10th Swiss Transport Research Conference* (2010)
30. M. Patriksson, *The traffic assignment problem: models and methods*. Courier Dover Publications (2015)
31. D. A. Hensher & K. J. Button (Eds.), *Handbook of transport modeling* (Emerald Group Publishing Limited 2007)
32. J. D. D. Ortuzar, L. G. Willumsen, *Modelling transport*. John Wiey & Sons Ltd., New Delhi (2011)
33. M. Wegener, *Overview of land use transport models*. Handbook of transport geography and spatial systems, 127-146 (Emerald Group Publishing Limited 2004)
34. S. P. Washington, M. G. Karlaftis, F. Mannering, *Statistical and econometric methods for transportation data analysis* (CRC press 2010)
35. R. Chrobok, O. Kaumann, J. Wahle, M. Schreckenberg, Different methods of traffic forecast based on real data. Eur. J. of Oper. Res., **155**, 558-568 (2004)

SHS Web of Conferences **58**, 01010 (2018) https://doi.org/10.1051/shsconf/20185801010
GLOBMAR 2018

Stochastic model of deteriorating cargo transshipment at port's terminal under irregular arrival of ships

Ludmiła Filina-Dawidowicz [1,*], and *Mykhaylo* Postan[2]

[1]West Pomeranian University of Technology, Szczecin, Faculty of Marine Technology and Transport, al. Piastów 17, 70310 Szczecin, Poland
[2]Odessa National Maritime University, Department of Management, Marketing, and Logistics, Mechnikov Str. 34, 65029 Odessa, Ukraine

Abstract. In the article the stochastic model is analyzed for deteriorating cargo transshipment taking into account the irregularity of arrived ships movement and cargo removing from warehouse with the regular surface transport. It is assumed that unloading rate of ships is given and fixed, but hold capacities of ships are randomly varying. During storage the cargo is subjected to deterioration with the given intensity. The problem of stationary probabilistic joint distribution of number of ships in terminal and quantity of cargo in warehouse is investigated. For this purpose the theory of Markov drift processes and methods of queueing theory are used. The problem of optimal value of deterioration rate is formulated and solved.

1 Introduction

The port terminals for transshipping and warehousing the perishable cargoes are the important links of transport-logistical chains in which transport streams of different kinds of transport are interacting. The effectiveness of such interaction is influenced by the regularity of transport units movement, the terms and cost of cargo delivery, as well as its safety. For perishable cargo transportation and warehousing, it is important to provide the optimal conditions for its storage between loading/unloading of transport units (TU). Therefore, these circumstances must be taken into consideration while port terminals design and operation.

Usually, for formal description of terminal functioning the combination of methods of the queueing theory (QT) and the storage (inventory) theory is used. For example, the significance of QT for project calculations lies in the fact that it makes it possible to predict variation of length of queues in a designed terminal and time of being in a queue of TU, that waits for beginning of loading or unloading for a remote prospect. Objectively existing variation and irregularity of movement of TU create difficult conditions for exploitation of production capacities of ports, that bring to necessity of reserves creation. The inventory theory makes it possible to give correct assessment of a level of expected amount of cargo

* Corresponding author: ludmila.filina@zut.edu.pl

in a warehouse of a terminal and to determine its necessary capacity, to give scientific grounds for a value of operation load on structural elements of moorage constructions and to improve its reliability. Description of a terminal in a form of one or another queueing system is also necessary for giving the scientifically grounded assessment of expected values of economic results and expenses that relate to exploitation of a terminal in the long term.

At the same time, while examining a problem of formal description of a port terminal in terms of the mentioned above theories, many unsolved tasks remain related to finding of the key dependence of the main characteristics of technological elements on characteristics of incoming streams of TU. Characteristics of technological elements that were calculated not in the proper way can bring to additional demurrage of TU and ships, as well as to losses connected with such a demurrage. The mentioned problem is important for a theory and practice of port terminals design and operation, but the theoretical difficulties can occur during it solving which require non-trivial special researches.

2 Literature review and problem statement

During the last decade, the international trade growth of perishable goods is observed that influences the increase of production/transportation of products, such as food, medicines etc. Transportation and warehousing of such goods using different transport modes (water, railroad) require maintaining special regimes (e.g. temperature conditions) in ports terminals and warehouses, etc. [1-3]. In other words perishable products logistics faces the deterioration problem. It can deal with products spoilage, physical depletion, gradual loss of qualitative properties of materials with the passage of time and above all storage conditions changing [4, 5]. Therefore, the great attention in theory of logistics and its applications is paid to the problem of deteriorating products flow control. The known models of inventory control theory cannot be applied directly in the process of perishable products production and transportation. In many situations arising in logistics management the adaptation and generalization of classical inventory control models is needed for the case of deterioration of perishable raw materials and finished product under prolonged transporting/warehousing. Recently, the inventory control models were reviewed by Williams & Tokar [6] and Bakker, Riezebos & Teunter [7], discussing main inventory system characteristics, i.e. price discounts, backordering or lost sales, single or multiple items, average cost or discounted cash flow, payment delay etc. In the works authored by Banerjee & Agrawal [8], Li, Zhang & Han [9] the inventory models for deteriorating products were proposed. It is naturally to suppose that the volume of perishable products deterioration depends on technical characteristics of refrigerating equipment and corresponding costs needed to support the special warehousing regimes [10, 11]. In the articles [12-14] the simple models for optimal lot sizing of perishable product based on generalization of the classical Wilson and Wagner-Whitin models were studied. However, in these works the possibility of warehousing regime control was not considered. Our paper aims to further develop the optimal inventory control models for perishable product, mainly, to develop the stochastic optimization model for the case of perishable product coming through the port's terminal considering deterioration process control at warehouse and taking into account additional investments directed for this rate decreasing. This idea was firstly mentioned in our previous works [10, 11, 15, 16].

It is known that nowadays while port terminals design the methods of simulation and operations research are widely used for giving the scientific grounds for capacity planning [17, 18] that make it possible to model a terminal work. In the literature dedicated to the modelling of port terminals operation, the ready models of queueing systems were used for the long time, which considered the specific character of port operation insufficiently

[19-21]. However, classical models of queueing theory were worked out mainly for solving the tasks of telecommunication system design, and for the most transport systems they are of little use. So, for example, they absolutely do not consider the factor of coordination of oncoming traffics in a terminal. For description of such traffics only a model of the stationary Poisson stream is usually used. At the same time, considering the present regulation of a transport operation (for example, in linear shipping), the TU streams, that come to a terminal, are not the Poisson ones but have some certain level of regularity.

The first researches that were dedicated to modelling of interaction of transport streams in a port in assumption of non-limited capacity of fronts of loading and unloading, were made in 70-80s of the last century. During the following two decades the accent in the researches was put on building and analysis of stochastic models of port terminals' work taking into consideration the finite capacity of loading and unloading fronts. In monograph [17] a level of research of this problem up to 2006 was analyzed, where stochastic models of transportation and storage systems are given with capacity limitation, that is built with the help of combination of methods of the storage (inventory) theory and QT. However, in it only relatively simple models of interaction of transport streams were analytically investigated in general, where one stream is regular, i.e. characterized by the constant rate of cargoes delivery (pickup). At the same time real transport streams have higher or lower level of irregularity and that is why they make interest for research of this more theoretically difficult case.

Among the latest researches in the field of modelling of interaction of transport streams in transit points a work [22] can be singled out, in which a model based on generalization of a classical model of Wagner - Whitin in the inventory theory is suggested for optimization of a coordination between a streams of ships (with the given schedule) and railway wagons. However, presented approach is efficient only for a case of fully controlled transport streams, the irregularity of TU arrival is not considered.

In the book [17] the analytical methods (with a help of a special class of Markov processes - Markov drift processes) are presented including enough researches of models of interactions (via a warehouse mainly) of transport streams in terminals for a case, when one of the TU streams (loaded and empty) is regular, and the other one is described by a model of the Poisson stream. In a case of an interaction of two irregular transport streams because of significant analytical difficulties within the Markov models, it was possible to investigate only the simplest case of interaction of two single TU. For solving the occurring mathematical difficulties and for getting the desired join queue-length distribution of TU and amount of cargo in a warehouse, it is necessary to make one or another simplifying assumptions, for example, related to unlimited capacity of one of the fronts of TU loading/unloading, unlimited capacity of a warehouse, etc. Above results are related to the usual cargo transhipment only, that is, to non-perishable cargo. In the article [11] the stochastic model of port's terminal was examined for transhipping the deteriorating cargo but under condition that one of the streams of TU is described by the model of compound Poisson process and another one is strong regular. That is why it is interesting to analyze more general case when the stream of ships is irregular but not obligatory is Poisson, for example, is finite or is described by the model renewal process. The analysis of such modelling case for deteriorating cargo is the aim of our paper.

3 Description of general modeling scheme of port's terminal

Let us consider the port's terminal that consists of one moorage and warehouse. The perishable cargo of one kind is delivered to a terminal with a single ship and immediately comes to a warehouse with rate W. The time of round voyage of the ship is random variable with the distribution function (d.f.) $A(t)$. The carrying capacity of ship is also random variable with the d.f. $G(x)$ and time of ship's unloading has the d.f. $B(t) = G(Wt)$. It is assumed that both random variables are mutually independent.

All cargo that was delivered to a warehouse is removed from warehouse with a help of surface transport (trucks, wagons) regularly with the constant rate $U < W$. The cargo in warehouse is subjected to deterioration with the rate $a\xi(t)$, where $\xi(t)$ is the amount of storing cargo at moment of time t (see Fig. 1).

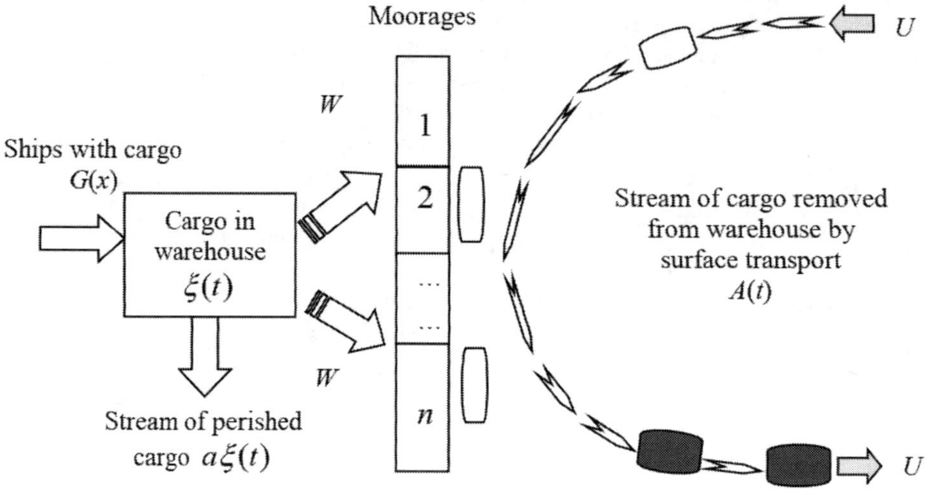

Fig. 1. Scheme of transport-warehousing system in port's terminal.

Remark. Our supposition that there are only single ship does not restrict the generality of terminal's model as a queueing system. If $A(t)$ is exponential distribution, then, as it is well known from the theory of queue M/G/1 with infinite waiting room [23], the periods of time when a server is busy and is idle form an alternative renewal process. Therefore, if we will consider this more general case we can put $B(t) = \Pi(t)$, where $\Pi(t)$ is d.f. of busy period. From queueing theory [24], it follows that the Laplace-Stieltjes transform $\pi(z)$ of d.f. $\Pi(t)$ satisfies the functional equation

$$\pi(s) = g\left(\frac{s+\lambda-\lambda\pi(s)}{W}\right), \operatorname{Re} s \geq 0,$$

where

$$g(z) = \int_0^\infty e^{-zx} dG(x), \operatorname{Re} z \geq 0.$$

Similarly it concerns the cases of one-server queueing systems of M/G/1 type with finite waiting room or with finite number of sources.

Let $\nu(t)$ be the number of ships that are in a terminal in the moment of time t, that is, $\nu(t)$ may take only two values: 0 or 1. In mathematical respect our task is to find the limit (with $t \to \infty$) distribution of a random vector $(\nu(t), \xi(t))$. From the physical point of view it means that we examine the work of terminal in a steady-state regime (or in equilibrium).

4 Finding the limit probabilistic distribution of inventory level

Due to above suppositions one can derive the following stochastic differential equation (with probability 1) describing the inventory level fluctuation in warehouse:

$$\xi'(t) = WI(v(t)=1) - UI(v(t)=0, \xi(t)>0) - a\xi(t), \tag{1}$$

where I(A) is the indicator of an event A. From (1), it follows that $\xi(t)$ is the random walk defined on the semi-interval [0, L), where $L = (W - U)/a$, i.e. 0 is sticky boundary and L is attracting boundary for this walk.

Let us denote $\{t_n\}$, n =1,2,... the sequence of the moments of time when process $v(t)$ changes its states and denote

$$F_k(x,t) = \mathbf{P}\{v(t)=k, \xi(t)\le x\}, \Phi_{kn}(x) = \mathbf{P}\{v(t_n)=k, \xi(t_n)\le x\}, 0\le x<L, k=0,1.$$

Our main goal is finding the limit distributions of process $(v(t), \xi(t))$ (assuming the existence of these limits).

$$F_k(x) = \lim_{t\to\infty} F_k(x,t),\ \ \Phi_k(x) = \lim_{n\to\infty} \Phi_{kn}(x), 0\le x<L, k=0,1.$$

For determination of the functions $\Phi_k(x), k=0,1,$ from (1) and by theorem of total probability one can derive the following system of linear integral equations:

$$\Psi_0(t) = \int_{+0}^{t} B(t-u)d\Psi_1(u) + \Psi_1(0)B(t),\ \Psi_1(t) = \int_{t}^{\infty}[1-A(h(u)-h(t))]d\Psi_0(u) + \Psi_0(t), t\ge 0, \tag{2}$$

where

$$\Psi_k(t) = \Phi_k(L(1-e^{-at})), k=0,1;\ h(t) = (1/a)\ln[1+\rho(1-e^{-at})], \rho = \frac{W}{U} - 1.$$

The equations (2) are the sequence of theorem of total probability. Furthermore, the functions $F_k(x)$ can be expressed through the functions $\Phi_k(x), k=0,1,$ by the relations

$$F_0(L(1-e^{-at})) = 2\alpha/(\alpha+\beta)\{\Psi_0(t) + \int_{t}^{\infty}[1-A(h(u)-h(t))]d\Psi_0(u)\}, \tag{3}$$

$$F_1(L(1-e^{-at})) = 2/(\alpha+\beta)\{\Psi_1(0)\int_0^t(1-B(u))du + \int_0^t\int_0^{t-u}(1-B(y))dyd\Psi_1(u)\}, t\ge 0,$$

where

$$\alpha = \int_0^\infty t dA(t) < \infty,\ \ \beta = \int_0^\infty t dB(t) < \infty.$$

Particularly, if

$$A(t) = 1 - e^{-\lambda t}, t \geq 0,$$

then from (2) we obtain the following loaded Volterra integral equation of the 2nd kind for determination of $\Psi_1(t)$:

$$[W + La(1 - e^{-at})]\Psi_1'(t) = \lambda W e^{-at} \int_0^t [1 - B(t - x)]\Psi_1'(x)dx + \lambda W \Psi_1(0)e^{-at}[1 - B(t)], t \geq 0, \quad (4)$$

For applications it is sufficient to know the mean and the variation of inventory level. For calculation of

$$\mathbf{E}\xi^j = \lim_{t\to\infty} \mathbf{E}\xi^j(t), j = 1,2,$$

immediately from (3), (4) the following formulae are valid (we omit the routing calculations)

$$\mathbf{E}\xi = L\{1 - \frac{2a}{1 + \lambda\beta}\Psi_1^*(a)[1 + \lambda \frac{1 - \beta(a)}{a}]\}, \quad (5)$$

$$\mathbf{E}\xi^2 = L^2\{1 - \frac{4a\Psi_1^*(a)}{1 + \lambda\beta}[1 + \lambda \frac{1 - \beta(a)}{a}] + \frac{4a\Psi_1^*(2a)}{1 + \lambda\beta}[1 + \lambda \frac{1 - \beta(2a)}{2a}]\},$$

where

$$\Psi_1^*(s) = \int_0^\infty e^{-st}\Psi_1(t)dt, \ \beta(s) = \int_0^\infty e^{-st}dB(t), \mathrm{Re}\, s \geq 0; \ \beta = g/W, g = \int_0^\infty x dG(x) < \infty.$$

Particularly, if

$$G(x) = 1 - e^{-x/g}, x \geq 0,$$

(the Markov model), then formula (5) can be reduced to the following one

$$\mathbf{E}\xi = L\{1 - \frac{2a\Psi_1^*(a)}{1 + \lambda\beta}[1 + \frac{\lambda\beta}{1 + a\beta}]\}, \quad (6)$$

where, as it follows from equation (4),

$$s\Psi_1^*(s) = \Psi_1(0)\{1 + \lambda\rho \int_0^\infty e^{-(s + a + 1/\beta)t}[1 + \rho(1 - e^{-at})]^{(\lambda/a - 1)}dt\}. \quad (7)$$

The constant $\Psi_1(0)$ may be found from condition

$$\lim_{s\to+0} s\Psi_1^*(s) = 1/2.$$

From this and (7) we find

$$\Psi_1(0) = \frac{1}{2}\{1 + \lambda\rho \int_0^{\infty} e^{-(a+1/\beta)t}[1+\rho(1-e^{-at})]^{(\lambda/a-1)}dt\}^{-1}. \quad (8)$$

In practice, parameter a is usually small. From the expression (7), (8) it may be shown that the following asymptotic formulae are valid:

$$a\Psi_1^*(a) \approx \Psi_1(0)\{1+\frac{\lambda\rho[2a+1/\beta-\rho(\lambda+a)]}{(2a+1/\beta-\lambda\rho)^2}\},$$

$$\Psi_1(0) \approx \frac{1}{2}\{1+\frac{\lambda\rho[1/\beta-\rho(\lambda+a)]}{(1/\beta-\lambda\rho)^2}\}^{-1}, \quad a \to 0. \quad (9)$$

The relations (9) are valid, for example, under condition

$$\lambda\beta\rho < 1$$

or

$$\lambda g < WU/(W-U).$$

Note that in equilibrium from equation (1) we obtain

$$\xi(t) = \int_{-\infty}^{t} [WI(\nu(\tau)=1) - UI(\nu(\tau)=0, \xi(\tau)>0) - a\xi(\tau)]d\tau, \quad -\infty < t < \infty. \quad (10)$$

(we moved up the initial moment to - ∞ and dropped out initial value $\xi(0)$). Applying the operator of mathematical expectation to both parts of (10) we obtain

$$\mathbf{E}\xi(t) = \int_{-\infty}^{t} [WF_1(\infty,\tau) - U(1-F_0(0,\tau)) - a\mathbf{E}\xi(\tau)]d\tau, -\infty < t < \infty. \quad (11)$$

After differentiation of both parts of equation (11) and accounting that $\mathbf{E}\xi(t)$ does not depend of time for stationary process we have

$$a\mathbf{E}\xi = WF_1(\infty) - U(1-F_0(0)). \quad (12)$$

Since

$$F_1(\infty) = \lambda\beta/(1+\lambda\beta),$$

from (12) we obtain finally

$$\lambda g/(1+\lambda\beta) = U(1-F_0(0)) + a\mathbf{E}\xi. \quad (13)$$

Relation (13) expresses so-called conservation law for transport-warehousing system under consideration. The expression in the right-hand side of equation (13) is the mean rate of cargo stream coming from ship to warehouse and in left-hand side is total intensity of cargo stream removed from the warehouse. The expression $a\mathbf{E}\xi$ is the average rate of deteriorated cargo stream removed from the warehouse.

From (13), one can find the stationary probability that there are no cargo in warehouse in arbitrary moment of time

$$F_0(0) = 1 - (\frac{\lambda g}{1+\lambda\beta} - a\mathbf{E}\xi)/U, \tag{14}$$

where $\mathbf{E}\xi$ is defined by (4)- (8).

5 Formulation of optimal inventory control problem

The results obtained above allow us to formulate and solve the problem of optimal inventory control for managers of port's terminal. In the models of optimal inventory control of perishable product it is expediently to consider the parameter a as control variable which is controlled by special investments directed into more modern refrigerating equipment [10, 15]. It is naturally to consider parameter a as a decreasing function of mentioned investments. Such function, for example, may have the following view

$$a(v) = a_0/(1+kv)^r \tag{15}$$

or

$$a(v) = a_0 \exp(-kv^r), \tag{16}$$

where a_0 is the value of parameter a when the above investments equals to 0, i.e. it corresponds to norms of natural perish under old technology of storage; k is a coefficient which characterizes the rate of cargo deterioration decreasing under positive value of investments directed into progressive storage technology; r is a positive parameter, $r \le 1$; v is value of investments that port's terminal managers direct in the beginning of planning horizon from own profit for decreasing of cargo deterioration.

Let us evaluate the average cost per time unit $\overline{C}$ related to functioning of port's warehouse in steady-state regime. If we take into account only the cost for storage, the loss from cargo perish, and investments for decreasing the deterioration of cargo, then the expression for $\overline{C}$ takes the form

$$\overline{C}(v) = (c_{st} + pa)\mathbf{E}\xi + v/T, \tag{17}$$

where c_{st} is holding cost for cargo unit and per time unit; p is price of product in good state; T is a planning horizon. In expression (16), it is assumed that parameter a depends of control variable v according to one of the dependencies (14), (15), and expression $\mathbf{E}\xi$ is defined by the formulae (6), (7).

Since under condition $v \to \infty$ the parameter $a\ (v) \to 0$ (see (15), (16)) and $\mathbf{E}\xi$ tends to a finite limit [17, 23], then $a\mathbf{E}\xi \to 0$ with the growth of v.

Taking into account the conservation law (13), (14) the objective function (17) will take the form

$$\overline{C}(v) = (c_{st} + pa)[\frac{\lambda g}{1+\lambda\beta} - U(1 - F_0(0))] + v/T. \quad (18)$$

The problem of parameter v optimal value finding that minimizes the function (18) is sufficiently complex computational procedure, since this control parameter v enters the integrals (see (6) - (8)), i.e. the objective function is not expressed through the elementary functions. Therefore, for this optimization problem solving a special computational algorithm is needed. In such a situation it is expediently to use some approximations, for example, asymptotic expansion under $a \to 0$ (see (9)).

6 Conclusion

The paper presents the development of the inventory control model for perishable cargo for the case of its servicing in port's terminal considering deterioration process control. It forms the next step in the development of inventory control theory considering irregularity of transport units movement and dependence of deterioration rate of cargo during its storage on the additional investments intended for this rate decreasing.

Presented stochastic model of port's terminal may be generalized in several directions, that will direct our future work. Two of them are noted below:

1) Refuse from the supposition of exponential distribution of inter-arrival periods of ships at terminal. In order to take into account more regular arrival of TU, it is possible to use the Erlang distribution or another kind of phase-type distribution [23, 24, 26]. But in a result, we must solve a more general system of integro-differential equation in respect of limit distribution of inventory level fluctuation $\xi(t)$. Another generalization may be based on application semi-Markov process for modeling the ships arrival.

2) Assume that the cargo is removed from warehouse irregularly, as well. The similar models (for non-perishable cargo) were studied in [17].

References

1. L. Chen, T. Notteboom, Distribution and Value Added Logistics in The Cold Chain Product Market with Application to the Role of Seaports. *5th ALRT Conference, Vancouver, June, 14–15,* (2012)
2. O. Laguerre, H.M. Hoang, D. Flick, Experimental Investigation and Modeling in the Food Cold Chain: Thermal And Quality Evolution. Trends Food Sc. & Techn., **29(2)**, 87–97 (2013)
3. Y. Zhang, X.D. Chen, An Optimization Model for the Vehicle Routing Problem in Multiproduct Frozen Food Delivery. J. of App. Res. and Techn., **12(2)**, 239–250 (2014)
4. J. Evans, Effects of Food and Beverages Storage, Distribution, Display and Customer Handling on Shelf Life. In: Kilcast, D., Subramaniam, P. (eds). *Food and Beverage Stability and Shelf Life*, Woodhead Publishing Limited, 273–300 (2011)
5. A. Kundu, P. Chakrabarti, T. Chakrabarti, An EOQ Model For Time-Dependent Deteriorating Items With Alternating Demand Rates Allowing Shortages By Considering Time Value of Money. Yug. J. of Op. Res., **23(2)**, 263–278 (2013)
6. B. D. Williams, T. Tokar, A Review of Inventory Management Research in Major Logistics Journals: Themes And Future Directions. The In. J. of Log. Manag., **19(2)**, 212-232 (2008)

7. M. Bakker, J. Riezebos, R.H. Teunter, Review of Inventory Systems with Deterioration Since 2001. Europ. J. of Oper. Res., **221(2)**, 275-284 (2012)
8. S. Banerjee, S. Agrawal, Inventory Model for Deteriorating Items with Freshness and Price Dependent Demand: Optimal Discounting and Ordering Policies. App. Math. Mod., **52**, 53-64 (2017)
9. Y. Li, S. Zhang, J. Han, Dynamic Pricing and Periodic Ordering for a Stochastic Inventory System With Deteriorating Items. Automatica, **76**, 200-213 (2017)
10. M.Ya. Postan, L. Filina-Dawidowicz, Optimal Inventory Control for Perishable Items Under Additional Cost for Deterioration Reduction. LogForum, **12(2),** 147-156 (2016)
11. M.Ya. Postan, L. Filina-Dawidowicz, Model of Stochastic Optimization for Deteriorating Cargo Inventory Control at Port's Terminal. In: "Dynamics in Logistics" – Proceedings of the 2018 International Conference LDIC2018, Berlin: Springer: 41−52 (2018)
12. B.P. Dash, T. Singh, H. Pattnayak, An Inventory Model for Deteriorating Items with Exponential Declining Demand and Time-Varying Holding Cost. Amer. J. of Oper. Res., **4**, 1−7 (2014)
13. R. Li, H. Lan, J.R. Mawhinney, A Review on Deteriorating Inventory. Eur. J. of Oper. Res., **3(1)**, 117−129 (2010)
14. N.H. Shah, Y.K. Shah, Literature Survey on Inventory Models for Deteriorating Items. Econ. Annals, **44**, 221−237 (2000)
15. M.Ya. Postan, L. Filina-Dawidowicz, Dynamiczny Model Optymalizacyjny Procesu Planowania Zapasow I Dostaw Zrożnicowanych Asortymentowo Produktow Szybko Psujacych się. In: Semenov, I.N., Wiktorowska-Jasik, A. (eds.) *Transport w Regionie Pomorza Zachodniego*, Szczecin, 13-20 (2013) (in Polish)
16. M.Ya. Postan, L. Filina-Dawidowicz, Dynamic Optimization Model for Planning of Supply, Production, and Transportation of Perishable Product, In: Suchanek M. (ed.) "Sustainable Transport Development, Innovation and Technology" − Proceedings of the 2016 TranSopot Conference, Berlin: Springer, 235−244 (2017)
17. M.Ya. Postan, *Economic-Mathematical Models Of Multimodal Transport*, Odessa, Astroprint (2006) (in Russian)
18. L. Kushnir, M. Postan, A method of determination of port terminal capacity under irregular cargo delivery and pickup. East.-Europ. J. of Enterp. Techn., **4(4/3)**, 30−37 (2016)
19. J. Wegierski, *Metody Probabilistyczne w Projektowaniu Transportu Szynowego*, Wydawnictwo Komunikacji i Łączności, Warszawa (1971)
20. J. Imakita, *Techno-Economic Analysis of the Transport System*, Saxon House, England (1978)
21. E.N. Voevudski, M.Ya. Postan, *Stochastic models in projecting of ports and in control of their operations,* Moscow, Transport (1987) (in Russian)
22. Yu. Kruk, M. Postan, Development and Analysis of Dynamic Optimization Model of Transport Flows Interaction at Port Terminal. East.-Europ. J. of Enterp. Techn., **1(3)**, 19–23 (2016)
23. B.V. Gnedenko, I.N. Kovalenko, *Introduction to Queueing Theory*. 3d Ed. Moscow, Kom Kniga (2005) (in Russian)
24. L. Kleinrock, *Queueing Systems: Problems and Solutions*, NY, J.Wiley&Sons (1996)
25. N.U. Prabhu, *Stochastic Storage Processes. Queues, Insurance Risk, Dams, and Data Communications*, 2nd Ed., Springer, Berlin (1998)
26. M.Ya. Postan, Yu.V. Kurudzhi, Modeling the Influence of Transport Units Movements Irregularity on Storage Level of Cargo at Warehouse. Acta Systemica, **12(1)**, 31−36 (2012)

SHS Web of Conferences **58**, 01011 (2018) https://doi.org/10.1051/shsconf/20185801011
GLOBMAR 2018

Impacts of vessel speed on bunker cost in short sea shipping: A cross-examination

Esa Hämäläinen[1], *Tommi* Inkinen[1,*]

[1]Brahea Centre, University of Turku, 20014 University of Turku, Finland

Abstract. Air emissions of shipping have decreased in SECA regions after 2015. Environmental impacts of regulation policies are identifiable and measurable, as a study from the port of Gothenburg recently exposed. In addition to emissions, it is important to understand how regulations and decisions, focusing on environmental sustainability, impact on the maritime business indicators such as transport costs and marginal revenues. When SECA decision was made, the oil and bunker prices were historically high. Prices were also highly volatile. Numerous studies estimated difficulties for shipping companies, and maritime transport dependent export companies. The oil price dropped dramatically in 2016 remaining at the same level till the early 2018. This empirical case study examines and simulates bunker price data in relation to different vessel speeds. The paper looks at how different speed and oil price combinations impact transport costs and export business. The results of this simulation can be summarized with the notion that the negative economic impacts of the oil price variation can be mitigated to some extent by using lower vessel speeds, thus slow steaming, on short sea shipping (SSS). However, variable relations are not straightforward and they require additional studies.

1 Introduction

The purpose of this study is to examine that how does slow steaming, together with four different bunker price options, impact on the essential key-figures of logistics. Original research data was obtained from a Big Data solution of a single paper mill (also [1]). This Big Data covers all the cost, product, logistics, invoice information, and allocations to market areas (destinations). The authors analyze how slow steaming combined with various (and increasing) bunker price variables would impact on the gross margin of the company. In the earlier studies, the authors [2, 3, 4] expected that bunker prices would even more than double, which did not happen.

Globally ship transportation (per a transported unit) has been considered to be the most environmentally friendly transport mode. Yang et al. [5] noted from a Taiwanese perspective that internal green practices and external green collaboration have positive impacts on green performance, which in turn helps to enhance company competitiveness. However, these arguments do not take into account the rising fuel costs that form a

* Corresponding author: tommi.inkinen@utu.fi

significant part of ship's operating costs, if new innovations and working solutions are not presented. Lindstad et al. [6] claim in their paper that emissions can be reduced up to 30% with a negative abatement CO2 cost per ton, if the existing fleet is replaced with larger vessels. Replacing old vessels may take as long as 25 years, so the anticipated reduction in emissions will be achieved gradually while the current fleet is being renewed. This argument is based on the assumption that the existing heavy fuel oils (HFO) are used as bunker fuels. Based on the above considerations, the main focus of this paper is to analyze the following two points:

- Conduct simulation calculations by using realistic bunker prices.
- Cross-examine, how bunker prices and lower speed (thus lower bunker consumption) have an impact on the economic key-performance indicators (KPIs) used by companies (such as gross margins and transport costs in €/Ton)?

2 About slow steaming and bunker consumption

Oil price skyrocketed to the all time high in 2008 (Fig. 1) and then dropped as dramatically. Currently, oil price has started to rise slowly, but reliable prediction of the level it will rise, and probably stay on, is difficult to estimate. In the cases, when oil price soars unexpectedly, shipping companies quickly realize and understand the need for new steaming solutions and vessel upgrades needed to be developed in order to lessen the impacts of high bunker costs on sea freights. Container shipping company Maersk have ordered the new large ships as large as 23 000 TEU and these have design speeds lower than the current ones (19 knots versus 25.5 for the Emma Maersk). The company calculated that they would operate at a fuel consumption of 50% less than the average industry, and also 20% more efficiently than the existing best. Slow steaming is practiced not only in the container market, although it may seem to make more sense due to higher container ship speeds as it is reported in every market. In December 2010, Maersk Tankers was reported to have their Very Large Crude Carriers (VLCCs) sailing at half the speed of a container ship. The design speed of 16 knots was reduced to speeds less than 10 knots on almost one third of its ballast legs, and between 11 and 13 knots on over one third of its operating days. For example, a typical round trip voyage from the Persian Gulf to Asia normally takes 42 days (at 15 knots laden and 16 knots in ballast). Maersk Tankers decreased ship speed to 8.5 knots on the ballast leg, thus increasing the round trip time to 55 days and saving nearly $400,000 off the voyage's bunker bill [7].

Globally there are different regions for different bunkers, so when a ship is entering a bunker-restricted region, they have to change their bunker fuel to cleaner and more expensive Marine Diesel Oil (MDO), or use other solutions, such as a combination of the HFO and scrubbers, which remove sulphur from bunker fuel. The effect of oil prices on shipping can be directly translated into increased bunker costs. Fuel costs represent as much as 25–60% of total ship operating costs, depending on ship and service types [8]. Among the Nordic export industries, the SECA decision is expected to increase discussions regarding new, alternative and relevant multimodal transport routes. These should be cost-efficient and capable of handling and regularly transporting large cargo volumes (see e.g. [9, 10, 11]). Accurate business forecasting in transportation is a real challenge, especially for the Finnish paper industry [3].

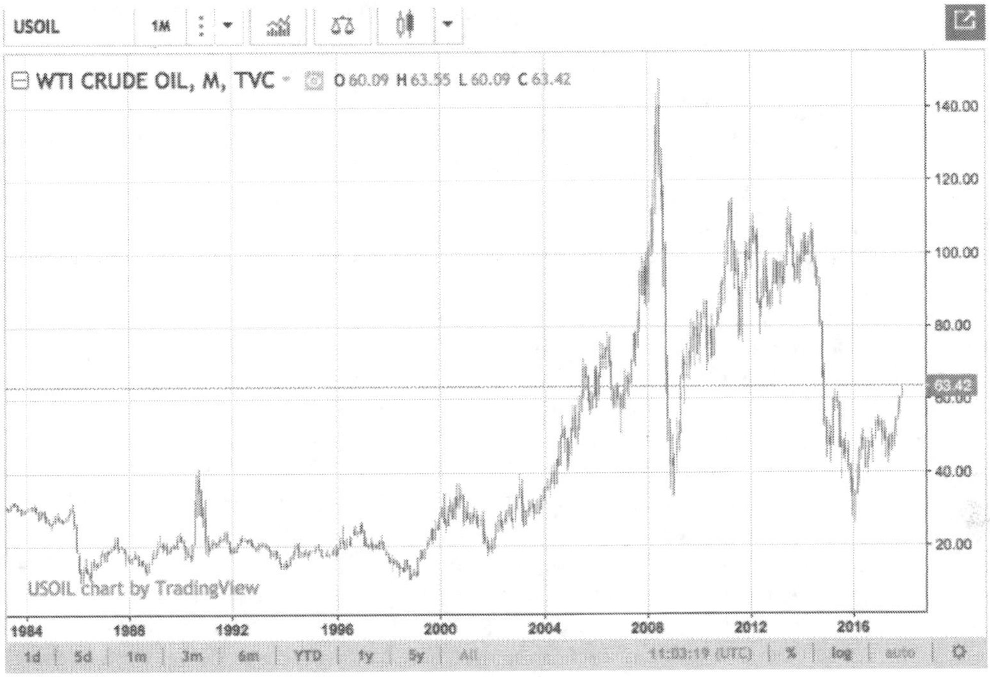

Fig. 1. Development of crude oil price since 1984.

3 Case data and the forecasting method

This study focuses on examining how slow steaming would impact bulk industry in the Baltic and North Sea region with different anticipated bunker prices. The relationship between slow steaming and the forecasted bunker costs was revealed by utilizing empirical data obtained from a paper mill located in Southern Finland and relatively close to an exporting port. The Nordic paper mills are normally integrated sites consisting of pulp processing, paper manufacturing, and a complicated mixture of converting, warehousing, transportation and cost management functions. This value chain covers the process functions from timber yard to end customers [2, 3, 4]. As background the case mill transports its final products mainly to the European markets (70%) and about 10% goes to the domestic customers.

The longitudinal data was obtained from the case mill's cost management SQL database (also [1]). This quantitative database stores the mill's production as well as economic, transportation and logistical information. Applied variables have been transformed into a single large dataset, based on packed net tons and Euro, and the parameters are calculated mainly in €/Ton, or in percentages, which gives the possibility to make relevant comparisons. The results of the calculations are shown in ratio to maintain business confidentiality, and the key-objective is to determine relevant value-added factors.

The method explaining how the research data was generated for forecasting calculations in order to simulate situation (with various bunker prices and various lower bunker consumption variables) is presented below:

1. From the acquired dataset, sea transportation costs were obtained. Transportation costs from the Nordic countries to Europe correlate significantly with oil prices,

and it was estimated that in 2017 the transportation costs will be slightly higher than at the level of 2014. Fig. 1 supports this assumption.
2. The data covers individual sea transport costs to four large European export countries.
3. The paper prices of the mill were compared with the prices received from the FOEX, which presents the latest market prices in Europe. The prices have stabilized at the level of 2009.
4. Variable and fixed costs were calculated and estimated from the values of 2008–2009, and it was forecasted how they will be in 2015 and onward. Additionally, there were discussions with the mill staff about the cost level. The mills have been able to stabilize manufacturing costs to match the level of 2009.
5. Notteboom et al. [8] have concluded that ship fuel costs, as a percentage of total daily operating costs, vary heavily from 25% up to 65% depending on ship type. In these forecasting calculations, 45% was used as a fuel cost-coefficient.
6. As to estimating total transport costs in 2017 with slow steaming, there was an added 10% extra costs driver; total sea freight costs in Euro per ton minus anticipated fuel costs of sea freights per ton = other operating costs per tons, and this figure was multiplied by 1.1 (10%) to take into an account the increased capital and warehousing costs in Euro per ton of cargo. As an example (100€-45€ = 55€ → 55€ *1.1 = 60.50€ → 45€+60.50 = 105.50). This 10% coefficient covers also other additional expenses, which will be charged from shippers including increases in salaries and vessels maintenance costs. These costs have a tendency to increase freight prices [12].

4 Findings and results

In Fig. 2, it is anticipated how the change of bunker costs (-25, 0, 25 or 50% of the existing cost) together with different bunker consumption estimations (0, -25 and -50%, these are based on slow vessel moving) have an effect on transport costs (%). Fig. 2 shows that if speed is optimal then the transport cost percent increases clearly linearly either if bunker price lowers from existing price by -25%, stays as 0 or increases by 25% or 50%. The authors want to expose the transport costs to four separate markets with previous described logistics variables. It is assumed that fuel costs are 45% of a ship's operating costs. The updated research data shows that if bunker price increases by 25% then transport costs percent per ton freight increases by 8% (see Fig. 2, Country 3). We have assumed that shipping companies add any rising fuel costs as such to freight prices, and therefore freight prices (transport cost from shipper's perspective) will certainly increase. Fig. 2 shows clearly that even bunker costs increases by 25% or even 50% transport costs go down by several percentages when ships start to utilizing lower speed as savings method when sailing.

 https://doi.org/10.1051/shsconf/20185801011

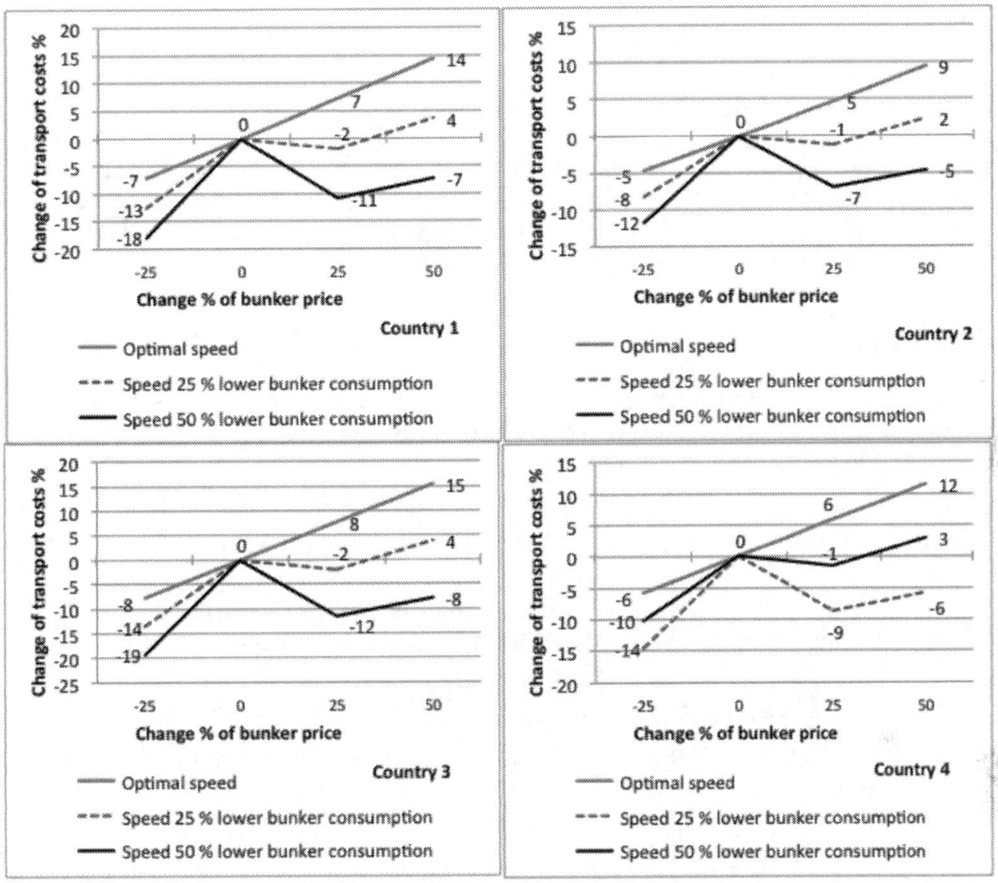

Fig. 2 Change of transport costs in % with four different fuel costs parameters: -25%, 0, 25% and 50% lower bunker consumption.

5 Discussion and conclusions

Before the SECA decision 2014 there were strong opinions stating that maritime transport will face on significant cost challenges because of the oil price development. At the same time also slow steaming was launched as a method to mitigate high oil price impacts. This study presents some anticipatory analysis with cross calculations between oil price development and slow steaming. The study exposes clearly that if bunker price increases, even up to 50% of the current level, there are means to mitigate this cost impact by sailing vessels slower, or even at the same speed but with larger cargo ships reducing the cost per unit. After 2015, when oil price decreased dramatically to their current levels, pressure to low steaming has decreased. However, oil prices in future are highly dependent on global politics and therefore sudden and drastic turns in oil/bunker prices are possible.

This study applied fresh data and presented different scenarios particularly for the Nordic export industry that is heavily reliant on SSS. The study expressed that cost savings by using slow steaming could be up to 25% depending on bunker cost development (see

[13]). The results also indicate how transport variables and their joint development impacts on gross margin, which is an important KPI in the paper industry. Transport cost has a direct impact on margins and the study argues that slow steaming in soaring price development is a viable method keeping transport costs under control.

This analysis addressed that there are variations between export markets due to different types of transport chains. The authors note that whole SSS industry, as well as logistics and supply chains, should work together if sudden changes occur in logistics business environment. Extrapolation results of this research should limit for transportation of large volumes bulky products. Usually every cargo group and sea length needs specific research on how slow steaming has an effect on freight prices. Impacts of slow steaming in SSS routes are clear and they vary according to different markets.

References

1. Hämäläinen, E. and Inkinen, T. 'How to generate economic and sustainability reports from Big Data? Qualifications of process industry', Processes, Vol. 5 No. 4: 64 (2017)
2. Hämäläinen, E. 'Cost efficiency of supply chain in a Nordic paper mill – A case study', International Journal of Management, Vol. 28 No. 3, pp. 945–958 (2011)
3. Hämäläinen, E. Economic geographical analysis of the Finnish Paper Industry. Annales Universitatis Turkuensis, SER A11 – TOM. 263. Turku. (2011)
4. Hämäläinen, E. 'Economic geographic characteristics in the Finnish paper industry – a case study', Fennia, Vol. 189 No. 2, pp. 63–75. (2011)
5. Yang, C-S., Lu, C-S, Haider, J. J. and Marlow, P. B. 'The effect of green supply chain management on green performance and firm competitiveness in the context of container shipping in Taiwan', Transportation Research Part E, Vol. 55, pp. 55–73 (2013)
6. Lindstad, H., Asbjørnslett, B. E. and Strømman, A. H. 'The importance of economies of scale for reductions in green house gas emissions from shipping', Energy Policy, No. 46, pp. 386–398 (2012)
7. TradeWinds (2010) 'Slow spur for Maersk VLCCs', TradeWinds Magazine, 13 December.
8. Notteboom, T., Delhaye, E. and Vanherle, K. Analysis of the Consequences of Low Sulphur Fuel Requirements. Report commissioned by European Community Shipowners' Associations (2010)
9. AECOM Rail Baltica Final Report: Volume I, AECOM Limited, Chelmsford (2011)
10. AECOM Rail Baltica Final Report: Volume II, AECOM Limited, Chelmsford (2011)
11. Hilmola, O.-P. 'Environmental and infrastructure payments and the future of road transports: case Tallinn-Warsaw', World Review of Intermodal Transportation Research, Vol. 4 No. 1, pp. 55 (2013)
12. Hämäläinen, E., Twrdy, E. and Inkinen, T. 'Cost aggregation in logistic chain', Journal of Open Innovation: Techhnology, Market, and Complexity, Vol. 3 No. 26, pp. 1–14 (2017)

13. Wiesmann, A. 'Slow steaming – a viable long-term option?', Wärtsilä Technical Journal, February (2010)

Research of the changeability of the vessels traffic intensity of inland waterways – case study for Lower Vistula

Patrycja Jerzyło[1], *Aleksandra* Wawrzyńska[2]

[1] Gdańsk University of Technology, Faculty of Civil and Environmental Engineering, Poland
[2] Gdynia Maritime University, Faculty of Navigation, Poland

Abstract. The article presents an analysis of the changeability of traffic intensity of inland waterway vessels in the Vistula Delta. The concept of traffic intensity was defined, traffic flows and disturbances were identified in the studied area. The characteristics of the inland waterway traffic model are presented. The results of the analysis indicate the need to carry out modernization of inland waterways for the needs of inland waterway transport: introduction of a traffic control system, unification of the navigational marking system and strict correlation of renovation/modernization works of hydrotechnical structures with a navigation period.

1 Introduction

The common transport policy pursued by the European Union aims at subordinating transport to the principles of sustainable development. This is to ensure real competition between the different modes of transport for environmentally friendly transport and the integration of freight transport corridors using two or more modes of transport. In light of these principles, inland waterway transport, which is environmentally friendly, safe and cheap, occupies an important place in the policy of sustainable development. Inland waterway transportation, providing crucial linkages between domestic and international shipping markets, plays a significant role in promoting the economic development of many countries [10].

Nowadays, the directions and trends of development, as well as the European Union's policy regarding inland navigation are aimed at adapting inland waterways to modern needs of shipping, forwarding and tourism and using inland infrastructure for intermodal transport as an element of handling and forwarding hubs.

The layout and length of inland waterways in Poland has remained at a similar level for years. The specificity of waterway infrastructure influences the factors shaping the demand for inland waterway transport. The navigational conditions translate directly into the basic

[1] Corresponding author: patjerzy1@pg.edu.pl

construction parameters of the rolling stock used for inland waterway transport, including the relatively low capacity of barges, as well as the volume of transport [6].

Traffic flow modeling is a mathematical record (structure of demand transport) based on the study of communication behavior in a separate territorial unit, which through the model of transport interactions is designed for traffic forecasting, study of traffic flow changes in the process of planning transport systems, optimization of transport networks, solving transport tasks and supporting decision-making in other transport problems related to transport systems [4].

The vessel traffic stream is described by a number of different parameters, of which the basic and additional parameters are distinguished [1]. The basic parameters are:

- intensity,
- density,
- speed.

In addition, the traffic flow can be described using the following quantities [1]:

- queue spacing,
- time between submissions of subsequent vessels,
- number of vessels traversed by a fixed point of the road at given time intervals,
- directional and generic structure of traffic.

Rivers have long been a communication route along with the development of the river's civilization in the natural state it was necessary to start regulating - to deepen and expand to allow the free movement of ever larger vessels. Poland in terms of nature has very good conditions for using inland transport.

Fig. 1 Transportation of non-normative cargo in 2017 [fot. P. Jerzyło]

Inland shipping, though heavily neglected in Poland, is slowly beginning to develop again (Fig. 1). A dense network of rivers and lakes favors the development of inland transport, however, such good conditions are used to a small extent. The advantage of inland waterway transport is low transport cost, as well as low environmental impact, which in today's world is one of the most important issues. The principle of sustainable development assumes such activities to maintain social, economic and environmental balance.

2 Shipping routes in the Vistula delta

One of the basic concepts in social sciences (particular in social economic geography), spatial management and economics is accessibility. First of all, it is the occurrence in the socio-economic space of at least two elements being, respectively, the source and destination of accessibility, which are accessible to each other. The second characteristic feature refers to the means of communication, acting as a carrier of connections connecting the previously mentioned points of space, overcoming its resistance in the form of socio-economic, political and administrative or natural factors. Such perceived availability implies the occurrence of two complementary notions of accessibility - in terms of transport and spatial approach [9].

The use of inland waterways through the Port of Gdansk offers opportunities for the development of the Polish economy by generating lower transport costs compared to land transport due to low transport prices, which result from high degressivity of unit costs, large freight loads over long distances, without significantly interfering in the natural environment [5].

To determine the potential transport accessibility of the Vistula Delta, use: analysis of available infrastructure in the form of ports (Tczew, Malbork, Elbląg, Gdańsk), floating fleet, which can conduct regular navigation in the surveyed area (Figure 2), location of large enterprises whose transport policy indicates for the interest in using transport navigation in logistics processes. The analysis of the condition of the existing waterway is also very important (Fig. 3).

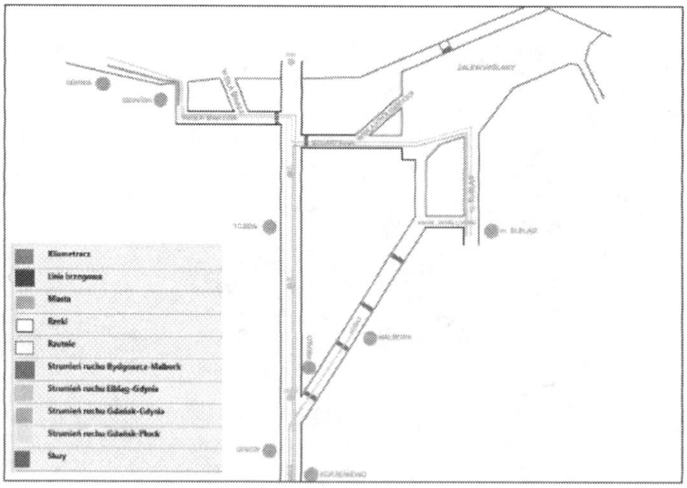

Fig. 2 Ship traffic jets in the Vistula delta between 2010 and 2016 [Own study]

Wisła from 684 km to Toruń, ie. 718 km, despite being classified to the Ib class of regional importance, does not meet the requirements of this class in terms of transit depth. Transit depths in relation to average water are 1.2 m in length. During low water depths, they fall below 1.0 m, often reaching not more than 0.6 ÷ 0.7 m. The guarantee required for this class of transit depth is only 5% the entire navigation period (ie several days a year). Wisła from Toruń km 718 to the Port of Tczew at km 910 remains normatively in class II of the waterway of regional significance. It meets the requirements regarding the width of the navigable route (30 m), the minimum radius of the bowels of the navigable route (300 m) and the minimum clearance under the bridges over (3.0 m). However, it still has a limited transit depth of up to 1.4 m. The guarantee of transit depth required for this class (1.8 m) is practically up to 10% of the entire navigational period. The Vistula River from Tczew at km 910 to the mouth of the Gulf of Gdansk remains in the third class of the waterway of regional significance. It meets the requirements regarding the width of the navigable route (40 m), the minimum radius of the bowels of the navigable route (500 m) and the minimum clearance under the bridges over (4.0 m). Does not meet the parameters of the Class III waterway with regard to the transit depth. Depths in the river's depth are limited to 1.6 m, compared to the 1.8 m required for this class [14].

Fig. 3 Regulatory buildings in the Vistula delta [fot. P. Jerzyło]

Nogat from the Vistula River to the estuary to the Vistula Lagoon, the length of 62.0 km remains in the 2nd class of the waterway of regional significance. The Nogat River belongs to the II class of a waterway. The section from km 0 + 000 to 38 + 650 is sewered (4 locks: Biała Góra, Szonowo, Rakowiec, Michałowo), Transit depth is 1.80 m. Section from km 38 + 650 to 62 + 000 - this is the navigable section of the river flowing along with limiting the transit depth to 1.60 m. On the sewered section it meets the requirements regarding the width of the navigable route (30 m), the minimum radius of the navigable water route (300 m), mucus parameters (length 65 m, width 9.5 m, depth on the lower threshold of 2.2 m) and a transit depth of 1.8 m [14].

The Szkarpawa River is 25.4 km long and belongs to the II class of a waterway with a restriction of the transit depth to 1.60 and the radius of the navigable route's curves up to 150 m. It meets the requirements regarding the width of the navigable route (30 m) and the parameters of the Gdańsk Głowa sluice (length 65 m, width 9.5 m, depth on the lower threshold 2.2 m).

The Martwa Wisła river is 11.5 km long and belongs to the Vb class of a waterway with a limited width of the lock sluice in Przegalin to 11.90 m. It meets the requirements regarding the width of the navigable route 50 m, the transit depth 2.8 m and the radius of the navigable route 800 m [14].

3 Acquiring data about unit traffic

Inland waterway is well known as an efficient and reliable transportation mode for economics, with the continuous expansion of trade volume, the traffic density has risen steadily, leading to maritime accident occurred frequently [11].

The information identifying the vessel floating in sea waters and identifying its traffic parameters can be obtained by means of the Vessel Traffic Service (VTS). VTS - these are services provided by competent administration to improve safety, navigation efficiency and environmental protection. This service must be able to interact and respond to traffic related situations taking place in the areas of VTS operations. Another system used to acquire data is the Automatic Identification System (AIS) [7].

In the case of obtaining information on the movements of units on inland waterways, data obtained from the river information system, the aim of which is to improve the safety of navigation, can be used [13].

Fig. 4 Transport of mass cargo in 2016 [fot. P. Jerzyło]

Thus the obligation to implement the River Information System (RIS) includes all inland waterways of international importance and the ports located on them, which connect with other roads of the same standard, i.e. from the 4th class of the waterway according to the classification United Nations Economic Commission for Europe (UNECE). The provisions of the RIS Directive, at the discretion of the Member States, may also be applied on inland waterways not covered by this obligation. However, on national inland waterways subject to RIS obligation, but not linked to the waterway network of another Member State, the requirements and technical specifications for a harmonized river information system set out in the RIS Directive are not mandatory but are only recommended. In Poland, most waterways have operational parameters of regional significance. Only less than 6% of their length meets the requirements of inland waterways of international importance (parameters of class IV and higher), i.e. those subject to the obligation to implement the RIS Directive. They are [13]:

- Dąbie lake to the border with internal sea waters - 9.5 km

River Odra from the town of Ognica to Przekopu Klucz-Ustowo and further as the Regalica river to the estuary to the Dąbie lake - 44.6 km

- the Western Oder River, which includes:

- from the weir in Widuchowa to the border with internal sea waters together with side branches - 36.6 km,

- Przekop Klucz-Ustowo, which connects the Odra Wschodnia River with the Western Odra River - 2.7 km.

- Parnica River and Przekop Parnicki from the Odra West River to the border with internal sea waters - 6.9 km.

Therefore, the part of the lower section of the Odra river from the town of Ognica to Szczecin is subject to the obligation to implement RIS, along with the Odra Wschodnia, the West Odra, the Dąbie Lake and other roads of the Szczecin water node. The total length of these waterways is 97.3 km. All sections have parameters of international significance and connect with each other. The requirement to develop technical specifications for a European river information system results from the need to build a harmonized, interoperable and open information system, available without discrimination for all system suppliers and users. For safety reasons, it was assumed that the technical requirements and specifications would be based on the work developed in this field by relevant international organizations, such as:

- International Sailing Association (PIANC),
- Central Commission for Navigation on the Rhine (CCNR),
- European Economic Commission of the United Nations (UNECE).

The framework structure of a harmonized river information system should enable system operators and users to achieve their goals by implementing a series of tasks related to inland waterway management, based on the information collected and sent to create the service [13].

The waterway in the Vistula Delta is not covered by any Traffic Control System, River Information System or Automatic Identification System. Data regarding inland waterway transport, unit traffic and transported cargo can be obtained from the Central Statistical Office and Regional Waterways Authority in Gdańsk and an observation method (Fig. 4) for determining the average speed of vessels. This information is obtained with a long time delay and can't be used to model the traffic flow of units in real time. Considering the current use of the navigable route in the Vistula Delta, it seems that this way of acquiring data is sufficient. In the case of an increase in traffic density and density of units moving along this route, it will be necessary to introduce a monitoring system in a continuous movement of units in the studied area.

4 The intensity of ship traffic

The model of ship traffic flows is called: the mathematical description the structure of demand transport based on the study of transport behavior in a separate territorial unit. The basis for consideration of models reflecting traffic flows is the measurement and archiving performed correctly.

To describe phenomena occurring on the road, which is the movement of vehicles, modeled in the areas of mathematics, physics and simulation models are used. The models described below are a combination of models:

- macroscopic model - it reflects the flow of motion streams in a certain section or area in which the disruption of a single stream affects the whole system, which causes a change in socio-economic factors. Examples of macroscopic models are: wave motion model, dispersion model.

- microscopic model - modeling of the movement of a single vehicle, expressed by relations between the unit and the person heading, between units and between the unit and its surroundings. Examples of microscopic models are: driving model with a leader, queue model.

On the basis of data provided by the Regional Economy Board in Gdańsk, the movement of tourist units in 2010 - 2014 in the area of the lower Vistula limited by locks: Przegalina, Gdańsk Głowa, Biała Góra, Czersko Polskie, Miejska in Bydgoszcz and Włocławek (Figure 5) was identified. Tourist units have been qualified for kayaks, sailing boats, rowing boats, motor boats and rafts.

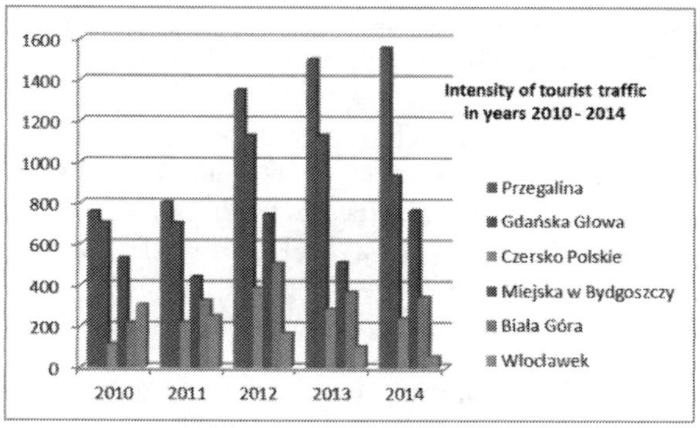

Fig. 5 Intensity of tourist traffic in 2010 - 2014 [Own study]

Analyzing the intensity of movement of cargo units, the following units were taken into account: motor barges, towed barges, pushed barges, tugs, pushers, passenger ships, ferries, icebreakers and wheelchairs (Fig. 6).

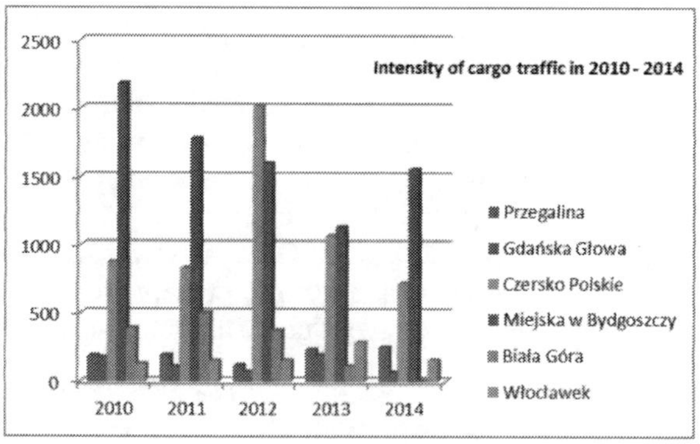

Fig. 6 Intensity of cargo traffic in 2010 - 2014 [Own study]

Intensity is a parameter of the unit traffic flow, specifying the number of vessels traverse the set point on the fairway per unit of time. Intensity is expressed by the formula [3]:

$$q = \frac{n}{\tau} \tag{1}$$

where:

q - the intensity of the unit traffic stream,

n - the number of units that traverse a given point,

τ - the time interval in which the units were registered.

In the studied area, one of the traverse points is the lock in Przegalina, and the unit of time 2014. Statistical analysis of the number of lockable vessels showed a large variation in the intensity of the ship traffic flow due to the month (Fig. 7). The largest number of units being flown took place in March, which was related to the icebreaking action on the Vistula river and increased number of icebreaker locks.

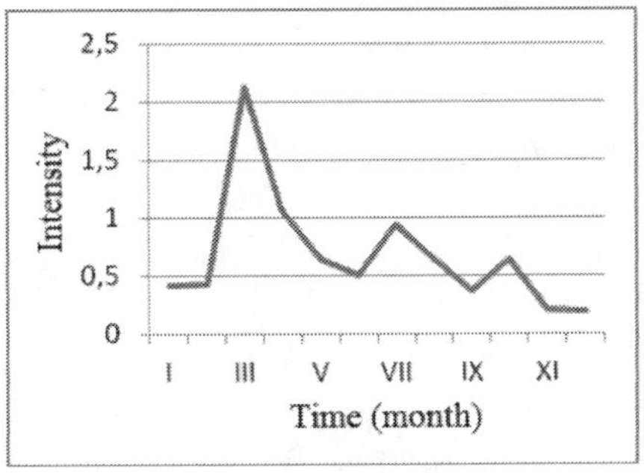

Fig. 7 Intensity of vessles traffic in particular months in 2014 [Own study]

5 Disturbances in the waterway network in relation to the identified traffic flows

Traffic modeling in transport networks on waterways is particularly complex. Such an area is a place of strong influence of various socio-economic, technical, organizational and economic factors. This affects the functioning of individual transport subsystems, which on one hand are characterized by distinct features, and on the other - depend on each other. This impact also applies to streams of traffic flowing through the transport network. In this situation even small changes in transport infrastructure may cause disturbances and changes in the traffic structure. The basic problem in modeling the parameters of the units' movement streams are difficulties in navigation, which more or less disturb the random character of the stream.

The basic purpose of modeling in case of disturbance analysis is proper identification of their causes and assessment of their impact level. An appropriate way of mapping the flow of traffic in the waterway network allows taking into account a number of factors that influence the achievement of reliable results. This is particularly important in traffic management, where the transfer of current, reliable and properly formulated information to network users can improve the functioning of the entire transport system [12].

Each user of the transport system realizes his journey within a certain time interval. From the point of view of modeling, the moment of commencing the displacement and the moment when it will be in certain places of the network are important. These can be places characterized by extremely arduous traffic conditions, as well as decision points, when the user has the opportunity to choose a way of moving. However, it is difficult to obtain such detailed information for each transport system user, although they would be very useful for precisely describing the flow of ship traffic flows.

The main components of the unit motion stream model are traffic flow parameters and probability distributions describing these parameters [8]. The basic problem in modeling the parameters of the units' motion streams are difficulties in navigation, which more or less disturb the random nature of the stream. A traffic disruption can be defined as a sequence of events leading to a state in which traffic can't be continued according to previous assumptions. In the examined model, the disturbances are, among others: waiting time for sluicing, slacking, low water level (insufficient depth) and temporary shut-off of locks and sections of the waterway. The purpose of such research is to identify places particularly susceptible to interference, to assess the impact of these disturbances on other elements of the transport system on waterways and their surroundings, and to select the appropriate traffic management tools to reduce the nuisances resulting from these disruptions.

Critical nodes of the network are the places of disruption, where the capacity reserves are stuck in liquid streams accompanying disrupted streams. Appropriate change of traffic structure in nodes may lead to improvement of the capacity and reduction of total time losses in the network. Therefore, it is important to accurately identify disturbance locations at the microscopic level [12].

Under real conditions, a single event may not cause a disturbance, and only the appropriate combination of several events will result in disturbances, requiring appropriate actions on the part of traffic management (Fig. 7).

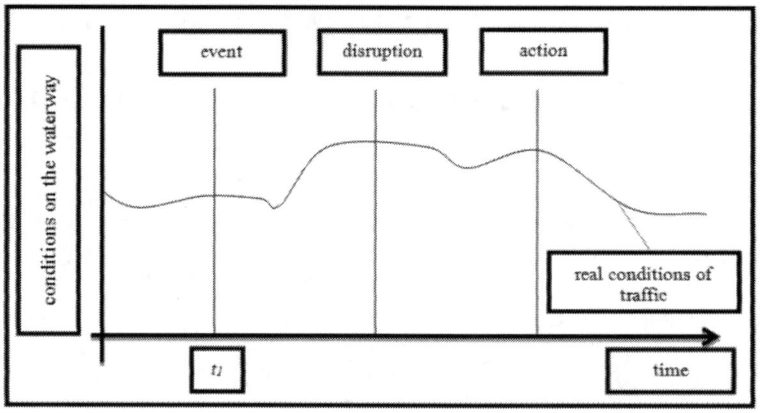

Fig. 7 Influence of disturbance on conditions occurring on the waterway [Own study]

Any disturbance may cause a violation of the prevailing conditions, and thus a decrease in the quality of the displacement. Therefore, an appropriate level of disturbance should be assumed, corresponding to a defined level of traffic loss, beyond which the effects of disturbances are felt by the network users. This level is strictly conditioned by the specificity of the area. Acceptance of specific traffic conditions and the related quality of movement is subjective and depends on both the user's characteristics, the period of the day, the type of travel, as well as the function and location of the elements of the transport network.

For the purpose of assessing traffic conditions in the network, a set of interference types has been defined:

$$R_z = \sum_{i=1}^{z} z(s,t,v,) \qquad (2)$$

where:

R_z - the size of the set, i.e. the number of disturbances accepted for analysis,
i – reference stream,
z – measure of the assessment - disturbances,
s – road,
t – time,
v –vessels in the disturbed traffic stream.

Interesting area characteristics calculated in this way include distribution of the number of disturbances in the network in relation to traffic flows. Measures of interference in the network are defined by the user for each network in the form of an acceptable drop in the speed of units, at which the movement of units is not considered to be disturbed yet. The distribution of the number of disturbances is determined for all (or selected) traffic flows globally in the area of the entire network as well as locally for each user-defined spatial unit in this network. It can be expected that there is a proper organization of traffic for a given network and is characterized by some statistical ordering (defined by the function of the density of the distribution or the composition of such functions as intensity, disturbances, length of queues, etc.).

The correct organization of traffic should lead to optimal use of all available infrastructure and organizational resources in a given network.

6 Summary

Inland shipping has been neglected for some time in Poland and is slowly beginning to develop again. A dense network of rivers and lakes is conducive to the development of inland transport, however, such good conditions are not yet fully used. The principle of sustainable development assumes such activities to maintain social, economic and environmental balance so that the next generations can benefit from what is now. It is worth considering the initiative of investing in inland transport in order to bring benefits not only to us, but also to future generations.

Traffic models are used in controlling ship traffic to predict the impact of control decisions on traffic efficiency. Adaptive cruise control systems must ensure that the measurement data can be processed in real time. Therefore, a compromise is needed between the accuracy of the vessel traffic model used and its computational complexity.

Current methods of ship traffic control are based mainly on macro and mesoscopic models. Such models include parameters defined for streams or groups of ships. They do not describe the movement of individual ships, and hence, do not allow full use of the information that can be obtained thanks to new traffic monitoring technologies (including sensor networks and video-detection).

Data concerning individual ships (eg location, speed, class, relation) are important and useful from the point of view of traffic control tasks.

The waterway in the Vistula Delta is not covered by any Traffic Control System, River Information System or Automatic Identification System. Data regarding inland waterway fleet, unit traffic and transported cargo can be obtained from the Central Statistical Office and the Regional Waterways Management in Gdańsk and an observational method for determining the average speed of vessels. This information is obtained with a long time delay and can't be used to model the traffic flow of units in real time. Considering the current use of the navigable route in the Vistula Delta, it seems that this way of acquiring data is sufficient. In the case of an increase in traffic density and density of units moving along this route, it will be necessary to introduce a monitoring system in a continuous of vessels traffic in the studied area.

References

1. Gaca S., Suchorzewski W., Tracz M., *Inżynieria ruchu drogowego. Teoria i Praktyka,* (Warszawa 2008)
2. Gucma S., Gucma L., Zalewski P., *Symulacyjne metody badań w inżynierii ruchu morskiego, Wydawnictwo Naukowe Akademii Morskiej w Szczecinie,* (Szczecin 2008)

3. Kasyk L., *Probabilistyczne metody modelowania parametrów strumienia ruchu statków na akwenach ograniczonych,* (Radom 2012)
4. Krych A., *Słownictwo kompleksowych badań i modelowania potoków ruchu, II ogólnopolska konferencja Naukowo-Techniczna: Modelowanie podróży i prognozowanie ruchu,* (Kraków 2010)
5. Główny Urząd Statystyczny, Urząd Statystyczny w Szczecinie, *Żegluga śródlądowa w Polsce w latach 2010-2013,* (Warszawa 2014)
6. Główny Urząd Statystyczny, *Transport wodny Śródlądowy w Polsce w 2014 r.,* Opracowanie sygnalne, (Warszawa 2015)
7. Pawlak M., Piaseczny L., *Modelowanie ruchu jednostek morskich dla określenia emisji związków toksycznych spalin,* Czasopismo Techniczne z.7-M/2008, (Kraków 2008)
8. Puszcz A. Gucma L., *Wpływ liczebności pomiarów na estymacje parametrów strumienia ruchu statków,* Proceedings of the XIV International Scientific and Technical Conference on Marine Traffic Engineering, (Szczecin 2011)
9. Wiśniewski Sz., *Dostępność transportowa Uniejowa – ujęcie regionalne,* Biuletyn Uniejowski, Tom 3, (Uniejów 2014)
10. Xie, Zhaoqing; Liu, Qing., *LSTM networks for vessel traffic flow prediction in inland waterway,* IEEE International Conference on Big Data and Smart Computing (BigComp), (Shanghai 2018)
11. Yan, X. P.; Wan, C. P.; Zhang, D.; et al. *Safety management of waterway congestions under dynamic risk conditions - A case study of the Yangtze River,* PPLIED SOFT COMPUTING Volume: **59** Pages: 115-128 (Published: OCT 2017)
12. Żochowska R., *Modelowanie potoków ruchu w sieci miejskiej dla potrzeb analizy zakłóceń,* Logistyka 4/2014, (Warszawa 2014)
13. [Online] Available: https:// www.ris-odra.pl/. [Accessed: 22-Feb-2018]
14. [Online] Available: http://rzgw.gda.pl/. [Accessed: 01-March-2018]

SHS Web of Conferences **58**, 01013 (2018) https://doi.org/10.1051/shsconf/20185801013
GLOBMAR 2018

Socio-economic impact of the International Waterway E60 on the Polish and Lithuanian coastal regions

Marcin Kalinowski[1] · *Rafał* Koba[1], and *Magdalena* Matczak[2]

[1]Maritime Institute in Gdańsk, Economics and Law Department, Długi Targ 41/42, 80-830 Gdańsk, Poland
[2]Maritime Institute in Gdańsk, Spatial Planning Laboratory, Długi Targ 41/42, 80-830 Gdańsk, Poland

Abstract. International Waterway E60 (IWW E60) is a sea-shore route running from Gibraltar to the North along European coast up to St. Petersburg then through the Baltic-White Sea Channel, then along the White Sea coast to Arkhangelsk. From the German-Polish border, along the Polish, Russian and Lithuanian coast to the Lithuanian-Latvian border, the route is 610 km-long and runs along 32 Polish municipalities and 4 Lithuanian regions. Determining the role of IWW E60 is important in the context of economic growth of municipalities and coastal regions, especially through the development of local seaports. Operations of the multifunctional port have a wide economic and social impact. Mutual interaction of functions can be observed on the example of fisheries, where in the port area fishing functions intermingle with the processing, tourism, storage and distribution function. The main socioeconomic benefits of E60 waterway development include the increase in employment, economic activity, generation of added value and improvement of transport infrastructure. In addition, ports influence attractiveness of the regions and create impulse for new jobs in the tourism industry. Until now there have been no attempts to make E60 route navigable or only on short sections, usually between two neighbouring ports. Despite the significant cross-border importance of this connection, no directions for its optimal development were defined. What is more, the route is not used. In connection with this, steps were taken to thoroughly investigate the problem and determine the possibility of its development as part of inland waterways network.

1 Introduction

The International Waterway E60 (IWW E60) is a sea-shore route running from Gibraltar to the North along the coasts of Portugal, Spain, France, Belgium, the Netherlands, Germany, Poland, Lithuania, Latvia, Estonia, Russia to the Sankt-Peterburg-Volga-Baltic waterway, then through the Baltic - Białomorski Channel, then along the White Sea coast to Arkhangelsk.

The E60 route on the Polish and Lithuanian sections runs along 32 Polish municipalities and 4 Lithuanian districts. In Poland, these are the following municipalities: Świnoujście, Międzyzdroje, Wolin, Dziwnów, Rewal, Trzebiatów, Kołobrzeg (urban and rural), Ustronie Morskie, Będzino, Mielno, Darłowo (urban and rural), Postomino, Ustka (urban and rural), Smołdzino, Łeba, Choczewo, Krokowa, Władysławowo, Jastarnia, Hel, Puck (urban and rural), Kosakowo, Gdynia, Sopot, Gdańsk, Stegna, Sztutowo and Krynica Morska. In Lithuania, these are the regions of: Klaipeda, Nering, Palanga and Silute. Polish municipalities have 1,070,032 inhabitants and occupy an area of 376 914 ha. In Lithuanian regions live 269,044 people and have an area equal to 332 200 ha. In total, the areas affected by the E60 waterway in 2015 were inhabited by 1 339 076 people and had an area of 709 114 ha. The following table provides detailed information on the study area.

Table 1. Data of communes and municipalities of the Polish and Lithuanian coast (2015)

Poland (year 2015)				
Region	**County**	**Commune**	**Population**	**Area, ha**
Zachodniopomorskie	Świnoujście	Świnoujście	41 152	19 7 1323
	Kamieński	Międzyzdroje	6 543	11 438
		Wolin	12 281	32 746
		Dziwnów	4 004	3 762
	Gryficki	Rewal	3 891	4 065
		Trzebiatów	16 565	1 025
	Kołobrzeski	Kołobrzeg (miejska i wiejska)	57 284	16 970
		Ustronie Morskie	3 663	5 698

	Koszaliński	Będzino	8 576	16 619
		Mielno	5 006	6 213
	Sławieński	Darłowo (miejska i wiejska)	22 016	28 966
		Postomino	7 036	22 686
Pomorskie	Słupski	Ustka (miejska i wiejska)	24 275	22 765
		Smołdzino	3 431	26 029
	Lęborski	Łeba	3 789	1 481
	Wejherowski	Choczewo	5 630	18 313
	Pucki	Krokowa	10 673	21 109
		Władysławowo	15 467	3 922
		Jastarnia	3 824	780
		Hel	3 531	2 172
		Puck (miejska i wiejska)	36 772	24 217
		Kosakowo	12 813	5 014
	Gdynia	Gdynia	247 478	13 514
	Sopot	Sopot	37 231	1 728
	Gdańsk	Gdańsk	462 249	26 196
	Nowodworski	Stegna	9 844	17 009
		Sztutowo	3 685	11 153
		Krynica Morska	1 323	11 601
		Total:	**1 070 032**	**376 914**
	Lithuania (year 2015)			
	County	**Municipality**	**Population**	**Area, ha**
	Klaipeda	Klaipeda	208 972	143 900
		Neringa	2 879	9 000
		Palanga	15 379	7 900
		Silute	41 814	171 400
		Total:	**269 044**	**332 200**
		TOTAL:	**1 339 076**	**709 114**

Source: Own elaboration based on GUS – Local Data Bank and Lithuanian Statistical Office

From the German - Polish border, along the Polish, Russian and Lithuanian coasts to the Lithuanian - Latvian border, this route is about 610 km long. Within the catchment area of this waterway, there are local seaports of the Polish and Lithuanian coast, such as: Kołobrzeg, Darłowo, Ustka, Władysławowo, Jastarnia, Hel, Svientoi - Butinge and two ports located on the Curonian Spit which connect with the E60 route through the International Waterway E70. The course of the E60 waterway is shown in Figure 1.

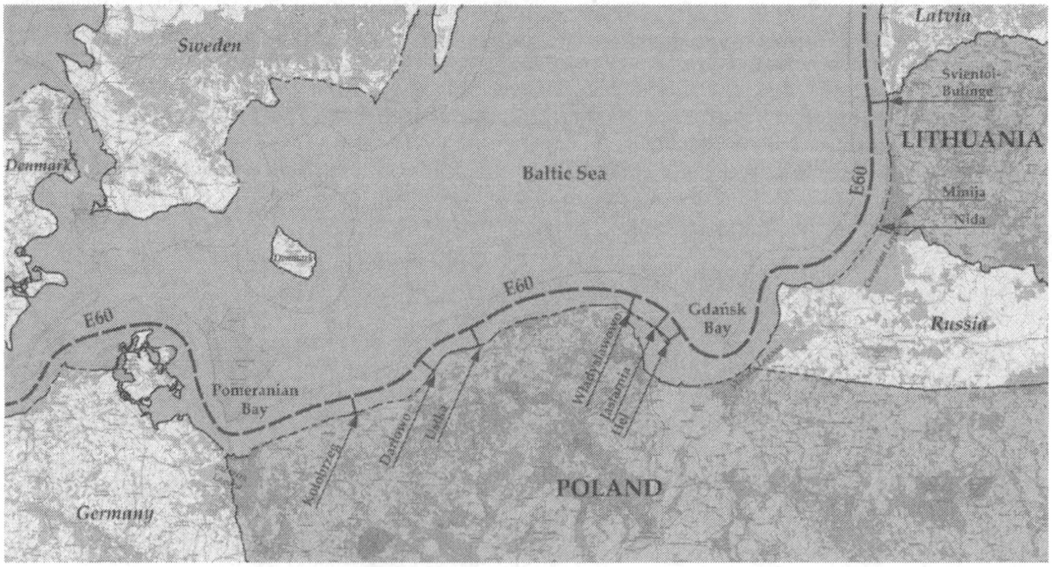

Fig.1. Seashore route E60 on the Poland - Lithuania section
Source: Own elaboration based on OpenStreetMap and *Inventory of Main Standards and Parameters of the E Waterway Network*, Economic Commission for Europe Inland Transport Committee United Nations, New York and Geneva 2012.

The E60 route has enormous tourist potential and can become a very important link between the local ports of Poland and Lithuania. All ports are located in areas of high tourist value, therefore the launch of passenger tourism on this route could attract many tourists.

Currently, regular ferry tourism along the IWW E60 does not exist. Regular lines in Poland connect only Kołobrzeg and Bornholm. Tourism between the local ports of Poland and Lithuania does not function due to legal and administrative constraints, difficult border crossing procedures, military zones closed to traffic and marine areas of landscape parks. The E60 route has big tourist potential, but there are many barriers that prevent its use. While it is possible to overcome barriers on the Polish and Lithuanian side, one should remember about the section of the Russian Federation existing along the Kaliningrad Oblast. In the territorial waters of this section there are both Russian military areas, which completely exclude any traffic of tourist vessels, as well as the area of the Curonian Spit landscape park, which is also a significant impediment. Therefore, the E60 route in its present form will be able to function only if close cooperation is established with the Kaliningrad Region and local ports on its coast.

Strengthening the role of IWW E60 is important in the context of economic growth of municipalities and coastal regions, in particular through the development of local seaports. Multifunction port operations have a wide economic and social impact. Mutual interaction of functions can be observed on the example of fisheries, where in the port area the fishing functions are mixed with processing, tourism, storage and distribution functions. The main socio-economic benefits associated with the development of the E60 waterway include increased employment, economic activity, the generation of added value and the improvement of transport infrastructure. In addition, ports affect the attractiveness of regions and create an impulse for new jobs in the tourism industry. Until now, no attempts have been made to navigate the E60 route or have been undertaken only on short distances, usually between two neighbouring ports. Despite the significant cross-border importance of this connection, no directions for its optimal development have been identified. Therefore, an attempt was made to examine the problem and determine the possibilities of its development, mainly in the context of the operation of local seaports. This article presents the synthesis of research results on the problem identified above.

2 Local ports as development centres

In the maritime policy of the state, seaports of fundamental importance for the development of the country's economy play an important role and they are privileged, especially when it comes to financing port investments. Countries such as Poland and Lithuania in their policies should also focus not only on these major ports but also link the development of maritime economy with the whole coast and take full advantage of the coastal location. Local ports are key in this respect. The standard and scope of services offered by these ports determines whether the use of the entire seashore will be limited to boarding houses and the beach, or whether it will be more comprehensive. This is particularly important due to the occurrence of such phenomena as:

- Environmental protection conditions make it necessary to limit the Baltic fishery, whose service should be one of the basic functions of local ports,
- A typical beach season at the Baltic Sea is very short,
- The growing wealth of the society creates pressure on a more varied offer for those who vacation by the sea,
- The open borders of the South Baltic countries create conditions for the development of short-sea passenger transport.

The proper development of local ports, which are still on the margins of the national maritime policy, will determine whether these factors will have a positive impact on the development of the coast.

The current European Union policy on integrated coastal zone management requires a comprehensive approach and recognition of all aspects of opportunities offered by coastal location for local communities on the entire coastline.

The issue of the possibility of developing local ports is also included in the broader spatial context, including not only their near-land (local and regional) terrestrial, but also maritime foreland, as the projected intensification of sea space use for various economic purposes should also result in new development impulses for local ports.

Research carried out as part of the SB LocPorts project allowed to identify barriers to the development of local ports, among which the most serious ones were:

- Economic barrier manifested in the lack of funds for financing port investments and other developmental plans,
- Legal and organizational barriers expressed in numerous restrictions imposed on local ports, which significantly hinder the development of port operations,

- Lack of efficient mechanisms of cooperation between stakeholders or those that should be interested in the development of local ports.

Based on the analysis, the possibilities of activating local ports were developed. One of the results of the analyses is the statement that the basic assumption of the local market economic recovery plan is to link the development of economic activity with the intensive and extensive use of existing or modernized port infrastructure. The type of activity that will include the broadest possible range of services based on and port infrastructure is of key importance. Local ports, due to their impact on the local and regional economy, should be treated as centers of economic development, i.e. centers that can indirectly significantly affect the national economy and therefore their importance cannot be undermined, especially in strategic national documents and laws regulating the seaports activities.

Taking advantage of the opportunities offered by the port requires the implementation of many investments, both in the field of port infrastructure facilities providing access to ports and ICT infrastructure that creates an efficiently functioning integrated information system, unified for all ports of the South Baltic Sea Region. In order for these investments to give the desired synergy effects, national authorities should create a clear vision of the development of the coastal regions and set directions for economic and functional-spatial development, which will allow integration of international, national and local social and economic interests. Obvious in this respect will be the beneficial impact of the waterway E60, which will undoubtedly create a strong impulse for economic development.

Formulating detailed proposals of actions that would allow to overcome, or at least reduce developmental barriers, requires in-depth research. It seems particularly important to undertake research on the place of local ports in regional policy and on the creation of regular maritime connections along the coast of the South Baltic countries, using the international waterway E60, which will undoubtedly have a positive effect on the activation of ports and regions in which they are located. It should also be borne in mind that the reduction of capital and thus infrastructure barriers will not be possible without full use of the possibilities offered by public-private partnership.

3 Local ports economic impact on the surrounding region

Poland and Lithuania, being member states of the European Union, the problem of cross-border cooperation between local ports could be solved at the EU level. However, current EU policy is targeted at the 300 largest European seaports and they have priority in granting funding for development. Local ports have a very difficult task.

The local ports of the Polish and Lithuanian coasts are located in poorly industrialized regions with high natural and tourist values. Therefore, cross-border cooperation between them should be implemented mainly on the basis of maritime tourism. However, to make this possible, many barriers have to be overcome. One of them is complicated and restrictive border procedures. The sea border crossing should provide adequate control, which must be efficient, fast, well-functioning and friendly to tourists.

In order for Maritime tourism to function, proper infrastructure and an integrated information system are needed. Uniform for all ports, providing tourists with information in a legible and easily accessible way. EU funds have a positive impact on the state of infrastructure in local ports, which is improving every year, but there is still much to do.

When the local ports of the Polish and Lithuanian coasts are fully adapted to passenger traffic, the next step should be to navigate the international waterway E60, running along the coast and connecting all the ports discussed.

The location of local ports is closely related to the development of economic functions of the immediate environment. The functioning of the port, in addition to certain financial effects, also has a fairly wide economic impact on the immediate surroundings. The main, though hard to measure, economic benefits include, among others, the possibility of reducing unemployment, economic activation of the region or an increase in tax revenues.

On the scale of the entire Polish coast, there is a great diversity between the communes where the ports and marinas are located. In the coastal areas there are numerous economic entities with different profiles and sizes. The spatial diversity of the number of economic entities is also relatively high, but their number decreases as the distance from larger agglomerations increases.

In economic activities in coastal communes, the largest share is market services, which include, among others, transport and warehouse management, as well as activities related to accommodation and gastronomy. An important area of activity is also industrial processing or fishing related activities. In the port cities, the largest number of companies deals with hotel services and conducts gastronomic activities. Other important activities are trade and repairs, business and real estate services, industry, construction and transport. The predominance of the tourist function is clearly visible.

As part of the sustainable development strategy, small sea ports have the role of economic activation centers for coastal regions. The state's regional and transport policy must be coordinated in order to avoid divergence of priorities, which would lead to the dispersion of financial resources from both structural and national funds.

Of particular importance among the factors determining the socio-economic development of cities and communes is the existing and anticipated transport infrastructure. Public transport is connected with the city's transport system, which should contribute to the development of areas through which transport routes run. The communication system and its efficiency to a large extent determine the possibilities of economic growth and create favourable conditions for the development of entrepreneurship. Entrepreneurship is one of the most important factors of socio-economic development and increase in competitiveness of coastal regions (other regions as well). Therefore, effective support for entrepreneurship should be an integral element of municipal and regional authorities. A positive example of this activity is the Kołobrzeg city support program for entrepreneurship, which assumes direct support through specific activities supporting local companies and indirect - through the implementation of a number of infrastructure investments that improve the functioning of the city.

The location relative to the back office and the foreground is an important factor in the implementation of the local port's transport function and economic activation. The meaning of the local port in the transport chain can be most generally expressed by the amount of cargo handled. In the case of large universal ports, the search for directions of service diversification led to the emergence of a distribution function, which is a synthesis of transport and commercial functions and resulted in the creation of logistics centers. At the basis of the development of the distribution function there is a need to improve the distribution of goods and services moved between production and consumption. In this context, in the era of increasing congestion in road transport and the tendency to implement modern logistic solutions and the concept of transferring cargo from roads to the sea, local ports have a potential chance of development as ports supporting large port centers. Therefore, local ports should activate the traditional transport function. Of course, it is necessary to create an appropriate infrastructure system.

4 Influence on local job market

Due to the complexity of their functions, seaports create large labour markets around them. The analyzed areas of local ports and their immediate surroundings create many employment opportunities. These ports generate employment in fisheries, processing and preservation of fish and fishery products. Besides, they are attractive to tourists in themselves. Thanks to this, they increase the attractiveness of the region and are an impulse to create new jobs in the tourism industry.

The coastal location and associated landscape features favour the development of tourism. The most popular form of tourism in the studied area is tourism, which results directly from the coastal location. For climate reasons, this type of tourism is limited almost exclusively to the summer season, and in addition largely depends on the current weather conditions. Active tourism, especially water tourism is also very popular. Regions offer conditions for sailing, windsurfing and kitesurfing enthusiasts. Tourism generates a lot of jobs, both in accommodation and gastronomy facilities, as well as recreation and entertainment facilities. In addition, tourists spending holidays in the vicinity of local ports benefit from services related to everyday life (hairdresser, beautician, car mechanic, doctor, etc.). This is a circumstance that may be an additional opportunity for industries not related to tourism.

Sea ports have always been a great job market, especially the so-called simple work. However, along with technological changes within particular port functions, and especially in the transport function, the importance of ports as a source of quantitative demand for labour decreases, analogously to that in industry. However, the importance of skilled work increases. From the point of view of the port-region relationship, it is difficult not to notice the relationship between the possibilities of port development and the possibilities of acquiring qualified staff.

In this connection, the port-region relationship has been observed in recent years. It is not the port that is the development stimulus of the city, but the development of the port depends on the impulses flowing from its surroundings. This applies to both education and accessibility of the port from the back office, which was a traditional "contact" of the city and port, as well as a new problem - joint port-urban activities for the optimal use of urban and port space.

Regardless of the size and significance of a given port, it is always an important component of development potential in the scale of the commune, region and country, and as part of the local development potential, each local port should be subject to local government policy shaped at the commune level.

The activity of the majority of local ports in Poland is mainly associated with fishing and, to a limited extent, with support for tourism and recreation. Few of the local ports form modest, though multifunctional, economic and spatial structures of both local and regional importance, capable of the typical operation of transport hubs.

The fishing industry is diverse, however, the fluctuations in its production in recent years appear in all types of activities related to it. Each area is also struggling with other problems, such as high maintenance costs, good environmental status, poorly developed port infrastructure or catch limits. This affects the condition of the fisheries sector, which more often tells related entrepreneurs to think about diversifying their activities.

In the Puck Bay area, employment in fisheries is characterized by a smaller share of young people (up to 30 years of age) than the average in Poland (8.8% to 13.5) and larger older people (over 60) (13.1% to 8) , 9%). In turn, in the years 2007-2013, the number of people working in the processing and preservation of fish and fishery products both in the Puck Land region and in the region remained at a similar level.

Among the employed in processing, women dominate all the time (59.6%). Current MIR estimates indicate that this area comprises 45.2% of employees on fishing vessels (at sea and on land) in the Pomeranian Region and ¼ of domestic employment. In 2014, the highest number of employees was in Władysławowo (217, i.e. 41.0% of the total) and in Jastarnia (208 people, i.e. 39.3%).

The coastal location of the Puck Land area and the unique environment and landscape values associated with it favour the development of tourism. The area of the Puck Land, as well as the entire Pomorskie Region, is considered one of the most attractive tourist regions in Poland. It is characterized by natural and cultural values, whose diversity and uniqueness on the national scale affect the high attractiveness of tourists visiting the region, but also for residents who positively assess their place of residence. The most popular form of tourism in the studied area is tourism, which results directly from the coastal location and landscape values of the region. Active tourism, especially water tourism is also very popular. The region offers excellent conditions for sailing enthusiasts, but there is no common and integrated offer of the entire region for sailors. In this region there are excellent conditions for windsurfing. The surroundings of Rewa and Jastarnia are very popular among kitesurfing fans. This information is also confirmed by the already mentioned LQ indicator. In 2014, it amounted to more than 5.5 for accommodation and food service activities. The scale of development of the sector is provided by data on the increase in accommodation facilities (between 2010 and 2014 by 327.6%, with the region average of 80.1% and the country's 37.2%), and consequently beds (respectively 53.1%, 21,1% and 13.8%), which means the regular development of the tourism sector and the steadily growing number of employed in tourism.

In the area of the Słowińska Grupa Rybacka, which includes the Ustka Port, there are 20 fish processing plants operate. Most of them have modern equipment and modernized or new infrastructure. MORPOL S.A., the largest fish processing plant in Europe and salmon in the world, has the largest share in production and employment. The health tourism is of major importance for the economic development of the SGR area, mainly Ustka Uzdrowisko, covering the City of Ustka and the Ustka commune part and, therefore, also a seaport. Ustka as one of 45 places in the country (in the Pomeranian Region - Sopot and Ustka) has the status of a health resort, which is an opportunity to extend the tourist season. The number of facilities providing accommodation services is above average, which reflects the tourist character of the region. In 2014, almost 137 thousand tourists rested in the SGR area that had at their disposal 11,500 beds in 131 facilities. The most developed accommodation base is located in the rural commune of Ustka, in which 52.7% of beds are concentrated, and their number, similarly to people using it, increases. 43% of accommodation is located in the City of Ustka, which is most often chosen by tourists for summer holidays (54.9% of the total number of guests.) The above data, similar to the surroundings of the local ports of the Bay of Puck, indicate a high level of employment in the tourism industry and its growing trend. Mostly it is connected with the seaport in Ustka and its immediate surroundings as well as the tourist attractions it creates.

In Darłowo Local Fisheries Group, within which the Port of Darłowo operates, there are 23 fisheries farms (some of them have many breeding and hatching centers scattered throughout the area) and 15 factories. In the fisheries sector (excluding processing) about 350 people are employed.

The type of business activity that is particularly important in the opinion of DLGR are tourist and tourism-related services. The nature potential is evidenced by the fact that legally protected areas occupy 17.1% of the area and there are more than 300 nature monuments on them. Tourism is one of the key potentials of the DLGR area due to the proximity of the Baltic Sea, attractive landscape, numerous lakes and tourist attractions. Data on the tourism services sector, however, indicate that the potential of the area is not fully used. According to the data at the end of 2013, there were 119 accommodation facilities in the DLGR area, half of them located in the city of Darłowo.

In the area of the Local Fishing Group "Morze i Parsęta", in which the Port of Kołobrzeg functions, in full-time fishing activity in 2013, 321 people were employed, including 250 at sea. With regard to data for the region in the area of SRLGD, over 29% of those working in fishing were employed, while in the case of nationwide data over 12.5%. According to the data contained in the RRW-19 forms, in 2014 in the municipalities belonging to SRLGD "Sea and Parsęta", there were 237 employees. The submitted forms concerned 44 units, or 64.4% of the size of the fleet registered in the municipalities of Kołobrzeg, Kołobrzeg and Ustronie Morskie commune at the end of 2014. Almost 80% of people working in the area of SRLGD "Morze i Parsęta" were employed in fishing units. This was a lower result than in most ports in Poland (average 87.3%).

Every year, over 400,000 people visit the region of SRLGD "Sea and Parsęta", mainly due to the sea port in Kołobrzeg. This number, thanks to the stays in the spa, spreads out throughout the year, with the highest volume of tourist traffic falling into the holiday season. This has an impact on the seasonality of employment in the area. This number is an additional, potential clientele of local enterprises. Of course, the expenses of tourists in the largest part are intended for accommodation, gastronomy, entertainment and recreation. However, it should be remembered that tourists staying in the area covered by LDS usually spend here at least a week.

In Lithuania the influence of the Šventoji seaport development on local job market depends very much upon the pace of reconstruction and marketing success. Altogether, it is expected that new 10 to 50 jobs can be created in Šventoji in maritime and ancillary services in relation to the seaport development. In assessing the prospects for the seaport development, two scenarios can be distinguished: a) if the port restoration is relatively slow, this growth could be restrained to 5-10 %; b) in the case if the economic indicators of the port will start to improve after the reconstruction in 2020, and over the next five years will improve steadily each year, this growth may reach 20 to 30 %.

The greatest demand for skilled and non-skilled labour force associated with the growth of the Šventoji seaport is related to the demand for catering to the visiting leisure boats and small-scale ferries in the summer season. According to statistics obtained from the existing Lithuanian and Latvian marinas, one can predict that the Šventoji seaport marina will be visited by 500-600 yachts and motor boats annually with foreigners comprising over 70 % of the visitors. During the first year of operation of the reconstructed seaport (first season) over 200 foreign boats can be expected arriving from Germany, Poland, Latvia, Russia, and the Nordic countries(Finland, Sweden, Denmark), if the information about the new marina will be successfully spread in the targeted foreign markets.

Pending upon successful development of Palanga resort marketing activities in the Nordic countries, we can expect at least 1-2 small ferry lines from these countries during the first 2-3 years of operation of the seaport, which would be particularly receptive to the skilled and unskilled labour force onshore. A significant increase in the labour force employed in the Šventoji seaport and the marina servicing the maritime and ancillary sectors can be facilitated by special events (e.g., an international Tall ship race), daily management of the Šventoji harbour, as well as by additional ground services in the leisure and informal education services (eg., a sailing club, a sailing school etc.). As will be mentioned below, there is a permanent demand for skilled labour force related to the maintenance of the fleet of service vessels of various institutions, i.e. the Lithuanian Navy, the Būtingė oil terminal, the Environment Protection Agency, and the State Border Guard Service. The impact of the Būtingė oil terminal on the local labour market is negligible since the staff is sufficient and it mainly lives in Palanga or Klaipėda.

As it was already mentioned above, currently there are 10 fishermen keeping their vessels in the Nida harbour. Due to the decline of commercial fisheries in the Curonian Lagoon, it is hardly feasible that the number of the fishermen can increase in Nida in the foreseeable future, although the number of berths specifically allocated to the small-scale fishing boats could be increased up to 40. On the other hand, increasing of the capacity of the dock for leisure boats can have some influence on the local labour market considering new port-related employment opportunities in direct and ancillary services. In spite of numerous restrictions and regulations related to the national park and World Heritage status, Neringa municipality is the leading one in Lithuania in terms of conditions for doing business. Hence, any upgrading of facilities for water tourism and leisure boating, particularly, opening of cross-border navigation to the Russian part of the Curonian Lagoon, might have a substantial positive impact on the local job market, especially in the summer season.

Due to the importance of water tourism and related services for the economy of the Minija village, in combination with rural tourism, angling and nature tourism services, and with addition of husbandry and small-scale ancillary fisheries, there is no structural unemployment in the Minija village. On the contrary, it attracts seasonal employees from as far as Kintai, the eldership centre, and Šilutė, the municipality centre. The seasonality of tourism in Minija is mitigated by the circumstance, that most of the facilities, including the marina, are closed in the low season and the remaining few open ones are able to survive catering to the needs of a few hundreds of ice-fishing and birdwatching enthusiasts visiting the Minija river delta in winter and in spring.

Most of the inhabitants in the low season either live on the earnings accumulated during the tourist season, or leave the village. This is a quite typical situation for seasonal touristic settlements worldwide. It is definitely not a sustainable state of affairs as it is prone to fast collapse if conditions of tourism change. However, for Minija, which is a remote village with very limited development alternatives it might be seen as a quasi-equilibrium state concerning the local job market. Bearing in mind that Minija is an ethnographic and architectural heritage village where any new investment and development plans must meet very strict conservation requirements, there is hardly any chance for the settlement to expand beyond its current limits and for the revival of any larger scale commercial fisheries in the Curonian Lagoon by local inhabitants, whose number and economic welfare is relatively stable, but their interests are not in fisheries anymore.

5 Barriers, opportunities and threats

The most important limitations of the possibility of developing local ports include, first of all, the **economic barrier** evidenced by the lack of funds for financing investments and other development plans.

The existing statutory solution, which provides entities managing local ports and sea harbours (or directly to municipalities) with the possibility of acquiring certain properties, at the same time imposing certain obligations on those entities requiring significant expenses, does not create any mechanism ensuring real regular income for these entities. The functioning solution does not guarantee a regular mechanism ensuring the acquisition of funds required for the construction, modernization and maintenance of infrastructure providing access to ports and harbors. The burden of constructing, maintaining and modernizing port infrastructure imposed on entities managing these ports and marinas in practice is impossible without state support.

Regardless of the management model, ports and marinas have very limited possibilities of obtaining inflows to cover high investment costs. The tasks and objectives included in local port development strategies require the owner to be included in the implementation. The ports themselves can meet current needs and strategic goals, however, they are not able to generate funds to cover even the most urgent needs.

The development of local ports is also not conducive to the system of tax burdens for entities managing ports and harbors, depending on their right of ownership or perpetual usufruct of real estate and buildings and structures located within the boundaries of ports and marinas. Admittedly, due to the existing actual ownership relations, in many specific cases these encumbrances do not constitute a development barrier for existing local ports and marinas within their present boundaries, however, the cases of limiting the development needs of these ports and marinas due to the burden of taking over further real estate they are more and more frequent.

Direct support by separated management boards of ports (eg Kołobrzeg or Darłowo) of economic entities that can undertake to implement part of the strategy is limited. Nevertheless, port management may create favourable conditions for economic development, attracting investors, acquiring aid funds from the European Union, as well as inspire entrepreneurship of port users.

The activities of local port and municipal authorities should focus on seeking funding for access. The activities of coastal communes that do not take into account the proper development of transport infrastructure will not bring the desired effects, and will additionally limit the possibilities of dynamic development not only of ports but also of other segments of the economy. With neglected transport infrastructure and difficult access to expressways and transport network of national and European scope, tendencies to economic marginalization of areas with neglected transport infrastructure are intensifying.

Investments in the development of ports are usually capital-intensive, in addition they are characterized by a long period of return. Neither maritime offices nor municipalities, in areas where local ports and sea harbours are located, have large resources. The situation does not seem to change quickly.

Port managing entities do not finance the infrastructure providing access to the port, through which the act means those that lead to the port or marina and the fairways located within the port or marina, including facilities, facilities and institutions connected with their functioning. These facilities are financed from the state budget within the limits provided for in the budget act. By virtue of other regulations, entities other than managing ports have the responsibility for the construction, extension, modernization and maintenance of infrastructure facilities providing access to the port from the land side, such as roads, railways, waterways. This is a state of affairs typical of European ports and can be treated as one of the forms of public aid for ports.

Location barriers arise both from the geographical location, concerning all ports, and the specificity of the Polish and Lithuanian coasts. The Baltic Sea has a peripheral position in relation to the main shipping routes, which limits the development of seaports, including to some extent local ports. In addition, the cold climate means that passenger and sailing are seasonal and can be practiced from late spring to early autumn.

In the case of Polish and Lithuanian harbours and marinas, the situation is exacerbated by the slightly diversified coastline, which does not encourage the development of cabotage, as is the case in Sweden, Finland or Denmark. In addition, in the case of many local ports and marinas located above the open sea, we are dealing with silting the entrance to the port. This limits the usable depth of the fairways and causes significant dredging costs. Even worse is accessibility from the land side.

The economic environment of local ports is not conducive to their development. They adjoin agricultural areas, and the small local industry does not generate any needs for maritime transport. This is accompanied by a decrease in catches, and hence a decrease in the number of cutters and fishing boats based in small ports and harbours.

Each port has its own specificity, which results in its chances and threats. The following analysis is an attempt to compile those features that may be common to all or most of the local ports, and so subject to a certain generalization.

Opportunities:

- A growing awareness that local ports are the main factor in increasing the attractiveness of coastal communes, favouring the expansion of their offer and the extension of the tourist season.
- An increase in the level of affluence of the society, which increases the demand for sea tourism and the practice of yachting.
- Gradual improvement of infrastructure on the coast.
- Gradual acquisition of experience by local government activists and the growing pool of qualified personnel.
- Progress in solving priority investment tasks of municipalities (sewerage, sewage treatment, schools, etc.), which, by their very nature, were assessed as more important than ports.
- The development of tourism and passenger traffic in sailing as a result of the increase in the standard of living in Poland and the improvement of the port infrastructure.

- Establishment of conditions for the development of other port functions.
- The possibility of using EU regional policy funds for the development of port infrastructure and access to ports from the sea and land.
- Development of cluster initiatives.
- A clear view of the major directions of port development in spatial development plans and development strategies of individual port municipalities.
- Linking investments in road infrastructure and tourist base of port communes with the development of ports.

Threats

- Weakness of links between local markets and abroad by the sea and a little varied shape of the coast,
- A short planning horizon, determined by the electoral calendar, at all levels of government.
- Lack of coherence of the activities of entities, on which the development of local ports, ie the government and all levels of local self-government depends.
- Spatial conflicts resulting from divergent interests in the coastal zone.
- Ecological fundamentalism, often with a corruption foundation.
- Capital limits.
- Poorly developed economic facilities of ports (mainly agricultural areas, low level of industrial development).
- Lack of the idea of part of port self-governments to use ports as important elements of the sustainable development of port centers.
- Too slow pace of improvement of the parameters of access infrastructure to ports from the land side.

6 Conclusions

Local ports, despite unfavourable conditions, can be an important factor in the activation of communes and coastal regions. For the proper development of local airports and their surroundings, a number of tools and legal and organizational instruments necessary to develop effective solutions are necessary. The development of local ports should at the same time be an integral part of the region, county and commune strategy.

Changes occurring in the structure of the national economy, including in the sectors related to the sea, make it necessary to change the destination of many port infrastructure facilities in local ports and to adapt them to new functions. In many cases, it is also necessary to modernize the infrastructure providing access to the port from the foreground (fairways, breakwaters, etc.) and from the back office (road and rail network).

Local ports and marinas are by definition the components of local infrastructure that provide the opportunity to take advantage of the opportunities offered by the coastal location. Hence, their functions are adapted to the forms of using the sea adopted by a given community. Basically, it should be mentioned here:

- providing access to a given section of the shore from the sea for cargo and passengers,
- handling of coastal and sea fishing using nearby benches,
- service of sport boats and various types of activities in the field of maritime tourism and recreation,
- providing conditions for the implementation of manufacturing activities related to the above-mentioned functions, such as:
  - storage and processing of fish,
  - construction and repairs of small marine units, mainly fishing and sports.

In spite of generally unfavourable factors determining development and barriers, local ports undergo transformation and gradually adapt to changed legal and economic conditions. This process is not fast and diverse, mainly due to differences in the potential and determination of local authorities. In each case, there is a certain range of development opportunities and forms of acceleration of activation.

The activation of local ports and marinas should be based on the following assumptions:

- business development should be related to intensive and extensive use of existing and / or expandable infrastructure comprising port / harbours and maximization of utility resulting from access to the sea and use of its resources.
- activities that meet the above-mentioned conditions include the traditional transport services sector (port services, freight transport), services related to tourism services, Baltic fishing and services based on knowledge and modern technologies.
- expenditures incurred on the revitalization of seaports and marinas are relatively high, as they are associated with the deepening of basins, port channels and fairways to ports, modernization of

breakwaters and wharves, construction of piers, shore protection, sea view terraces, promenades, etc. On the other hand, the types of activity developed on the basis of existing / modernized infrastructure facilities are usually too small as to the size and production time of infrastructure use, so that entities (public, private) could obtain any economies of scale, and local effects have significant effects positive feedback between the conducted activity and the increase in value added. The key from an economic point of view is therefore the development of activities, which will include the broadest possible range of services / production based on the available port / harbour infrastructure. In this way, economies of scale (not achievable in most local seaports) will be replaced by the benefits of the scope. The provision of a wide range of activities or the development of several types of production will increase the use of capital-intensive infrastructure, which will reduce the costs of its production engagement, and customers (tourists, recipients of products / services) will receive a bundle of goods and services in a given town.

- the activation of small ports and harbours according to the economies of the scope has a chance to be successful, provided that:
  - business development will be based on the idea of locally created clusters, i.e. complementary chains and complementary services and types of production performed by cooperating private and public entities (local government and state),
  - local clusters will develop functional relationships with such centers in other cities to the widest extent possible, and thus obtain additional benefits arising as a result of running and developing activities in the network of economic connections (network economies).

In creating clusters, it is necessary to initiate local authorities and create a platform that facilitates the partners' agreement and sometimes the creation of a public-private enterprise.

The activation of local ports and marinas finds their justification in theory and practice to stimulate the socio-economic development of small and large economic systems. All European Union programs, whether related to sectoral policy (transport policy), industry policy (maritime policy) or horizontal policies (cohesion, sustainable development, competition, growth and employment), as well as regional development strategies developed by the European Union (Strategy for the development of the Baltic Sea Region) are built largely on the assumptions presented above, which will undoubtedly facilitate public entities implementing the strategies of local development constructed in this way to access Community funds. It is also significant that the current attempts to develop local maritime communities based on other premises did not bring expected results in most cases.

Practical ways of activating local seaports / marinas / sea centers with the use of infrastructure, boil down to:

- development of maritime sports and tourist clusters and passenger navigation,
- development of fishing, recreational angling and mariculture,
- development of commodity production and traditional transport and reloading services,
- development of services related to knowledge and modern technologies.

Referring to the first of the above-mentioned items and in view of the fact that the Polish and Lithuanian local seaports discussed in this article are located in areas of high natural and tourist value, the development of sea tourism can be an important impulse for the development of ports and their surroundings. Attempts should be made to navigate the IWW E60 that runs along the coast. This trail has enormous tourist potential and can become a very important link between local and national ports both for Poland and Lithuania. The launch of passenger tourism on this route would undoubtedly be very interesting and would attract many tourists, thus supporting the activation of the regions. To this end, it is necessary to carry out detailed research in this area and to develop a plan of action to revive the local ports and their surroundings through the creation of regular passenger navigation along the coastal route E60.

References

1. *Analiza stanu infrastruktury w portach rybackich i przystaniach pod kątem dalszych potrzeb inwestycyjnych,* Morski Instytut Rybacki, Gdynia 2012
2. H. Bierndgarski: *Małe porty prą do przodu,* Polska Dziennik Bałtycki, Baltexpo 2009, 1 września 2009
3. Cz. Christowa: *Uwarunkowania funkcjonowania i rozwoju polskich portów morskich* W: *Funkcjonowanie i rozwój polskich portów morskich w świetle zapisów "Zielonej Księgi" polityki morskiej Unii Europejskiej,* VII Międzynarodowa Konferencja Naukowa "Porty Morskie 2007", Pod red. K., Chwesiuka, Prace Monograficzne Akademii Morskiej w Szczecinie, Szczecin 2007
4. Formy aktywizacji rozwoju średnich i małych portów morskich polskiego wybrzeża. Modele zarządzania małymi portami – rekomendacje dla Polski. Kancelaria Senatu Biuro Analiz i Dokumentacji, Dział Analiz i Opracować Tematycznych, Warszawa 2010

5. *Formy aktywizacji rozwoju średnich i małych portów morskich polskiego wybrzeża,* Kancelaria Senatu Biuro Analiz i Dokumentacji, Listopad 2010
6. A.S. Grzelakowski: *Formy i kierunki aktywizacji rozwoju małych portów morskich w UE* W: *Dwa lata Polski morskiej w Unii Europejskiej,* XXII Sejmik Morski, Świnoujście-Ystad, Zeszyty Morskie 2007
7. W. Kuszewski, *Identyfikacja potencjalnych warunków determinujących rozwój portów lokalnych w aspekcie niebieskiego i zielonego wzrostu,* Gdańsk 2017
8. W. Kuszewski, *Przegląd prowadzonych badań nad niebieskim i zielonym rozwojem portów lokalnych,* Gdańsk 2017
9. Lokalna Strategia Rozwoju na lata 2014-2020 Darłowskiej Lokalnej Grupy Rybackiej, Darłowska Lokalna Grupa Rybacka, Darłowo 2015
10. Lokalna Strategia Rozwoju na lata 2014-2020, Północnokaszubska Lokalna Grupa Rybacka 2012
11. Lokalna Strategia Rozwoju Obszarów Rybackich, Stowarzyszenie Lokalna Grupa Rybacka Pradolina Łeby, Gniewino 2013
12. Lokalna Strategia Rozwoju Słowińskiej Grupy Rybackiej 2014-2020, Słowińska Grupa Rybacka 2016
13. K. Luks: Polityka wobec portów lokalnych na przykładzie Portu Elbląg, 2009
14. J. Miszczuk: Problemy małych i średnich portów morskich z punktu widzenia administracji państwowej, Szczecin.
15. P. Nowaczyk: Małe porty morskie w strategiach rozwoju gmin nadmorskich. Przedsiębiorczość i Zarządzanie Wydawnictwo SAN, tom XVI, zeszyt 4, część I, s.57-72.
16. Plan rozwoju portu w Helu, Zarząd Portu Morskiego Koga 2009, http://www.porthel.home.pl/files/plan_rozwoju_portu_w_helu_2009.pdf
17. Program rozwoju polskich portów morskich do roku 2020 (z perspektywą do 2030 roku), Ministerstwo Transportu, Budownictwa i Gospodarki Morskiej, Warszawa, Lipiec 2013
18. Rocznik Statystyczny Gospodarki Morskiej, Główny Urząd Statystyczny, Szczecin 2016
19. Strategia Rozwoju Gminy Miejskiej Łeba na lata 2014-2025, Listopad 2014.
20. Strategia Rozwoju Portu Morskiego Darłowo, Zarząd Portu Morskiego Darłowo, Darłowo 2007
21. Strategia rozwoju portu morskiego Kołobrzeg, Zarząd Portu Morskiego Kołobrzeg 2010, http://www.zpm.portkolobrzeg.pl/303,strategia-rozwoju-portu
22. Strategia Rozwoju Portu Morskiego Kołobrzeg, Zarząd Portu Morskiego Kołobrzeg Sp. z o.o., Kołobrzeg 2010
23. Strategia rozwoju portu morskiego w Ustce do roku 2021, Business Mobility International Sp. z o.o., Ustka 2007, *www.christowaczeslawa.pl/.../2007-strategia-rozwoju-portu-morskiego-w-Ustce.pdf*
24. Strategia Rozwoju Ziemi Puckiej na lata 2016-2025, Stowarzyszenie Północno Kaszubskie Lokalna Grupa Rybacka, kwiecień 2016.
25. Strona internetowa: http://ustka24.info/przebudowa-wejscia-portu-ustka/,
26. Strona internetowa: http://www.portalmorski.pl/morze-inne/bezpieczeństwo-i-srodowisko/39735-szykuje-sie-gigantyczna-przebudowa-wejscia-do-portu-w-ustce,
27. Strona internetowa: http://www.portalmorski.pl/porty-morskie/port-kołobrzeg,
28. Strona internetowa: www.gp.pl/strefa-biznesu/wiadomosci/z-regionu-a-wladze-ustki-maja-koncepcje-nowego-portu-wizualizacja,10172840.
29. W. Szczurek, Działalność gospodarcza gmin w portach morskich, Fundacji Rozwoju Uniwersytetu Gdańskiego, Gdańsk (2002)
30. Ustawa z 20 grudnia 1996 r. o portach i przystaniach, Dz.U.2002, nr 110, poz. 967, z późn.zm.
31. Zarząd Portu Morskiego Darłowo (2007), Strategia rozwoju portu morskiego Darłowo - dokument, http://www.port.darlowo.pl/pl/?div=dokumenty [28 wrześnie 2014].
32. J. Zieziula, *Identyfikacja czynników determinujących wielkość obrotów przeładunkowych w małych portach morskich na przykładzie Portu Darłowo,* Uniwersytet Szczeciński
33. J. Zieziula, P. Nowaczyk: Identyfikacja czynników determinujących wielkość obrotów przeładunkowych w małych portach morskich na przykładzie Portu Darłowo, Szczecin.

SHS Web of Conferences **58**, 01014 (2018) https://doi.org/10.1051/shsconf/20185801014
GLOBMAR 2018

Container terminals competitive factors – results of an international survey

Adam Kaliszewski[1,3], *Arkadiusz* Kozłowski[2], *Janusz* Dąbrowski[1], and *Hanna* Klimek[1]

[1]University of Gdańsk, Faculty of Economics, Institute of Maritime Transport and Seaborne Trade, Poland
[2]University of Gdańsk, Faculty of Management, Department of Statistics, Poland

Abstract. Container terminals require continuous benchmarking of competitiveness factors. This paper discusses factors of competitiveness with a main focus on feeder and mixed terminals (in which both feeder and transshipment services are provided). Method statement includes an economic literature review with respect to competitiveness factors and a container port expert survey process to achieve a list of most important factors among global container port operators. Experts were selected from different countries and private container terminals. A questionnaire was developed and sent online to them. Questions required both a ranking of factors as well as assigning monetary values to critical variables. Responses were analyzed using a pre-established strategy. Overall, experts ranked first the role of market related competitiveness factors related to service quality, price level and adaptability to the changing market environment. Intermodal links and general hinterland connections factors were more often mentioned than factors relating to nautical accessibility or other factors. Both assignment of monetary values as well as ranking have produced mutually coherent results. The paper does not claim to be definitive; it aims to provide an update of competitive factors hierarchy and highlights issues for further port development.

1 Introduction

Container terminals are functioning within a framework of constant change resulting from globalization of supply chains, changing alliances among shipping lines and competition for better access to hinterland markets. As year 2017 resulted in more vessel capacity consolidation among shipping lines community, ports worldwide face a problem of shrinking direct customer base and frequent changes to the port compositions of the main loops of major world trade routes. This trend continues in

[3] Corresponding author: kaladam@wp.pl

2018 as Japanese carriers (K Line, MOL and NYK) form ONE alliance from 1st April. This situation increases operational risk that the less competitive ports can suddenly be left out without any container volumes, while it is not possible to dispose of surplus assets or excess debt in the short term. Thus, a careful monitoring of competitiveness parameters is needed. Looking from a single point of view, in a given container terminal, it is not often evident what should be done in order to increase competitiveness. In this paper C-level executives share their views regarding current level of competitive determinants, as well as give insights into hierarchy of these factors. By aggregation of these individual and independently collected views from various corners of the world, to a certain extent, individual biases are smoothed out and a clearer picture emerges of what matters most in this industry [1].

2 Competitive factors in port literature

Parola *et al.* [2] reviewed economic journals for a 32-year period (1983-2014) searching for words 'competitiveness', 'selection', 'choice' in combination with the term 'port'. Based on results from top 25 journals and their impact factor a total of 39 drivers have been reported. The top ten drivers in decreasing order of relevance are following: port costs, hinterland proximity, hinterland connectivity, port geographical location, port infrastructures, operational efficiency, port service quality, maritime connectivity, nautical accessibility and port site.

Haezendonck *et al.* [3] have proposed 25 determinants related to both cost related determinants as well as qualitative non-costs elements that have been studied during a survey of port experts for port of Antwerp, in the Hamburg-le Havre port range. These determinants include: 1) port infrastructure, 2) port superstructure, 3) port labor (human capital), 4) logistical techniques and communication systems, 5) intra-port competition, 6) inter-port competition, 7) intra-port co-operation, 8) extra-port co-operation, 9) client-relationships in the port, 10) client-relationships outside the port, 11) local government intervention, 12) regional government intervention,13) (supra)national government intervention, 14) port supporting services,15) activities maritime accessibility, 16) shipping, 17) transshipment, 18) warehousing, 19) value added logistics, 20) manufacturing, 21) activities by agents, forwarders etc., 22) distribution activities within port, 23) road transport, 24) rail transport, 25) inland navigation. These determinants were tested during a field survey with 75 face-to-face interviews.

Tongzon [4] proposed eight determinants of competitiveness with a view of a logistic hub centered around a seaport, with port of Singapore being a world class example of a well-functioning system in the South East Asia. These determinants include: 1) port (terminal) operation efficiency level, 2) port cargo handling charges, 3) reliability, 4) port selection preferences of carriers and shippers, 5) the depth of the navigation channel, 6) adaptability to the changing market environment, 7) landside accessibility, 8) product differentiation.

Esmer *et al.* [5] investigated into non-price competition strategies of port sector competitiveness factors. The research was intended to reveal non-price competition strategies prevailing in the Turkish cargo terminals handling more than 100,000 TEUs or 500,000 tons of bulk and general cargo. There have been no respondents outside Turkey. Results of the Esmer survey indicated five groups of non-price factors that support business strategy of Turkish ports: customer care, service customization and bundling, service expansion, service diversification and auxiliary services.

The competitiveness determinants of seaports have been directly and indirectly subject of academic research, including Burns [6], Defilippi [7], Fan *et al.* [8], Haralambides [9], Heaver *et al.* [10], Magala and Sammon [11], Ng [12], Notteboom and Rodrigue [13], Wan and Zhang [14], Yap and Notteboom [15], Yuen *et al.* [16].

In light of literature review, there is a high degree of commonality among competitiveness determinants. Thus, for the purpose of expert survey, a set of 20 factors has been proposed in Table 1 that incorporates most of above findings.

Table 1. The competitiveness factors applied to the expert survey.

No	Competitiveness factors	Area of competitiveness
1	Terminal charges (THC), price, rebates and other financial incentives	Terminal market offerings
2	Scope of terminal services and logistic value-added services	
3	Level of container terminal service quality (speed, reliability, availability, security, non-discriminatory access, eco-friendliness)	
4	Port community system (serving port clients, other stakeholders as well as inside container terminal)	Terminal resources
5	Terminal's ability to serve mega container vessels (TEUs +18k)	
6	Intermodal transport availability in the container terminal (by rail, inland waterways and roads)	
7	Private ownership of terminal	Terminal management
8	Partial ownership of a terminal by shipping lines	
9	Terminal's adaptability to the changing market environment	
10	Level of harmony in management-labour-government relationships (no strikes, conflicts and others)	
11	Corporate Social Responsibility (incl. business ethics, respect of natural environment and involvement with local communities)	
12	Terminal operations respecting natural environment protection laws	
13	Port's reputation, public relations and marketing	
14	Port's nautical accessibility	Terminal accessibility
15	Maritime connectivity (frequency of shipping services)	
16	Hinterland connection (road and rail networks, inland waterways)	

17	Port authority charges, price and pricing strategies	External business environment
18	Fast customs and admin clearance of cargo, incl. port's regulations and customary duties	
19	Shipping lines concentration level (M&A, alliances) and changes in shipping lines' preferences	
20	Supportive government active in promoting ports and logistics transport policies	

3 Method statement

In contrast to Esmer *et al.* [5] this survey was intended for an international usage. The target group contained experts currently working in container terminals, mostly managed by multinational port operators from private sector. With a view to shorten data collection process across different time zones, instead of face-to-face or telephone interviews, an online survey was conducted. Experts have been selected among C-level managers of international port operators. This survey is of explanatory nature, with main purpose of learning what factors matter most to container terminal competitiveness level. First question asked respondents to rank among 20 factors in order of decreasing importance. The second question aimed to differentiate responses in light of monetary weight assigned to each of the factors. As human nature responds in a more cautious way once money is at stake, especially taking into account behavioral effects of endowment and loss aversion with respect to money that respondents already have [17]. In this survey they were asked to expense USD 100 among 20 factors. Then respondents were asked to specify terminal type (feeder, mixed with transshipment over 10% and pure transshipment terminal with min. 80% transshipment share).

The online survey has been conducted from 15th February 2018 to 14th March 2018, mainly among executive level members of LinkedIn port and container terminal online communities. Respondents have provided a total of 68 rankings of competitiveness factors, including 39 responses regarding monetary weights. Within that figure 30 respondents have chosen to provide full responses. In order to map an approximate location of respondents based on their computer (or smartphone) IP numbers a freegeoip.net service have been consulted to search the geolocation associated to cities along with other relevant information like time zone, geographical latitude and longitude. Thus, a total of 68 responses related to respondents located in 35 countries, incl. Africa, Asia, Europe and Americas.

4 Results of the survey

4.1 Statistics of respondents and container terminals

The survey was anonymous thus 30 exact international container terminals could be identified out of total 68 respondents that ranked competitive factors. This sample represents a widespread geographic range of expert opinions, also including voices from less developed countries.

Survey respondents represent voices from container terminals in Russia (3 terminals), Poland (2 terminals), China (2 terminals), Italy (2 terminals), United States (2 terminals) and terminals in Brazil, Cameroon, Croatia, DR Congo, France, Germany, Iraq, Madagascar, Mexico, Panama, Philippines, Qatar, Republic of Korea, Republic of Lithuania, Slovenia, Sri Lanka, Sweden, United Kingdom and Vietnam.

Among above mentioned terminals 14 of them represent feeder terminals with gateway containers, 13 mixed terminals (with both gateway containers and min. 10% transshipment share) and 3 pure transshipment terminals (min. 80% transshipment share). Thus, feeder and mixed terminals dominate among sampled respondents. The ownership of these terminals is private, as no managers representing public bodies have been selected for this survey. It helps with a more focused and homogenous interpretation of the results. In effect this survey captures the voice of the private sector container terminal operators, with more common management and business approach. The total number of experts who have participated in this research, as per country affiliation, is presented in the Table 2.

Table 2. Container terminal experts per country of reply.

No	Country	Respondents	Percentage	Cumfreq.	Cumperc.
1	United States	6	8.82%	6	8.50%
2	Poland	5	7.35%	11	15.85%
3	Brazil	4	5.88%	15	21.74%
4	Russia	4	5.88%	19	27.62%
5	Croatia	3	4.41%	22	32.03%
6	Germany	2	2.94%	24	34.97%
7	Italy	3	4.41%	27	39.38%
8	Sweden	3	4.41%	30	43.79%
9	United Arab Emirates	3	4.41%	33	48.21%
10	United Kingdom	3	4.41%	36	52.62%
11	Belgium	2	2.94%	38	55.56%
12	China	2	2.94%	40	58.50%
13	Mexico	2	2.94%	42	61.44%
14	Pakistan	2	2.94%	44	64.38%
15	Philippines	2	2.94%	46	67.32%
16	Republic of Lithuania	2	2.94%	48	70.26%
17	South Africa	2	2.94%	50	73.21%
18	Cameroon	1	1.47%	51	74.68%

19	Colombia	1	1.47%	52	76.15%
20	Denmark	1	1.47%	53	77.62%
21	France	1	1.47%	54	79.09%
22	Georgia	1	1.47%	55	80.56%
23	Iraq	1	1.47%	56	82.03%
24	Ivory Coast	1	1.47%	57	83.50%
25	Madagascar	1	1.47%	58	84.97%
26	Netherlands	1	1.47%	59	86.44%
27	Panama	1	1.47%	60	87.91%
28	Papua New Guinea	1	1.47%	61	89.38%
29	Republic of Korea	1	1.47%	62	90.85%
30	Singapore	1	1.47%	63	92.32%
31	Slovenia	1	1.47%	64	93.79%
32	Spain	1	1.47%	65	95.26%
33	Sri Lanka	1	1.47%	66	96.74%
34	Ukraine	1	1.47%	67	98.21%
35	Vietnam	1	1.47%	68	100.00%
	Total	68	100.00%		

4.2 Statistics of survey questions

The respondents have provided replies by ranking each of the factors on a position from 1 to 20. Note that individual replies have been averaged and a global ranking is shown in Table 3.

Table 3. Ranking of competitiveness factors.

Factor of competitiveness	Average rank	Group letter based on Least Significant Difference test
Level of container terminal service quality (speed, reliability, availability, security, non-discriminatory access, eco-friendliness)	4.8	a
Terminal charges (THC), price, rebates and other financial incentives	7.2	b

Intermodal transport availability in the container terminal (by rail, inland waterways and roads)	8.4	bc
Hinterland connection (road and rail networks, inland waterways)	8.8	bc
Scope of terminal services and logistic value-added services	8.8	bcd
Port's nautical accessibility	9.8	cde
Maritime connectivity (frequency of shipping services)	10.0	cdef
Terminal's ability to serve mega container vessels (TEUs +18k)	10.5	def
Terminal's adaptability to the changing market environment	10.9	efg
Fast customs and admin clearance of cargo, incl. port's regulations and customary duties	11.0	efg
Port community system (serving port clients, other stakeholders as well as inside container terminal)	11.3	efgh
Port authority charges, price and pricing strategies	11.5	fghi
Shipping lines concentration level (M&A, alliances) and changes in shipping lines' preferences	12.8	ghij
Level of harmony in management-labour-government relationships (no strikes, conflicts and others)	12.9	hij
Port's reputation, public relations and marketing	12.9	ij
Supportive government active in promoting ports and logistics transport policies	13.2	jk
Partial ownership of a terminal by shipping lines	14.9	kl
Terminal operations respecting natural environment protection laws	15.6	l
Corporate Social Responsibility (incl. business ethics, respect of natural environment and involvement with local communities)	15.8	l
Private ownership of terminal	15.8	l

Above result can be compared with results from the second question regarding allocation of USD 100 into twenty competitive factors. Respondents have not been limited to the way this allocation is done. Values from all responses are counted and compared to the total USD amount for all factors. As responses are consistent the ranking in Table 4 is similar to the ranking in Table 3. In both approaches respondents have indicated that terminal offering (including service level, level of prices and access to hinterland) matter most. Other factors in this sample carry less importance. Perhaps, if sample comprised of more pure transshipment terminals, such issues as ability to serve mega vessels TEUs +18k and nautical access would have gained importance. Softer factors like social corporate responsibility, government support and environment protection are not yet crucial for achieving quarterly and annual volume and profit targets, thus lower positions reported.

Table 4. Ranking of competitiveness factors based on assignment of nominal USD 100 amount per each respondent.

Factor of competitiveness	Average USD amounts assigned to each factor	Group letter based on Least Significant Difference test
Terminal charges (THC), price, rebates and other financial incentives	14.15	a
Level of container terminal service quality (speed, reliability, availability, security, non-discriminatory access, eco-friendliness)	9.85	ab
Terminal's adaptability to the changing market environment	7.33	bc
Scope of terminal services and logistic value added services	6.85	bcd
Hinterland connection (road and rail networks, inland waterways)	6.49	bcde
Intermodal transport availability in the container terminal (by rail, inland waterways and roads)	5.72	bcde
Terminal's ability to serve mega container vessels (TEUs +18k)	5.44	bcde
Port's nautical accessibility	5.23	bcde
Maritime connectivity (frequency of shipping services)	4.82	cde
Fast customs and admin clearance of cargo, incl. port's regulations and customary duties	4.62	cde

Port authority charges, price and pricing strategies	4.44	cde
Shipping lines concentration level (M&A, alliances) and changes in shipping lines' preferences	3.85	cde
Port community system (serving port clients, other stakeholders as well as inside container terminal)	3.67	cde
Partial ownership of a terminal by shipping lines	3.62	cde
Level of harmony in management-labour-government relationships (no strikes, conflicts and others)	2.92	cde
Private ownership of terminal	2.67	cde
Port's reputation, public relations and marketing	2.64	cde
Corporate Social Responsibility (incl. business ethics, respect of natural environment and involvement with local communities)	2.44	cde
Supportive government active in promoting ports and logistics transport policies	1.97	de
Terminal operations respecting natural environment protection laws	1.67	e

The order of factors in Table 3 and Table 4 results from the average ranking position and the assigned USD amount respectively. However, as the number of respondents was limited, thus the classification obtained does not have to be the same as it would have been obtained had the whole population of container terminal managers worldwide been examined. In order to verify if, on the basis of the obtained results, it is possible to infer in some measure about the importance of particular factors in view of port managers, appropriate statistical tests have been performed. First, it was checked if the ranking performed by the respondents gives grounds for stating that certain competitiveness factors were significantly statistically more frequently selected at higher places than others. Table 5 contains the results of the Friedman test, checking if the average ranks for at least one pair of factors are significantly different from each other.

Table 5. Results of Friedman test.

Test statistics	**Df**	**p-value**
$Q = 316.25$	19	<0.001

The P-value for the test statistic (assuming the asymptotic chi-square distribution of the Q statistic) is practically equal to zero, which means that there are at least two such factors that have significantly different average positions in the ranking. In order to check how many and which factors have significantly different average positions in the ranking, a post hoc analysis was performed using the Least Significant Difference (LSD) criterion. The results of this analysis are presented in the last column in Table 3. Factors that do not have the same letter are significantly different from each other, while the same letter marked is not significantly different from each other. The first factor is assigned the letter "a" and it has no other factor, which means that the average position in the ranking for this factor is significantly higher than for any other factor. The next letter "b" is assigned to the next four factors, which means that the average ranks for these factors are not significantly different from each other. Groups of factors created as a result of this analysis are not disjoint, e.g. the next group "c" contains three factors from the group "b". Nevertheless, on the basis of post-hoc tests, groups of more important and less important factors can be distinguished in the opinion of port managers. This ranking of grouped factors is shown in Figure 1.

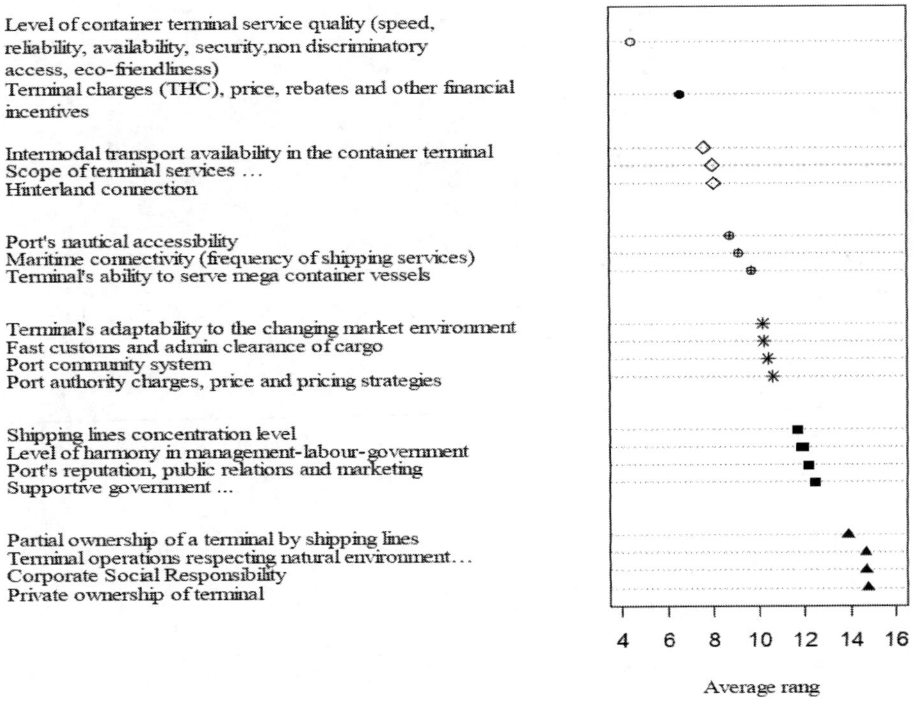

Fig. 1. Ranking of competitive factors.

In the next step, it has been tested if the average USD amounts that the respondents allocated to each factor are significantly different. To take into account the larger differences that could have occurred between the allocated amounts than between ranks, this time the analysis of variance (ANOVA) is applied. The results of the basic ANOVA test can be found in Table 6.

Table 6. Analysis of Variance Table.

	Df	Sum Sq	Mean Sq	F value	p-value
Respondent	38	5.1	0.14		
Factor	19	6504.5	342.34	9.8	<0.001
Residuals	722	25214.1	34.92		

The value of the test statistic F is very large, which results in practically zero p-value. This means that the average amounts for at least two factors are significantly different from each other. As in the previous case, a post hoc analysis was carried out. LSD, with the Bonferroni adjustment for p-values, was selected as a criterion for comparing the means. The results of this analysis are presented in the last column in Table 4. Analogously to the previous post hoc analysis, factors with the same letter are not significantly different from each other.

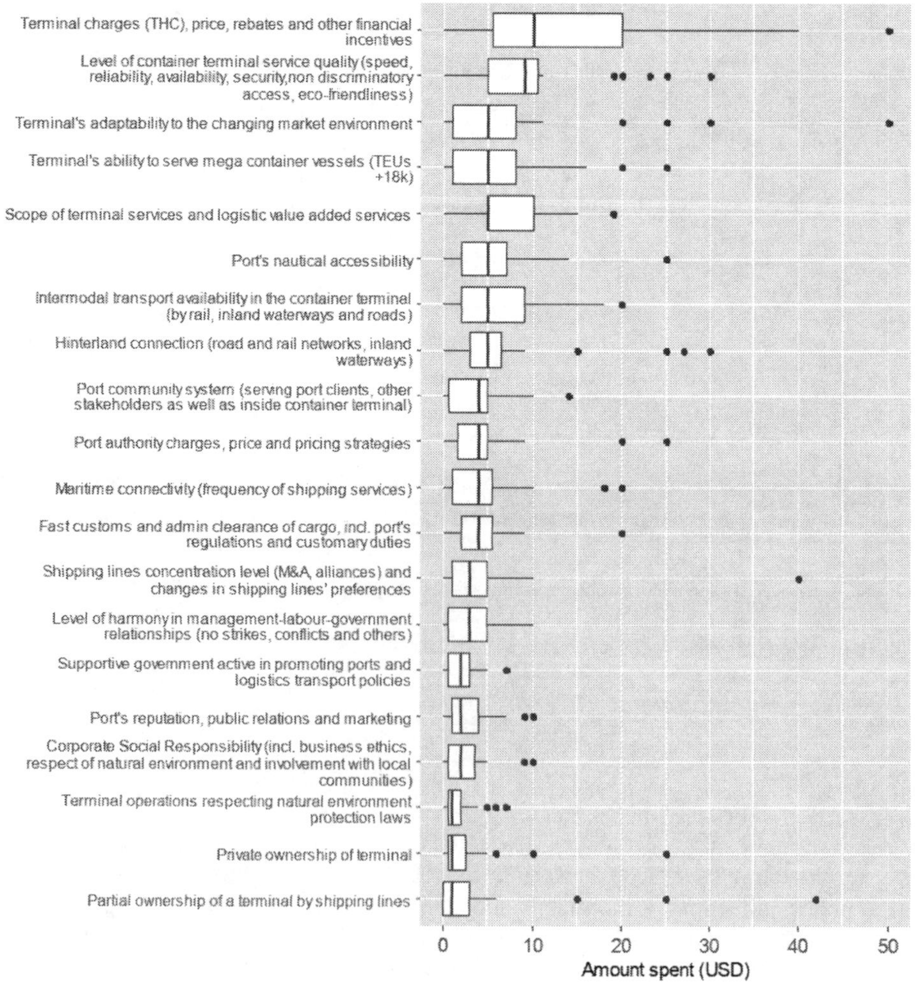

Fig. 2. Boxplots of USD assigned to each factor, sorted by median.

As mentioned earlier, the order of the factors by the amount of money allocated is similar to that resulting from the ranking, however the differences between the means of the allocated amounts are not as visible as between average rankings. For example, in the analysis of groups of factors by assigned dollar amounts, the group "c" includes as many as 16 out of 20 factors, and only five groups of factors (letters from "a" to "e") were distinguished. However, in the analysis of the rankings, these groups were as many as 12 (from "a" to "l"). These results are in part due to the smaller number of responses to the question about monetary amounts than to the question about ranking. In addition, quite a large variation in the amounts allocated and their skewed distributions (see Figure 2) also have an impact on the lower statistical power of this test.

5 Discussion and conclusions

Various factors describing the competitiveness of seaports have been described in the literature so far, some of them have greater or lesser importance in the role they play in shaping port competitive capacity and resulting competitive position achieved. Despite previous attempts to organize them, there has been no study so far that would indicate the rank of individual competitiveness factors of marine container terminals worldwide. A total of 68 experts being top managers of maritime container terminals from 35 countries took part in this survey.

The results of this research indicate that in the opinion of these C-level managers operating in various seaports of the world, the most important factors affecting the competitiveness of these terminals are: level of container terminal service quality, which on average received the highest position in the ranking, and "terminal charges (THC), price, rebates and other financial incentives", for which the average amount of money assigned was the greatest. Further, intermodal transport availability in the container terminal, hinterland connection, as well as scope of terminal services and logistic value-added services are considered important factors of competitiveness. On the other hand, for the least important factors managers recognized those related to caring for the natural environment and corporate social responsibility.

In order to get a full picture of the situation, and in particular to verify the opinions of port experts, it is necessary to conduct a similar survey among customers of container terminals, i.e. shipping lines and shippers. The results of such research will help in the development of a ranking of the most important factors of competitiveness of marine container terminals, taking into account the opinions of all participants of the port services market for containers. The conclusions from the survey, presented in the form of recommendations to interested entities (e.g. owners of container terminals, ministries shaping transport and maritime policy of the countries concerned), will be used to take steps to strengthen the competitiveness of marine container terminals in the port services markets. Such research should be carried out regularly in order to take into account the changing market situation, business environment of terminals, and in particular technical and technological progress in sea and land container transport as well as consolidation activities of container shipping lines.

References

1. J. Surowiecki, *The wisdom of crowds: why the many are smarter than the few and how collective wisdom shapes business, economies, societies and nations* (2004)
2. F. Parola, M. Risitano, M. Ferretti, E. Panetti, Transport Reviews, **37**, 1 (2016)
3. E. Haezendonck, J. van den Broeck, T. Jans, J Prod Anal, **36**, 2 (2011)
4. J. Tongzon, Maritime Economics & Logistics, **9**, 1 (2007)
5. S. Esmer, H. Nguyen, Y. Bandara, K. Yeni, The Asian Journal of Shipping and Logistics, **32**, 1 (2016)
6. M.G. Burns, *Port managemet and operations* (2015)
7. E. Defilippi, Maritime Economics & Logistics, **6**, 4 (2004)
8. L. Fan, W.W. Wilson, D. Tolliver, Maritime Economics and Logistics, **11**, 4 (2009)
9. H. Haralambides, International Journal of Maritime Economics, **4**, 4 (2002)
10. T. Heaver, H. Meersman, F. Moglia, E. van de Voorde, Maritime Policy & Management, **27**, 4 (2010)
11. M. Magala, A. Sammons, Maritime Economics and Logistics, **10**, 1-2 (2008)
12. K.Y.A. Ng, Maritime Economics & Logistics, **8**, 3 (2006)
13. T.E. Notteboom, J.P. Rodrigue, Maritime Policy and Management, **32**, 3 (2005)
14. Y. Wan, A. Zhang, Journal of Transport, Economics & Policy, **47**, 1 (2013)
15. W.Y. Yap, T. Notteboom, Maritime Policy & Management, **38**, 5 (2011)
16. A. Yuen, A. Zhang, W. Cheung, Research in Transportation Economics, **35** 1 (2012)
17. D. Kahneman, *Thinking. Fast and Slow* (2011)

SHS Web of Conferences **58**, 01015 (2018) https://doi.org/10.1051/shsconf/20185801015
GLOBMAR 2018

How to measure the economic value of ecosystem functions and processes and link such value to the MSP?

Ilona Kamińska[1,*], *Tomasz* Zarzycki[1] , *Jacek* Zaucha[2] and *Dorota* Ciołek[3]

[1]University of Gdansk, Faculty of Oceanography and Geography, Institute of Oceanography, 80-309 Gdansk, Poland
[2]University of Gdansk, Faculty of Economics, Department of Macroeconomics, 80-309 Gdansk, Poland
[3]University of Gdansk, Faculty of Management, Department of Econometrics, 80-309 Gdansk, Poland

Abstract. The paper aims at answering questions on implications of the economic valuation of regulatory ecosystem service in maritime spatial planning. The biogeochemical processes that occur in marine sediments such as denitrification, contribute indirectly, though significantly, benefits to human wellbeing. Such direct and indirect benefits are called ecosystem services. According to the Common International Classification of Ecosystem Services (CICES V5.1) for the Integrated Environmental and Economic Accounting, the case study: ecosystem service lies within the: Section - Regulation & Maintenance (Biotic); Division - Transformation of biochemical or physical inputs to ecosystems; Group - Mediation of wastes or toxic substances of anthropogenic origin by living processes and Class - Bio-remediation by micro-organisms, algae, plants, and animals (Code 2.1.1.1). Marine sediments are economically evaluated mainly as sources of providing services (sand and gravel extraction or space for potential industrial use). Their regulatory function is often neglected in such studies. Therefore, we employed the methodology from the field of ecological economics and assessed the economic value of denitrification of the Gulf of Gdansk. This paper suggests the possible implications of using such data in MSP and discusses the competition sea uses in order to determine the final benefits for society.

1 Ecosystem services and MSP

Marine spatial planning is considered as a tool of ecosystem based management [1] and requires the most recent baseline maps of ecosystem components. Usually data layers on biomass, species abundance and/or richness (biodiversity) are used, though they are not straightforward indicators of an area value. In several cases more comprehensive approaches have been used, such as Biological Valuation Maps – BVM [2, 3]. BVMs

* Corresponding author: oceika@ug.edu.pl

integrate biological features such as abundance of species with additional criteria such as fitness consequences and rarity, though no social dimension is incorporated in such approach.

The goal of marine spatial planning is to spatially and temporarily organize maritime uses so there is no conflict between each other and between them and the marine ecosystem. These activities and uses are possible due to the flow of direct and indirect benefits from ecosystem resources, as well as functions and processes that are called ecosystem services [4, 5]. Consequently, mapping of marine ecosystem services is an essential issue for preparing environmentally and societally relevant plans for the use of marine resources [6]. Mapping of ecosystem services allow one to integrate biological and ecological valuation maps with social and economic aspects of marine ecosystems.

1.1 Ecosystem services provided by marine ecosystems

Marine and coastal ecosystems provide a wide range of goods and services that are fundamental to human well-being. Ecological processes, products and outcomes arising from the interactions of organisms with their environment and their activities in the ecosystem are translated into the terms of goods and services in order to increase the understanding between the science and decision makers or society [7, 8]. Despite the fact that the goods and services concept is not free of criticism [9], research in this field has become widely accepted and has developed rapidly in the last decade. Many different methods of categorisation of ecosystem services have been proposed in the literature [5, 10-15]. The classification introduced by Beaumont [4], following Millenium Ecosystem Assessment [5] and Hein [15] proposes the division of goods and services into four categories: (i) production services – products obtained from the ecosystem; (ii) regulating services – the benefits obtained from the regulation of ecosystem processes; (iii) cultural services – the nonmaterial benefits which people obtain from ecosystems; and (iv) supporting services – the ones which are necessary for the production of all other ecosystem services, but do not yield direct benefits to humans. Except the four abovementioned categories, an additional group of 'the option use value' services was proposed by Beaumont [4], so that all Total Economic Value (TEV) categories are represented. The detailed definitions of particular goods and services can be found in the earlier referred work of Beaumont [4]. In addition, several case studies, according to this framework, including the Gulf of Gdansk, were preliminarily analysed in terms of the presence of goods and services provided by marine living organisms as a proxy of marine biodiversity.

More recent approaches to classification of ecosystem services started to appear in the literature and policy such as The Economics of Ecosystems and Biodiversity TEEB [16], and the EU-level proposal for the CICES – Common International Classification of Ecosystem Services (http://www.cices.org) developed for the purpose of the EU Biodiversity Strategy implementation by the Mapping and Assessments of Ecosystem Services working group [17, 18].

Several approaches became more national or regional in terms of geographical scope, such as the UK National Ecosystem Assessment [19] and the UK National Ecosystem Assessment Follow On [20]; Swedish Environmental Protection Agency [21] for the Baltic Sea and Skagerrak; HELCOM [22], [23] etc. that address marine ecosystem services.

The Common International Classification of Ecosystem Services (CICES) developed from the work on environmental accounting undertaken by the European Environment Agency (EEA). The idea of a common international classification was elaborated, due to the recognition that if ecosystem accounting methods were to be developed and comparisons made, then some standardisation in the way we describe ecosystem services was needed

[24]. In addition to the need for the standardization in the context of environmental accounting as well as work on mapping (including spatial visualisation), as well as valuing ecosystem services more generally would benefit from more systematic approaches to naming and describing ecosystem services.

1.2 Regulatory ecosystem services – removal of nitrogen

Eutrophication is one of the main threats to the biodiversity of the Baltic Sea and is caused by excessive load of nutrients to the marine environment. Eutrophication is driven by a surplus of nitrogen and phosphorus in the sea. Nutrient over-enrichment causes elevated levels of algal and plant growth, increased turbidity, oxygen depletion, changes in species composition and nuisance blooms of algae [25].

The role of microorganisms and other biota in removal of nitrogen as well as ecosystem functions and processes contribute to regulatory services such as mitigation of eutrophication. For the purpose of presented research, only the part of the regulatory services provided by marine processes was analysed. One of the key benefits from marine sediments is the removal mechanism for nitrogen from aquatic systems which mitigates eutrophication. According to CICES this particular service can be classified as: *section* – regulation and maintenance (Biotic); *division* – transformation of biochemical or physical inputs to ecosystems; *group* – mediation of wastes or toxic substances of anthropogenic origin by living processes and *class* – bio-remediation by micro-organisms, algae, plants, and animals (CICES Code 2.1.1.1).

2 Methodology – economic valuation of ecosystem services

A number of methodologies have been developed to attempt to quantify the benefits of different ecosystem services [26-29]. Similarly to private market goods, all methods of economic valuation of ecosystem goods and services are found in the theoretical principles of welfare economics. Changes in welfare are being observed through the investigation of people's willingness to pay (WTP) or willingness to accept compensation (WTA) for changes in the level of provision of a particular good or service [30-31]. Some valuation techniques are based on actually observed behaviour data which are known as 'revealed preference' techniques. They include methods that intercept values indirectly from behaviour in surrogate and existing markets, which are assumed to have a direct relationship with the ecosystem good or service being assessed [32-34]. In contrast to the ones described above, other techniques generally known as 'stated preference' are based on rather hypothetical than actual behaviour data. Questions describing hypothetical markets or situations regarding changes in the quality of service provided are asked. The value is inferred from the obtained people's responses [32-34]. Certain techniques can be applied on a broad scale, for many components of TEV, some are applicable to specific issues or impacts, and others are adapted to particular data sources [35, 36].

2.1 Economic valuation of regulatory ecosystem services – removal of nitrogen

The goal of presented research was to estimate the economic value of mitigation of the effects of eutrophication processes – regulating ecosystem service of the Gulf of Gdansk using a combination of two non-market goods valuation methods, i.e. the Replacement Cost Method (RC) and the Contingent Valuation Method (CVM). For the purpose of the study it was assumed that services balancing the effects of eutrophication are linked with regulating nitrogen through denitrification in sediments. The use of a set of direct and indirect methods of economic valuation enabled us to verify the research hypothesis which assumes that the value of mitigating the effects of eutrophication services, estimated on the basis of the replacement cost, is higher than the value expressed by social preferences.

The initial estimation of the economic valuation of removal of nitrogen and mitigation of eutrophication processes using the Replacement Cost Method has been divided into four stages: (I) Identification and quantification of the service regulating the effects of eutrophication understood as balancing the amount of nitrogen through denitrification in sediments; (II) Indicating a substitute for the provided ecosystem service; (III) The economic valuation of mitigation of eutrophication; (IV) Conducting the research using the Contingent Valuation Method necessary for verifying the estimated value.

It is worth remembering that the Replacement Cost Method cannot be assessed as a credible measure of the economic value without performing comparative research using the Contingent Valuation Method – CVM [37-39]. Consequently, the estimation of the economic valuation of balancing the eutrophication effects had been divided into the following research tasks: (I) The analysis of additional substitutes for an ecosystem service; (II) Constructing two valuation scenarios and conducting a Focus Group Interview; (III) Conducting pilot research and verifying the scenarios; (IV) Conducting the main research – estimating the socio-economic value of an eutrophication effect balancing service using the CVM; (V) The statistical analysis of survey results; and (VI) The analysis of value changes resulting from the declared preferences in connection with a reduction in the amount of nitrogen in the Gulf of Gdansk and aggregation of the results for the Polish population.

The Contingent Valuation Method, until recently was considered the most controversial method of the environmental valuation. It is used for a broad spectrum of problems since it enables estimating both use and non-use values. The Contingent Valuation Method is based on two assumptions. Firstly, consumers know best what is best for them, and secondly, the market participants are able to reasonably and consciously determine their preferences. This method attempts to translate people's opinions to monetary values. The technique tries to determine the value that people attribute to particular ecosystem goods or services. The way to achieve this goal is to examine peoples' opinions through interviews. Surveys are conducted among a representative group of consumers.

In everyday life it often happens that we realise the real value of something, only when we lose it. The basis for valuation is therefore a fact or even a possibility of losing the value [32]. Consequently, the valuation is expressed by the willingness to cover the costs. Subsequently, the respondents are asked to clearly state how much they are willing to pay for ecosystem services (WTP: Willingness To Pay).There is no market where the services balancing the effects of eutrophication would be the subject of trade. In consequence, the scenario presenting a particular situation in which a respondent is supposed to provide a particular WTP value, should contain enough information and ought to be the most probable. As a rule, the scenario used during the Contingent Valuation Method research informs respondents about the hypothetical, but sometimes real situation of the deteriorating environment.

It was assumed that respondents behave identically to the real market, they are aware of decisions taken, and their consequences, and finally understand the presented situation. Nowadays, most economists agree that the results obtained by this method can be reliable under certain necessary conditions. This method, like any other tool, must be used with full awareness of its shortcomings and limitations.

The survey will took into account the specificity of the examined service, i.e. balancing the effects of eutrophication services in the Gulf of Gdansk ecosystem and the NOAA's recommendations [40]. The process of designing the survey assumes that: (i) the situation in terms of which the respondent makes the transaction will be thoroughly described, (ii) factsheet illustrating the effects of eutrophication, used in order to familiarise the surveyed with the valuation scenario, was tested in the pilot research, (iii) the used format of questions was a question about willingness to pay – WTP, (iv) the respondents were informed about the existence of substitutes for the services mentioned in the valuation scenario, (v) following questions about participation in the transaction (yes / no), a question about the reasons for taking a particular decision followed (vi).

The survey consisted of 11 questions plus a section providing information on socio-demographic characteristics of respondents, as for instance age, sex, residence, education, occupation, number of persons constituting the household and monthly net household income. The main survey was carried out at 1000-person representative sample of adult Polish citizens in 2014. Interviews were conducted face-to-face, and performed by the trained interviewers. The representative sample being tested is a random-quota sample whose composition corresponds to the population according to the following features: voivodeship, place of residence (city or village), gender and age. The sizes of individual features were calculated on the basis of the Local Data Bank of the Central Statistical Office. Stratified sampling was used. In randomly chosen locations, the interviewers selected the interviewees ensuring the nationwide representativeness of the sample due to the above-mentioned features.

3 Results – economic valuation of regulatory ecosystem services – removal of nitrogen

3.1 Quantification of the regulating service, the amount of nitrogen removed through denitrification in sediments

Ecosystem processes responsible for removal of nitrogen from the water column (denitrification) and mitigating the effects of eutrophication roughly utilise 20410 tons of nitrogen (41). This corresponds to ca. 15% of the total supply of nitrogen from external sources.

3.2 Indicating a substitute for the provided ecosystem service

The second stage aimed at indicating the artificial substitute for the service provided by the environment, complying at the same time with the condition that the substitute should be the cheapest one [39]. Consequently, the biological wastewater treatment process on the example of the company Saur Neptun Gdansk S.A. was selected. The company is responsible for the collective water supply and sewage disposal and serves about 500, 000 inhabitants from the area of the city of Gdansk and Sopot.

3.3 The economic valuation of mitigation of eutrophication by using Replacement Cost Method

Annually, the sewage treatment plants of the discussed company receive 33 640 228 m^3 of wastewater containing 2 795,5 tons of nitrogen [42]. The wastewater is biologically treated, which costs about 3.38 PLN/m^3. In the sewage treatment process the nitrogen content is reduced by 85.3% which corresponds to the removal of 2 132,8 tons of nitrogen via process of denitrification. As it was mentioned in point 3.1, the Gulf of Gdansk is responsible for the removal of 20 410 tons of nitrogen [41]. It takes all phases in biological wastewater treatment (incl. primary mechanical treatment) to perform the nutrients reduction, therefore it can be assumed that the total cost of treatment per $1m^3$ of wastewater can be used in the nitrogen reduction. The estimated value of service that is being discussed is approximately assessed at 1 088 099 231 PLN per year. And removal of one tonne of nitrogen costs about 53 312,06 PLN annually.

3.4 Verification of RCM results by conducting Contingent Valuation Method (CVM)

The results of the survey allowed mainly to estimate the economic value of the ecosystem service, i.e. is the denitrification process occurring in the Gulf of Gdansk ecosystem. However, the wide range of information obtained in the study made it also possible to find an answer to the question how the declared WTP depends on various sociological, economic and demographic factors. In order to investigate the impact on the WTP value of these variables, the approach used by Ressurreição [43] was applied. An econometric model explaining the natural logarithm of the WTP amount declared by the respondents was estimated [44].

Based on the results of the survey, we can determine the average WTP value declared by respondents, not protesters. This is 21.14 PLN per person with a 95% confidence interval (17.33; 24.95). Taking into account the average income of respondents, this would be about 0.78% of the average monthly salary.

When interpreting the results of the model, it should be stated that the income of respondents had a high positive impact on the declared value of WTP, the impact is statistically significant at the lowest level of significance regardless of the specification of the model. This means that people with higher incomes are more inclined to get involved in financing the analysed project. A positive statistically significant impact of the level of education has been demonstrated. People with higher education are more likely to incur costs related to the protection of the Gulf of Gdansk. It also turned out that among the surveyed people, urban residents declared smaller amounts than the residents of rural areas, but assuming the same level of income, education, age and number of members in the family.

Analysing the impact of environmental awareness of respondents, it was noticeable that the more negative effects a respondent experienced, the greater the amount they were willing to declare. In addition, knowledge of each of the natural regulatory functions of the Gulf of Gdansk contributed to a greater tendency to co-finance - the most important function turned out to be ‘maintaining the gas and climate balance’. It was also shown that a respondent's region of residence was significant for the declared WTP amounts. People living in the following voivodeships: Masovia, Podlaskie, Silesia, Warmia-Masuria, Greater Poland, and Lubusz declared lower values than the national average. Holy Cross Province and West Pomerania - on the national average level. The other voivodeships indicated a greater tendency to co-finance the analysed project

The non-protesters group [45] constituted 69.8% of the respondents. Therefore, treating the sample on which the survey was conducted as a representative sample, we may assume that a similar part of the entire adult Polish population would be willing to incur such a one-off payment to finance technological solutions replacing the lost natural self-cleaning functions of the Gulf of Gdansk. Data from the Polish Statistical Office shows that in 2013, the population above 20 years of age was 34,482,874 people. If 69.8% of those people paid 21.14 PLN then the amount collected would be 508 819 978 PLN. Taking into account the 95% confidence interval, this would be between 417,210,334 PLN and 600,429,622 PLN.

4 Discussion – how to link the economic value of ecosystem functions and processes to the MSP?

The use of a set of direct and indirect methods of economic valuation enabled us to positively verify the research hypothesis which assumed that the value of mitigating the effects of eutrophication services, estimated on the basis of replacement cost, is higher than the value expressed by social preferences. The results of the survey made it possible first of all to estimate the economic value of the ecosystem service, i.e. is the denitrification process taking place in the Gulf of Gdansk ecosystem, according to the respondents' opinions. The average value of participating in the costs of this service was declared by respondents at the level of 21.14 PLN per person with a 95% confidence interval (17.33; 24.95). The generalised total value of WTP for all residents of Poland is 508 819 978 PLN, which means about 87 thousand PLN per km^2 of the Gulf of Gdansk sediments [44]

Compared to the value of the service calculated on the basis of replacement costs, the removal of 20,410 t of nitrogen means an expenditure of approx. 1 088 099 231 PLN. In simplified terms, denitrification services are provided by sediments cost-free. Theoretically, the use of sediments for bioremediation may give rise to an alternative cost (e.g. a ban on erecting structures located on the bottom). In Polish conditions in internal and territorial waters it is not possible to build wind farms, and mariculture in the Gulf of Gdansk is not a viable alternative to seabed. There are also no studies on external costs generated by such sediments. Therefore, the zero value of the alternative cost was assumed. In view of the above, it is enough to divide the amount by the surface the Gulf of Gdansk sediments (5,852 km2 of sediments) to obtain the value of the spatial rent of 185.9 thousand PLN per km2 of sediments for 1 year. The WTP value per ha of sediments is significantly lower than the previously calculated replacement costs (87,000 PLN per km2). The survey used the elicitation format, i.e. one-off payment, which makes it very difficult to compare with the substitution costs results by a particular substitute for nitrogen removal on land expressed in PLN per year. The value declared by the respondents should therefore be 'sufficient' to solve the problem of eutrophication with this one payment. This means a significant underestimation of the value of ecosystem services by Polish citizens. The degree of supply of organic pollutants to the Baltic is decreasing, but unfortunately to a small extent, so the Gulf of Gdansk function is needed all the time. Despite these differences, the way to the practical use of the study results is not closed.

This study shares knowledge on the economic value of maritime services (in this case regulatory services) to the maritime policy. This knowledge can be used to support actions for the sustainable use of the seabed and its mineral resources, complementing the existing bottom maps, indicating selected areas of the seabed important from the viewpoint of the economic development and protection of ecosystem services. Maps supplemented with data on economic and ecological value of ecosystem services will be the starting point for spatial planning in maritime areas and proper management of seabed areas. The estimated economic value of maritime services can be used as a tool to raise environmental awareness among maritime decision-makers and users, and this may result in fewer conflicts between the economic exploitation of the sea and the needs of the environmental protection.

To make the marine spatial plans deciding upon conflicting uses, the monetary value of various benefits might provide a useful insight and guidelines. Nonetheless, at the same time they can bias planning towards quantifiable benefits only. Many research questions are still pending. What form of the results of economic valuation is needed to support a knowledge based (evidence based) decision in marine spatial planning? To what extent will the understanding of space requirements change if we introduce economic values to MSP?

Acknowledgments

The paper presents the outcomes of the research financed by the Polish National Science Centre under the project: 'Economy of maritime space' 2015/17/B/HS4/00918 as well as 'Economic valuation of balancing the effects of eutrophication processes - regulating ecosystem services of Gulf of Gdansk in the context of promoting sustainable development' granted by decision number DEC-2012/05/N/HS4/00960. We would like to also thank the company Saur Neptun Gdańsk for sharing the data with authors.

References

1. EU Directive 2008/56/EC of the European Parliament and of the Council of 17 June 2008 establishing a framework for community action in the field of marine environmental policy (Marine Strategy Framework Directive) (2008)
2. S. Derous, M. Austen, S. Claus, N. Daan, J.C. Dauvin, K. Deneudt, J. Depestele, N. Desroy, H. Heessen, K. Hostens, A.H. Marboe, A.K. Lescrauwaet, M. Moreno, I. Moulaert, D. Paelinckx, M. Rabaut, H. Rees, A. Ressureição, J. Roff, P.T. Santos, J. Speybroeck, E.W.M. Stienen, A. Tatarek, R. Ter Hofstede, M. Vincx, T. Zarzycki, S. Degraer, *Building on the concept for marine biological valuation with respect to translating it to a practical protocol: Viewpoints derived froma joint ENCORA-MARBEF initiative*. Oceanologia **49**, 4 (2007)
3. T. Zarzycki, *Ecological and Socio-Economic Valuation of Marine Biodiversity of the Gulf of Gdańsk*, PhD thesis, Faculty of Oceanography and Geography, University of Gdansk (2012)
4. N.J. Beaumont, M.C. Austen, J.P. Atkins, D. Burdon, S. Degraer, T.P. Dentinho, S. Derous, P. Holm, T. Horton, E. van Ierland, A..H. Marboe, D.J. Starkey, M. Townsend, T. Zarzycki, *Identification, definition and quantification of goods and services provided by marine biodiversity: Implications for the ecosystem approach*, Marine Pollution Bulletin **54**, 253–265 (2007)
5. Millennium Ecosystem Assessment. Island Press, Washington DC (2005)
6. C.M. Holmlund, M. Hammer, *Ecosystem services generated by fish populations*, Ecol. Econ. **29**, 253–268 (1999)
7. E.M. Borgese, *The economics of the common heritage*, Ocean and Coastal Management **43**, 763–779 (2000)
8. J.M. Wesławski, E. Andrulewicz, L. Kotwicki, E. Kuzebski, A. Lewandowski, T. Linkowski, S.R. Massel, S. Musielak, K. Olanczuk-Neyman, J. Pempkowiak, H. Piekarek-Jankowska, T. Radziejewska, G. Różynski, I. Sagan, K.E. Skóra, K. Szefler, J. Urbanski, Z. Witek, M. Wołowicz, J. Zachowicz, T. Zarzycki, *Basis for a valuation of the Polish Exclusive Economic Zone of the Baltic Sea: Rationale and quest for tools*, Oceanologia **48**, 1, 145–167 (2006)
9. L. Lewan, T. Söderqvist, *Knowledge and recognition of ecosystem services among the general public in a drainage basin in Scania, Southern Sweden*, Ecol. Econ. **42**, 459–467 (2002)
10. R. Costanza, R. d'Arge, R.S. de Groot, S. Farber, M. Grasso, B. Hannon, K. Limburg, S. Naeem, R. O'Neill, J. Paruelo, R. Raskin, P. Sutton, M. van den Belt, *The value of the world's ecosystem services and natural capital*, Nature **387**, 253–260 (1997)

11. D. Pimentel, C. Wilson, C. McCullum, R. Huang, P. Dwen, J. Flack, Q. Tran, T. Saltman, B. Cliff, *Economic and environmental benefits of biodiversity*, Bioscience **47**, 11, 747–757 (1997)
12. K.C. Ewel, R.R. Twilley, J.E. Ong, *Different kinds of mangrove forest provide different goods and services*, Global Ecol. Biogeogr. **7**, 83–94 (1998)
13. F. Moberg, C. Folke, *Ecological goods and services of coral reef ecosystems*, Ecol. Econ. **29**, 215–233 (1999)
14. R.S. de Groot, M.A. Wilson, R.M.J. Boumans, *A typology for the classification, description and valuation of ecosystem functions, goods and systems*. Ecol. Econ. **41**, 3, 393–408 (2002)
15. L. Hein, K. Koppen, R. de Groot, E. Ierland, *Spatial scales, stakeholders and the valuation of ecosystem services.* Ecol. Econ. **57**, 209–228 (2006)
16. *The economics of ecosystems and biodiversity: mainstreaming the economics of nature: a synthesis of the approach, conclusions and recommendations of TEEB* (2010)
17. J. Maes, A. Teller, M. Erhard, C. Liquete, L. Braat, P. Berry, B. Egoh, P. Puydarrieux, C. Fiorina, F. Santos, M.L. Paracchini, H. Keune, H. Wittmer, J. Hauck, I. Fiala, P.H. Verburg, S. Condé, J.P. Schägner, J. San Miguel, C. Estreguil, O. Ostermann, J.I. Barredo, H.M. Pereira, A. Stott, V. Laporte, A. Meiner, B. Olah, E. Royo Gelabert, R. Spyropoulou, J.E. Petersen, C. Maguire, N. Zal, E. Achilleos, A. Rubin, L. Ledoux, C. Brown, C. Raes, S. Jacobs, M. Vandewalle, D. Connor, G. Bidoglio *Mapping and Assessment of Ecosystems and their Services. An analytical framework for ecosystem assessments under action 5 of the EU biodiversity strategy to 2020*. Publications office of the European Union, Luxembourg (2013)
18. MAES, *Indicators for mapping ecosystem services*. 2nd MAES Working Paper. European Union, doi: 10.2779/75203 (2014)
19. UK National Economic Assessment (UK NEA) *The UK National Ecosystem Assessment*. Technical Report. UNEP-WCMC, Cambridge (2011)
20. K. Turner, M. Schaafsma, M. Elliott, D. Burdon, J. Atkins, T. Jickells, P. Tett, L. Mee, van S. Leeuwen, S. Barnard, T. Luisetti, L. Paltriguera, G. Palmieri, J. Andrews, *UK National Ecosystem Assessment Follow-on*. Work Package Report 4: Coastal and marine ecosystem services: principles and practice. UNEP-WCMC, LWEC, UK (2014)
21. T. Söderqvist, L. Hasselström, *The economic value of ecosystem services provided by the Baltic Sea and Skagerrak*. The Swedish Environmental Protection Agency Report 5874 (2008)
22. HELCOM, *Ecosystem Health of the Baltic Sea 2003–2007*: HELCOM Initial Holistic Assessment. Balt. Sea Environ. Proc. No. **122**. (2010)
23. B. Hasler, *Marine Ecosystem Services: Marine Ecosystem Services in Nordic Marine Waters and the Baltic Sea - Possibilities for Valuation,* Nordic Council of Ministers, (2016)
24. R. Haines-Young, M.B. Potschin, *Common International Classification of Ecosystem Services (CICES) V5.1 and Guidance on the Application of the Revised Structure*. Available from www.cices.eu CICES (2018)
25. HELCOM, *Sources and pathways of nutrients to the Baltic Sea*. Baltic Sea Environment Proceedings No. **153** (2018)
26. M.M. Hufschmidt, D.E. James, A.D. Meister, B.T. Bower, J.A. Dixon, *Environment, natural systems, and development: an economic valuation guide*. Baltimore, EUA: Johns Hopkins University Press (1983)
27. J. Braden, C. Kolstad, (Eds.), *Measuring the Demand for Environmental Quality*. Elsevier Science Pub., New York (1991)
28. W.M. Hanemann, Preface. In: *Pricing the European Environment*, S. Navrud (ed.), Scandinavian University Press, Oslo (1992)

29. A.M. Freeman, *The measurement of environmental and resource values*. Washington: Resource for the Future (1993)
30. J.A. Dixon, L.F. Scura, R.A. Carpenter, P.B. Sherman, *Economic Analysis of Environmental Impacts*. London: Earthscan Publications Ltd. (1994)
31. W.M. Hanemann, *Willingness to pay and willingness to accept: How much can they differ?* American Economic Review **81**, 3, 635–647 (1991)
32. J. Shogren, J. Hayes, *Resolving differences in willingness to pay and willingness to accept: A reply*. American Economic Review **87**, 241–2 (1997)
33. T. Żylicz, *Ekonomia wobec problemów środowiska przyrodniczego*, PWN, Warszawa (1989)
34. T. Żylicz, *Ekonomia Srodowiska i Zasobów Naturalnych*, PWE, Warszawa (2004)
35. *Integrated Coastal Zone Management: Review of Progress in Selected OECD Countries*, OECD, Paris (2007)
36. S. Pagiola, P. Agostini, J. Gobbi, C. de Haan, M. Ibrahim, E. Murgueitio, E. Ramírez, M. Rosales, J.P. Ruíz, *Paying for biodiversity conservation services in agricultural landscapes*. Environment Department Paper No. **96**. Washington: World Bank (2004)
37. L.A. Shabman, S. Batie, *Economic Value of Natural Coastal Wetlands: A Critique*. Coastal Zone Management Journal, **4**, 231-247 (1978)
38. A.M. Freeman, *The Measurement of Environmental and Resource Values – Theory and Methods*. 2nd Edition. Washington, D.C.: Resources for the Future (2003)
39. S. Sundberg, *Replacement costs as economic values of environmental change: A review and an application to Swedish sea trout habitats*. Swedish Environmental Protection Agency (2004)
40. K. Arrow, R. Solow, P. Portney, E. E. Learner, R. Radner, H. Schuman, *Report of the NOAA Panel on Contingent Valuation*, National Oceanic and Atmospheric Administration (1993)
41. B. Graca, Dynamika przemian azotu I fosforu w strefie kontaktu wody z osadem dennym w Zatoce Gdańskiej. Wydawnictwo Uniwersytetu Gdańskiego, Gdańsk (2009)
42. SNG, Saur Naptun Gdansk, pres. comm. (2014)
43. A. Ressurreição, J. Gibbons, P.T. Dentinnho, M. Kaiser, R. Santos, G. Edward-Jones, *Economic valuation of species loss in the open sea*, Ecological Economics, **70**, 4, 729-739 (2011)
44. I. Wrześniewska, D.Ciołek, T. Zarzycki, Modelowanie ekonomicznej wartości denitryfikacji jako jednej z usług ekosystemowych Zatoki Gdańskiej z wykorzystaniem WTP, Journal of Management and Finance **13**, 4/2 (2015)
45. J. Halstead, A. Luloff, T. Stevens, *Protest Bidders in Contingent Valuation*, Northeastern Journal of Agricultural and Resource Economics **21**,160-169 (1992)

Spatial Evolution of the European Container Ports' System in Perspective of the Location Theory

Karolina A. Krośnicka[1*]

[1]Department of Urban Design and Regional Planning, Faculty of Architecture, Gdańsk University of Technology, Narutowicza Street 11/12, 80-233 Gdańsk, Poland

Abstract. The maritime container terminal is no longer a spatially coherent object. Functionally it ends, where their most external components are located. The process of location splitting of container terminals (ger. *Standortspaltung*) can be treated as next stage of their discrete growth. The new container facilities are being built to improve the containers' flow, passing from port terminals to cities situated in their hinterland and vice versa. The external components of container terminals have a very diverse program, are functionally complex, and due to advanced technologies of information and logistics they are interconnected into one system. The structure of the functional bindings of a maritime container terminal could be compared to the dendritic shape of a neuron, the kernel of which is a terminal, and the arms are transportation corridors ended with distant land intermodal terminals. The physical feature of this system is the logistics landscape with vast areas and large cubature. The aim of the paper is to present the graphical model explaining the changes in distribution and hierarchy of container terminals' external elements in relation to the network of cities and transportation.

1 Evolution of the container system

The UNCTAD report [1 p. 4], same as the work of Rodrigue, Comtois and Slack [2], distinguishes three phases of the evolution of the container system. These are in turn: containerization of the maritime transport system, containerization of the land transport system and the development phase of intermodal operations. This paper [see also: 3] proposes adding to the mentioned phases the fourth one: the phase of integration of the container system. These stages are clearly reflected in changes of the spatial structure of the container system. Although they do not have sharp chronological boundaries, they can nevertheless be given a contractual time frame. The next phases of the spatial evolution of the container system are described below and are shown in Figure 1.

The stage of development of the maritime container subsystem (1956-1970) is the period of crystallization of the maritime network. Land transport connections of container terminals were based mainly on the existing general cargo distribution system. In this phase the first terminals serving containers were erected. Their appearance shaped the later course

* Corresponding author: karkrosn@pg.edu.pl

of shipping routes. In the initial period, the implementation of innovative container infrastructure entailed an increase in the cost of shipping companies, as the port authorities fearing the lack of investment depreciation, passed on the cost of adapting quays for ship-owners [4, pp. 7, 136].

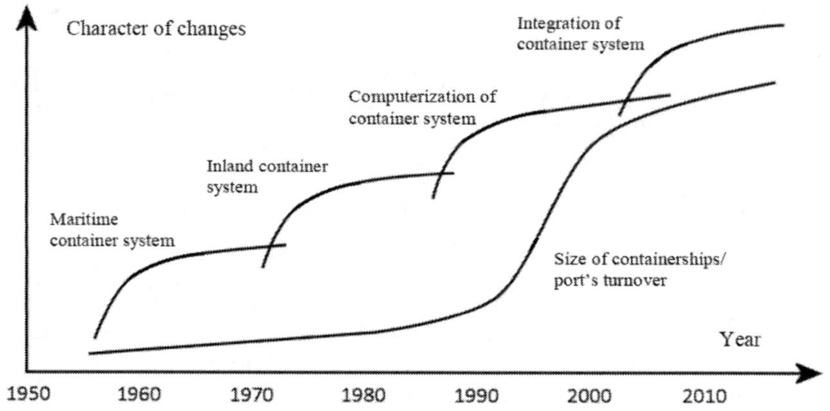

Fig. 1. Stages of the world's container system development [3, p. 69].

In the phase of development of the inland container subsystem (1970-1985/1988) a division into main container ports and feeder ports took place. Transport corridors running from the main container ports inlands have developed and transport connections parallel to the shoreline appeared. In the area directly adjacent to container terminals (up to 5 km), in vicinity of large inland cities, and in the transportation nodes (usually situated about 300 km from sea terminals) the new logistic objects, functionally connected with container terminals, have been created. In this phase the development of intermodal operations and the containerization of the inland transport system took place (1970s - the end of the 1980s). This stage consisted in improving the efficiency of the land-based container system, as well as in compacting the transport and logistics connections (highways) and in the development of regional inland transport hubs. In many cases, the organization of the land transport chain was taken over by ship-owners. This phenomenon Krasucki and Neider [4, p. 136] described as "the descent of ship-owners to land". As the route of the load could be organized using a single transport document, it became possible to introduce combined container transports. Containerization has forced the railway network to adapt to the operation of heavier trains, with a higher frequency of commodity trains. A new type of commodity railway stations operating at the interface of at least two modes of transport, also emerged – inland railway container terminals were created (intermodal terminals). In the 1980s the differentiation between container hubs and feeder ports has deepened, and the gradual transformation of the global shipping routes took place.

In the third phase of development (1985/1988 - 2001/2003), sometimes referred as the "logistics platform of trade", the container system was computerized. An integrated electronic data transmission system has started to develop, which has allowed port services to extend beyond the immediate port borders. The result of the development of port info-structure strengthened the links between the port and ports' users, as well as between the port and the port city. Container transport has proliferated. At the same time, the phenomena of main-porting (development of hub ports) and of multi-porting (development of many small feeder ports) occurred. In the foreland of some container hubs gate-way terminals started to be created. Deepwater container terminals of the off-shore type have appeared in most of ports [5]. At the ports' hinterland the multimodal door-to-door transport and inland terminals have developed. Some of the functions were derived from the terminals into the urban area of the port city (up to 50 km). Inland transport corridors extended perpendicularly to the shore (over 300 km), and the second-line logistic connections appeared. In the area of further hinterland (150-300 km) a network of logistics centers has been developed. The development of the logistics system has led to the expansion and overlapping of hinterlands of larger container ports.

According to Lu and Chang [6, p. 155] at the beginning of the 21st century the container system entered the fourth phase of development (2001/2003-). For this phase according to Monios [7, p. 37] two parallel tendencies are characteristic: enlargement of port facilities by creating inland terminals integrated with the maritime ones (port regionalization), and integration of logistics chains. At the same time, following the concept of Robinson [8] and Woo et al. [9] that the port has now lost its monopoly position and has become a place of dynamically changing logistic links. So far, seaports competed with each other mainly in terms of the scale of transport, quality of services, diversification of activities and prices for services. At present, it is not ports that compete with each other – it is integrated logistics supply chains, in which ports are only one of the components, and not any more an independent transport node [10, pp. 155, 157, 168]. The fourth phase of the container system development consists in the possibility of providing transport services within the whole available land connection network and various loading and unloading ports, using various means of land transport. Therefore, now the transition from the model of "determined supply chain" to "flexible road selection" (synchro-modal transport) take place. This concept is, among other things, the result of treating the supply chain as a warehouse space (the concept of traveling warehouse) and possibility of taking decisions on the means of transport on an ongoing basis during the transport process [11, p. 570]. For this reason, the management of seaports in various ways supports and coordinates the creation of investments in their hinterlands, developing and strengthening the functioning of transport corridors and logistics supply chains, including inland transshipment terminals, logistics centers and dry ports. During the integration of the container system, in addition to strengthening the previously mentioned trends, the construction of automated deep-water terminals and dry ports took place, both in the zone of the inactive port city region (at a distance of 50 to 150 km from the terminal) and in the zone above 300 km from the terminal. The transport corridors of the hinterland were also extended and reoriented. The networking of transport and popularization of synchro-modal transport took place. In China the phenomenon of clustering dry ports was observed [6] . Notteboom and Rodrigue [12, p. 303] also noted, that the inland terminal may now be an island formation in the hinterland of another container port, intercepting the cargo in favor of its "own" port. Thus the port's hinterland may be discontinuous. According to Notteboom and Rodrigue [12, p. 306] local and national authorities are increasingly interfering in the process of shaping the logistics network trying to regulate the location of individual logistic facilities, including through fiscal incentives. Roso and others [13], relying on Swedish experience, believe that dry ports are a manifestation of a conscious strategy of port or regional authorities. Other authors however suggest, that the process of regionalization is imposed on ports by external forces, and that so far these activities are not coordinated enough to be considered as strategic management of the logistic network in the region or country [14, p. 41], [12, p. 307].

The trends that currently appear, which may be the next phase of containerization, are the automation of sea and land transport (automatic vesels and cars) and blockchain technology (the so-called dispersed register). The tests using blockchain technology have already been carried out by the port of Antwerp, Rotterdam and by Møller - Mærsk.

2 Model of spatial evolution of the container system in terms of location theory

The paper presents a model of the evolution of the Europe container system in terms of the location theory, taking into account in particular the development of Polish container ports (Fig. 1.). In Europe, changes within the container system were different than in Asia or North America. In Europe, for historical reasons, a regular network of large cities (as for a European scale) is complemented by a dense network of small and medium-sized cities, which results in a more even distribution of transport corridors at the port cities' hinterlands. In Asia, large cities are concentrated in the coastal zone, hence port cities are also the main producers and recipients of cargo so the inland transport network is relatively underdeveloped. Significant distances between the larger cities of North America demand long-distance transport corridors, so-called land bridges, which achieve dominance over other, shorter transport connections [15, slide 24].

The model was based on research methodology, joining the following methods:

- Review of both historical and contemporary literature concerning container terminals operations and their spatial configuration;
- Spatial analysis of both historical and contemporary container terminals' plans and configuration (including the distribution of logistics' objects co-operating with container terminals) based on cartographic material (plans, maps, and aerial photos) and study visits;
- Spatial analysis of urban network evolution in the light of W. Christaller central places theory;
- Analysis of cargo turnover and operation technologies of container terminals based on statistic data, professional literature, as well as study visits and interviews with terminals' employees.

The presented model was built on the basis of Rodrigue and Notteboom's [16, pp. 19-29] and Wang and Decruet's [17] models. The spatial evolution of the container system has been the subject of many researchers' work. One of the first studies was created by Taaffe and others [18, p. 505] proving that the concentration of cargo in selected seaports (defined as the level of annual turnover) leads to the creation of priority corridors in the transport network of a hinterland. Based on the Taaffe's and others model [18] a five-stage Rimmer's model [19] was created, describing the development of ports along with the hinterland. In the 1980s Hayuth [20] introduced the concept of dominant ports and land load centers, which expansive development took place by gradually taking over the hinterlands of other ports. This phenomenon led to the concentration of cargo in a small number of strongly developed hub ports and monopolization of the main transport corridors. Barke [21], continuing Hayuth's considerations, additionally took into account the relocation of port functions to the undeveloped areas of port cities with efficient transport and communication links with the port. Van Klink [22] stated that the new role of the port was to take over logistic control over the inland network, and to integrate the work of spatially independent facilities. He also suggested enlarging the range of port logistics receipts in the subsequent phases of port development from the local scale (port city), through the regional one, to the macro scale (transport network). Van Klink [22] thus demonstrated the importance of the cooperation strategy between the port support entities and the inability to operate the port as an independent transport facility.

Fleming and Hayuth [23] and then Ducruet [24] supplemented the concept of "centrality" (the creation of gate-way ports being an access to hinterland) with the concept of "intermediacy". According to the concept of intermediacy, hinterland is spatially discontinuous (the logistics network is reduced to linear elements related with islands of inland distribution centers and transshipment/gate-way ports). Therefore the vast direct hinterland's market is no longer an indispensable factor in the concentration of large cargo streams.

Slack and Wang [25], Notteboom [26], Frémont and Soppé [27], Monios and Wilmsmeier [28] noticed the coexistence of cargo concentration trends in specific ports and cargo de-concentration in their hinterlands. As the next stage, following the phenomenon of cargo de-concentration in ports' hinterlands, Notteboom and Rodrigue [29] isolated the phase of regionalization. The existence of the regionalization phase was partly questioned by Rimmer and Comtois [30, p. 38], who considered it a continuation of the decentralization phase. Rodrigue and Notteboom [16, pp. 19-29] have synthesized previous models of the development of the port system and presented their own model, in which they distinguished six successive phases. Rodrigue and Nottebom [16, pp. 19-29] as well as Wang and Ducruet [17, pp. 58-69] referred to the model, indicating the last four phases of the model (two concentrations and two de-concentration) as those, that actually involve the evolution of the container system, recognizing the first two phases of Rodrigue and Notteboom model (2010) as the period of containerization experiments.

The six phases of the containerization process, mentioned in the Rodrigue and Notteboom model [16], of which the first four phases were described by Taaffe, Morrill and Gould [18], Hayuth [20], Barke [21], are therefore as follows:

1. an even distribution of ports and their hinterlands in the coastal space,
2. penetration and gradual capture of hinterlands by other ports,
3. formation of strong inland and sea connections (transport corridors) and gradual centralization (concentration of cargo mass in specific ports), which results in division of ports into hubs and feeder ports,
4. phase of full port centralization and creation of a hub-and-spoke service system,
5. phase of decentralization and appearance of large deep-water transit terminals in the foreland of ports (called: transshipment hubs, intermediate hubs, offshore hubs, gate-way terminals),
6. regionalization of both the foreground and the hinterland.

Since the concentration of cargo in a given container port occurs parallel with the spatial dispersion of functional elements of container terminals, with regard to the spatial evolution of the container system, it is not entirely accurate, that Rodrigue and Notteboom (2010, pp. 19-29) as well as Wang and Ducruet (2012, pp. 58-69) separate the phases of concentration and de-concentration from each other.

In reference to the above-described stages of the container system evolution (Figure 1), and on the basis of analysis of technologies used in container terminals, spatial layout and road-rail network serving terminals, four phases of spatial evolution of the European container system have been separated [3]. These phases are presented by the model (Fig. 2, Tab. 1) of the evolution of Europe's container system in terms of location theory:

- **Phase 1** (1956-1970)
  - almost every container terminal had both its direct ocean connections and feeder links
  - in large terminals feeder connections from smaller terminals were concentrating
  - the terminals had inland transport connections reaching the border of the region (up to 150 km) and the distribution system of bigger cities (about 300 km).
- **Phase 2** (1970-1985/1988):
  - concentration of cargo in selected ports has led to the creation of large hub-type container terminals
  - the number of container terminals in the ports has increased - container multi-terminal ports occurred
  - ocean connections were existing almost exclusively in container hubs
  - a line of transport links parallel to the shoreline has been created

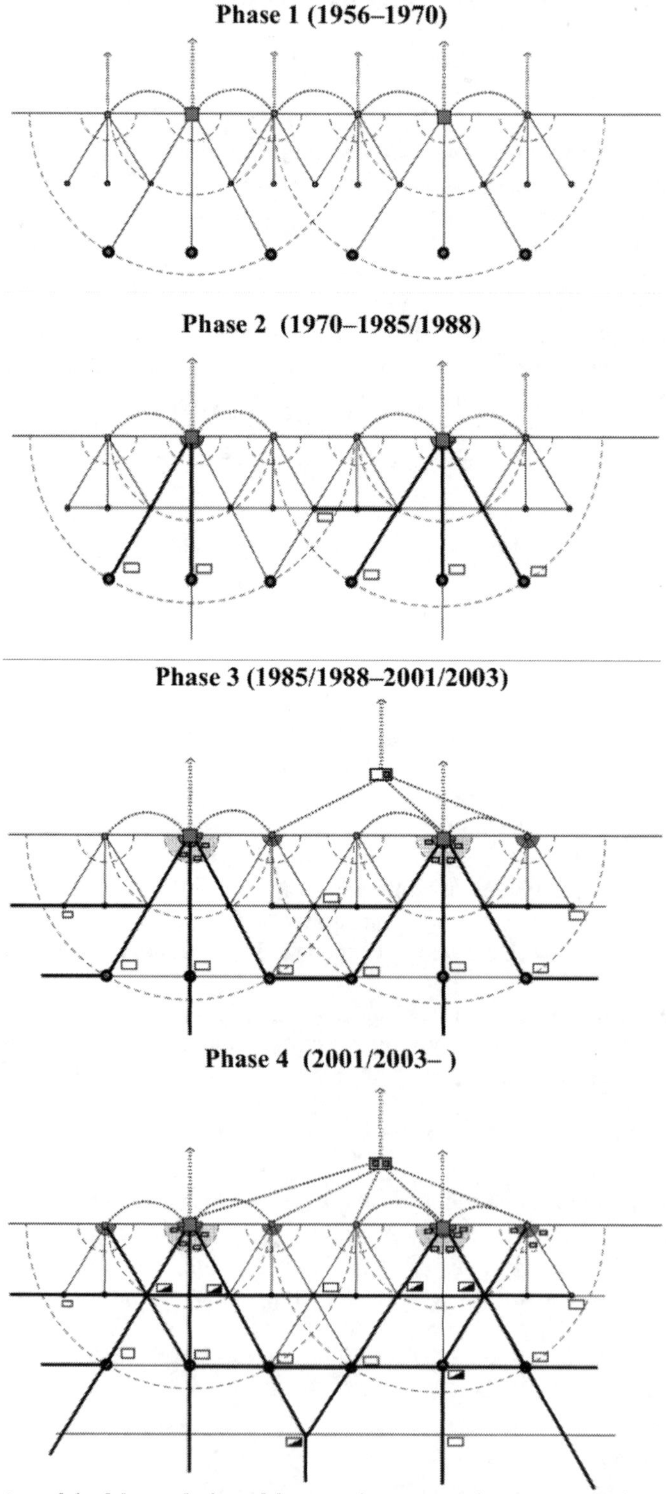

Fig. 2. A model of the evolution of the container system in Europe. Explanations at Table 1. [3, p. 73-75].

Table.1. Explanations to Fig. 2.

Nodes and objects:	**Zones of hinterland:**
City	Direct (up to 5 km)
Container terminal	Metropolitan (up to 50 km)
Gate-way terminal in foreland	- Regional (up to 150 km) - Further hinterland (up to 300 km) - Far hinterland (over 500 km)
Inland logistic centres	
Dry port	
External objects associated with container terminals in the metropolitan zone (up to 50 km)	

Transportation connections

Ocean connections	Inland transport corridors
Feeder connections	Inland secondary connections

Source: [3, p. 73-75]

- the hinterlands of smaller container ports were captured by parallel transport connections and by the inland transport corridors of large terminals, which in result transformed smaller ports into feeder-oriented ones
- container terminals began to expand external functions in the area directly adjacent to them (up to 5 km)
- strong inland transport corridors have developed, which connected container hubs with their further hinterland in the further hinterland (about 300 km from the terminals), at the nodes of transport corridors and in vicinity of inland metropolises, the distribution and logistics centers of these cities were created, functionally connected with container sea container terminals.

• **Phase 3** (1985/1988-2001/2003)
- in the crossings of shipping routes the gate-way terminals were established, concentrating cargo of many container terminals/ports in their foreland
- the largest ports have maintained ocean connections, while the rest served as terminals/feeder ports for hub or gate-way terminals
- big ports have built new deep-water container terminals of the third generation
- some of the functions were derived from container terminals outside (outsourced), into the area of the port city agglomeration (up to about 50 km)
- the inland transport corridors of larger container ports have been extended and strengthened (up to over 300 km)
- the network of logistic centers has expanded in the area of further facilities (150-300 km).
- transport links parallel to the shore (also of the second line) have been established and strengthened, so in result the hinterlands of the larger container ports began to overlap.

• **Phase 4** (2001/2003-)
- expansion of gate-way terminals on ports' forelands

- concentration of cargo in hub ports, which in addition to direct ocean connections, have feeder connections to smaller ports/terminals and to gate-way ports
- erecting in larger container ports/terminals the new deep-water fourth generation automated terminals
- construction of dry ports in the previously not active zone of the port city region (50-150 km).
- construction of dry long-haul ports (in the area above 300 km)
- as a result of mutual capturing of ports' hinterlands, the extension, expansion and partial reorientation of the transport corridors in the hinterland takes place. The system of transport connection obtained the surface character, and enabled flexible route selection (synchro-modal transportation).

The last phase (Figure 2, phase 4) of the development of the container system consists in networking of the cargo flow, and in the reorganization of hinterlands' coverage ranges of individual container ports. It can therefore be considered as an activity that at least theoretically equalizes the development opportunities of container ports by unifying transport accessibility from these ports to the hinterland. In practice, however, it leads to the expansion of the hinterland of these container ports, which have the most extensive system of functional arms in the hinterland. In this situation, one should agree with van Klink [31, 22], that currently the port has lost an ability to function as an independent transport hub. The new task of port management (or container terminal operators) has therefore become to control the availability of the land network and to integrate the work of spatially independent logistic facilities in the hinterland. Such an approach requires close cooperation of port authorities and terminals, both with the transport and logistics entities in the hinterland, as well as with the territorial self-governments leading their spatial policy.

Spatial de-concentration of container terminals consisting in the construction of their "booms" should be considered as a method of strengthening the position of a given container port and its functional area, and an attempt to concentrate the cargo on its main area. Concentration of cargo within a given supply chain is accompanied by spatial de-concentration of objects related to the flow of containers through this chain. This approach to the issue explains why Ducruet et al. [32, p. 359] ascribe the causes of spatial concentration to the agglomeration's activities, while the basis of the process of de-concentration is seen as the operation of global forces.

Christaller's communication principle (Ger. Verkhersprinzip) says that the most favorable location for settlement centers occurs when the largest number of towns in the cheapest way is communicated with the most important centers [33, pp. 63-64]. Hence, it can be concluded that the phenomenon of gradual transformation of polycentric settlement systems into node-band systems is a natural consequence of this principle.

In the theory of Christaller [33] and Lösch [34] the grid of the central centers system was determined by determining the temporal availability assuming a homogeneous, isotropic space, where the range assigning the area to the center was 21 and 36 km (one-day pedestrian/horse accessibility). The shape of Christller's hexagonal mesh [33, pp. 63-64] should still be treated as an extremely efficient transport solution with a very favorable structure index [35, pp. 34-35]. But modern technological possibilities and the real speed of transport (lorries, railways) have meant that temporary accessibility from large port cities should now be designated for other zones than was the case in the original theory. The coverage of these zones is determined by the actual accessibility of the car: within 1 hour (50 km), 4-6 hours - round trip within 1 day (approximately 300 km) and 1 day (over 500 km). The region of the port city of 150 km (1.5-2 hours of travel) should also be taken into account, where there are strong functional links with the central region of the region. These zones also have their fixation in the location of cities in the hinterland - the first line of cities are medium-sized cities located within a radius of 150 km from the port city, the next line is larger cities located within 300 km of the port cities.

In fact, the regular grid of accessibility is distorted by the resistance of a heterogeneous space [36, p. 185]. In the case of gate-way cities [37, pp. 269-285], the network significantly disturbs the course of far-reaching transport corridors. Therefore, Christaller's theory should be supplemented with the theory of system-network theory, developed by Hohenberg and Hollen-Lees [38]. It refers to the open (not self-sufficient) and anisotropic space and assumes that the settlement structure is a system of nodes (vertices), whose hierarchy develops mainly through the transport system and far-reaching contacts and interactions. Therefore, the regular Christaller's hexagon in the case of harbor metropolis is disproportionately stretched along the main transport routes leading to the back of the port city. The theory of network-systems indicates, that the role played by port cities in the settlement structure depends on the nature of transport infrastructure (i.e. the system of highways, railway lines, inland waterway connections, reloading and distribution and logistics centers).

Analysis of the spatial development of the container system authorizes the thesis, that the reorganization of container ports is one of the signs of their adaptation to the dynamically transforming hierarchical settlement system [33] into a mesh of an order of magnitude higher than the previous one - in the arrangement of metropolitan connections. As is known, the organization of the settlement structure is, as a rule, hierarchical, which results from the principle of concentration and accessibility of areas [2, p. 14]. The high cost of building the transport infrastructure means, that communication links are first established for the most important locations [2, pp. 14-15]. Therefore, the market areas of the centers are also hierarchical [36, p. 198], e.g. container ports facilities. What's more, Christaller [33, p. 15] pointed to the fact, that central places do not have to be cities at all, although they are intuitively associated with such a concentration [39, pp. 14-16], [40, p. 30]. Therefore, large complexes of logistic and distribution buildings, i.e. large dry ports located in strategic nodes of the transport network, could also be considered as central places so understood. Along with the expansion of transport corridors, perpendicular to the shoreline, corridors parallel to the shore are gradually being built, and transshipment centers are developed at the intersection of both directions, what leads to deepening and widening the range of hinterlands of selected container ports.

3 Conclusions

Concentration of cargo in a given port is accompanied by the development of a network of external facilities at the near and further hinterland of the container terminal (this phenomenon is called "Standortspaltung" by Schätzl [41]) to increase the retention of the hinterland and to increase the cargo flow through the terminal. Output of subsequent functions beyond the basic outline of the terminal (spatial de-concentration of container terminals) takes place respectively: in neighboring areas and in the distant hinterland, in the area of the port metropolis, in the port city region, and in its distant hinterland.

The effect of progressing globalization is the integration of the land and sea subsystems of each from container ports into one coherent (so-called seamless) supply chain, in which the flow of cargo and information is fast, fluent and highly automated. Currently, the process of integration of logistics chains of individual container ports into one global delivery system is underway, based on synchro-modal transport (flexible choice of route and mode of transport in real time). This process changes the current understanding of the concept of port's hinterland (the hinterland is no longer continuous), and remodel the previous hierarchy (structure of connections) of individual port and land agglomerations (metropolises), and thus the entire settlement structure. At the same time, the reorganization of the container system is also partly the result of the dynamic transformation of the settlement system. Modern possibilities of movement and the associated cardinal change in the temporal accessibility of the areas caused that the cell of the settlement structure increased by an order of magnitude in relation to the canvas of the original theory of central places [33]. From the point of view of transport, the main layout of the settlement system is currently a network of metropolitan connections with a radius of around 300 km. The port's metropolises, as gate-way cities [37], according to system-network theory [38], have a distorted netting, extending along the main transport corridors leading to the hinterland. In addition, according to the Christallerian communication principle [33], polycentric settlement structures are gradually transforming into node-band systems.

References

[1] UNCTAD (eds), *Development and improvement of ports. The principles of modern management and organization* (TD/B/C.4/AC.7/13), pp. 1–80 (1992)
[2] J.-P. Rodrigue, C. Comotois, B. Slack, *The Geography of transport systems*, 2 ed., Routledge Taylor and Francis Group, London-New York, pp. 1–352 (2009)
[3] K. A. Krośnicka, *Przestrzenne aspekty kształtowania i rozwoju morskich terminali kontenerowych*, Wydawnictwo Politechniki Gdańskiej, Gdańsk, ISBN 978-83-7348-684-3, pp. 1-233 (2016)
[4] Z. Krasucki, J. Neider, *Konteneryzacja w transporcie międzynarodowym*, PWN, Warszawa, pp. 1–131.
[5] A Ashar., 2002, *Revolution Now*, Containerization International, January, pp. 56–59 (1986)
[6] J. Lu, Z. Chang, *The Construction of Seamless Supply Chain Networks: The Development of dry Ports in China*, [in:] Bergqvist R., Wilmsmeier G., Cullinane K. (eds), *Dry Ports – A Global Perspective. Challenges and Developments in Serving Hinterlands*, Ashgate Publishing, Routledge, pp. 155–172 (2013)
[7] J. Monios, A functional analysis of dry port systems: the case of Spain, presentation at the Conference on Intermodal Strategies for Integrating Ports & Hinterland, Edinburgh (2010)
[8] J. Robinson, Global and world cities: a view from off the map, International Journal of Urban and Regional Research, **26(3)**, pp. 531–554 (2002)
[9] S.-H. Woo, S.J. Pettit, A.K.C Beresford, Port evolution and performance in changing logistics environments, Maritime Economics & Logistics, **13(3)**, pp. 250–277 (2011)
[10] J. Lu, Z. Chang, *The Construction of Seamless Supply Chain Networks: The Development of dry Ports in China*, [in:] Bergqvist R., Wilmsmeier G., Cullinane K. (eds), *Dry Ports – A Global Perspective. Challengesand Developments in Serving Hinterlands*, Ashgate Publishing, Routledge, pp. 155–172 (2013)
[11] J.G de Gijt, M.L. Broeken (eds), *Handbook Quay Walls*, Balkema – Proceedings and Monographs in Engineering, Water and Earth Sciences, CRC Press, pp. 1–718 (2005)
[12] T. Notteboom, J.-P. Rodrigue, Port regionalization: towards a new phase in port development, Maritime Policy and Management, **32(3)**, pp. 297–313 (2005)
[13] V. Roso, J. Woxenius, K. Lumsden, The dry port concept: connecting container seaports with the hinterland, Journal of Transport Geography, **17(5)**, pp. 338–345 (2009)
[14] J. Monios, G. Wilmsmeier, Giving a direction to port regionalization, Transportation Research Part A, 46(10), pp. 1551–1561 (2012)
[15] J.-P. Rodrigue, T. Notteboom, *The Future of Containerization: Box Logistics in Light of Global Supply Chains*, conference presentation, IAME, Greece, July 2007, pp. 1–25 (2007)
[16] J-P. Rodrigue, T. Notteboom, Foreland based regionalization: Integrating intermediate hubs with port hinterlands, Reaserch in Transport Economics, **27(1)**, pp. 19–29 (2010)

[17] C. Wang, C. Ducruet, New port development and global city making: emergence of the Shanghai-Yangshan mulilayered gateway hub, Journal of Transport Geography, 25, pp. 58–69 (2012)
[18] E.J. Taaffe, R.L. Morrill, P.R. Gould, *Transport expansion in undeveloped countries*, Geographical Review, 53(4), pp. 503–529 (1963)
[19] P.J. Rimmer, The search for spatial regularities in the development of Australian seaports 1861–1961/62, Geografiska Annaler, **49(1)**, pp. 42–54 (1967)
[20] Y. Hayuth, Containerisation and the load center concept, Economic Geography, **57(2)**, pp. 160–176 (1981)
[21] M. Barke, *Transport and trade*, Oliver & Boyd, Edinburgh, pp. 1–249 (1986)
[22] H.A. Klink van, The port network as a new stage in port development: the case of Rotterdam, Environment and Planning A, **30(1)**, pp. 143–160 (1998)
[23] D.K. Fleming, Y. Hayuth, Spatial characteristics of transportation hubs: centrality and intermediacy, Journal of Transport Geography, **2(1)**, pp. 3–18 (1994)
[24] C. Ducruet, Port city relationships In Europe and Asia, Journal of International Logistics and Trade, **4(2)**, pp. 13–35 (2006)
[25] B. Slack, J.J. Wang, The challenge of peripheral ports: an Asian perspective, GeoJournal, **56(2)**, pp. 159–166 (2002)
[26] T. Notteboom, *The peripheral port challenge in container port systems*, [in:] Leggate H., McConville J., Morvillo A., *International maritime transport: perspectives*, Routledge, pp. 173–188 (2005)
[27] A. Frémont, M. Soppé, *Northern European range: shipping line concentration and port hierarchy*, [in:] J. Wang, T. Notteboom, D. Olivier, B. Slack (eds), *Ports, cities, and global supply chains*, Ashgate, Alderschot, pp. 105–120 (2007)
[28] J. Monios, G. Wilmsmeier, The role of intermodal transport in port regionalization, Transport Policy **30**, pp. 161–172 (2013)
[29] T. Notteboom, J.-P. Rodrigue, Port regionalization: towards a new phase in port development, Maritime Policy and Management, **32(3)**, pp. 297–313 (2005)
[30] P.J. Rimmer, C. Comtois, China's container-related dynamics, 1990–2005, GeoJournal, **74(1)**, pp. 33–50 (2009)
[31] H.A. Klink van, Creating port networks: the case of Rotterdam and the Baltic region, International Journal of Transport Economics, **14(3)**, pp. 393–409 (1997)
[32] C. Ducruet, S. Roussin, J.C. Jo, Going West? Spatial polarization of the North Korean port system, Journal of Transport Geography, **17(5)**, pp. 357–368 (2009)
[33] Christaller W., *Die Zentralen Orte in Süddeutschland. Eine ökonomisch-geographische Untersuchung über die Gesetzmässigkeit der Vorbereitung und Entwicklung der Siedlungen mit städtischen Funktionen*, Jena, pp. 1–340 (1933)
[34] Lösch A., *Gospodarka przestrzenna – teoria lokalizacji*, PWE Warszawa (oryginał: Lösch A., Economics of Location, New Haven 1954), pp. 1–407 (1963)
[35] W. Ratajczak, *Analiza i modele wpływu czynników społeczno-gospodarczych na kształtowanie sieci trans-portowej*, PWN Warszawa-Poznań, pp. 1–140 (1980)
[36] P. Śleszyński, Dostępność czasowa i jej zastosowania, Przegląd Geograficzny, **86(2)**, pp. 171–215 (2014)
[37] A.F. Burghardt, A hypothesis about gateway cities, Annals of the Association of American Geographers, **61 (2)**, pp. 269–285 (1971)
[38] P.M. Hohenberg, L.H. Lees, *The making of urban Europe, 1000–1950*, Harvard University Press, pp. 1–398 (1985)
[39] E. Nowosielska, Teoria Christallera – prawda i mity (w sprawie nieporozumień pojęciowych), Dokumentacja Geograficzna **3**, pp. 1–83 (1992)

[40] E. Siemianowska, O zastosowaniu niektórych modeli teoretycznych w badaniach nad ośrodkami miejskimi w średniowieczu, Archaeologia Historica Polona **22**, pp. 27–54 (2014)
[41] L. Schätzl, *Wirtschaftsgeographie 1 – Theorie*, 6th Edition, UTB, Paderborn, pp. 1–280 (1996)

SHS Web of Conferences **58**, 01017 (2018) https://doi.org/10.1051/shsconf/20185801017
GLOBMAR 2018

The long arm of the sustainable port

Bartłomiej Laskowski*

Abstract. As there has been a lot written regarding sustainable transportation methods, there is still some opportunity to enhance the perspective on the economic advantage of the zero-emission supply chain. This has the potential of an accelerated adaptation in a cost to serve delivery system. The paper will introduce into the discussion the concept of working with nature, to develop durable structures and its capability of delivering value on its interactive environment or face the natural selection process. The functionality of the Port will be revisited in contrast of today's confinements of operation, while creating an opportunity to strengthen its perspective on its deliverables and reassess today's market. Further it will reflect whether sustainable transportation planning is still a choice or a competitive requirement.

1 Research Goal

To describe the delegating role of ports on the sustainability development in the port's hinterland access by inland waterway corridors

1.1 Methodology

The research is a qualitative analysis of prescriptive methodologies on sustainable development, offering a problem definition, diagnosis, design and implementation, concluded by the research results.
Limitation of research is the lack of quantitative data available for the application of some of the resent management methods, therefore the limitation is an evaluation of applying some of these methods to a corridor's social development.

- The problem definition is introduced in chapter, the function of a port.
- The diagnostic of issue identification and interventions is introduced in the chapter, Justifying the business purpose.
- The design to addresses the issues and application of the interventions is elaborated in the chapter, Equilibrium in institutional relationships.

* Corresponding author: Bartlomiej.Laskowski@outlook.com

- In the application and opportunity chapters, the implementation of the changes and engaging intervention participation are discussed.

2 The function of the Port

The traditional function of the port is that of connecting means of transition.

Yet the classical models of measuring the ports value, is that of revenue streams and its effectiveness is depicted by how much throughput it can accommodate in a specific period.

However, Ports are connecting much more than just modalities, thus the emphasis should be more on the term transition. Although the competitiveness that attracts foreign direct investment can be measured by how well connected throughput streams are. This measuring model is still evolving.

Today's challenge is to be able to reduce the cost to serve. This attracts a foreign entities investment and improves the regional economical capability to deliver productivity. This offers opportunities to reflect on the standard.

Companies are continuously challenging their throughput streams on their involved modalities, by identifying the optimal supply stream of production to its consumer. Thereby, allocating cost-efficient throughput along the available modal split. In this perspective, the key would be flexibility to mitigate between the available modality options of the market one is trading in. Balance its market shocks, risks and maximize present market advantages, allowing the market forces to synchronize its modal split utilization. This can be referred to as, synchro modal optimization of the available transport portfolio.

In this perspective, a sustainable port is one that offers efficiency on the supply chain as the entry and exit point of an overall economy. It should be integrated with a transportation modality that offers energy efficiency on the work performed. Thus, a sustainable infrastructure mechanism with the port having a delegating entry and exit position to that system. A transfer point from the sea to the land as an exchange chokepoint.

The article Chokepoints in Corridors: Perspectives on bottlenecks in the European transport network, by Patrick Witte et al. (2012) elaborates that in Europe a lack of thorough understanding of bottlenecks and their cumulative effects on each other is lacking in the transport network.

"The findings from Patrick Witte's research have indeed identified that cumulative accelerator affects exist in the current conduct of logistical interaction between the user groups and planning bodies of the infrastructures. It also identified that the interactions and effects between these user groups are far wider and more complex than the traditional dimensions of infrastructure and spatial structure perspectives, as such the subject of chokepoints or bottlenecks deserves in today's fast pace environment a wider and more complex recognition to the fields it is interacting with." [1]

He points out the interaction between the realms of infrastructure, spatial structure, governance structure and economical structure through interaction of their sub dimensional categories, have the ability to cumulate and accelerate effects of bottlenecks.

Witte's concept offers the possibility for issue identification on trade corridors.

Understanding the issue to be exposed is crucial, as it is said by Copernicus:" To know that we know what we know, and to know that we do not know what we do not know, that is true knowledge."

As previously stated Ports have a connecting function, as such they have the ability to connect interaction with its surrounding potential.

"The main challenge today is to implement and develop methods that improve the environment and strengthen the economy simultaneously by making optimal use of the available space." Waterman (2015)

Such a space is not merely reserved for connecting the tangible throughput streams.

This can also be exposed in the Ports ability to connect interdependencies of corridors and even the ability to intertwine with the economic entanglement by creating shared values and with that connecting human capital that can drive the knowledge economy. A system that is perfectly mature where its boundaries of optimal utilization have been reached, does not offer a driving competitiveness, it drives only an idle state of progress stagnation.

2.1 Problem definition

The problem definition is the ports immature market potential from improving the environment and strengthening the economy, by making optimal use of the available space in it's hinterland network.

3 Justifying the business purpose

A perfect system that has had its potential already exposed, will not create growth or progress. This is left to a system with build in failures at critical points to create continues iterative regeneration on comfortability, in order to generate process resilience and counter stagnation of progress.

Such a system can create assimilation into its surrounding circumstances of the micro economies of Ports and enable durable competitive investments. This iterative development of resilience fosters commitment in harmony of its macroeconomic opportunity spectrum.

Porter and Kramer have described in their articles, The Big idea creating shared value and measuring shared Value (2011) the concept of a self-reflecting system. Their shared value concept describes a different type of business conduct. One where social and community challenges and are part of the professional business conduct. After all, "strategy theory holds that to be successful, a company must create a distinctive value proposition that meets the needs of a chosen set of customers." According to this, not only the economic needs but also social requirements, specify the composition of a market. Their work reflects the cooperate social responsibility applications in perspective to creating shared value. Today it is understood by the CSR concept that social or community issues are not the main concern of a classical company. In general, the view persists that external institutions outside of a company, force an internal financial strain on the business conduct. A lack of awareness of available concepts creates the perception that social costs are externalities, that are only an additional general cost of running a business. Governmental institutions are also not supportive in developing a modern perception. They see business to have a short-term orientation, with the perception to internalize social costs on companies by using taxes as an instrument. Having only one instrument that is typically associated with a punishing function in order to attempt internalizing social costs, companies are driven to take a defensive stand against the social responsibility approach, experiencing it as to only generate additional costs. This leads unfortunately to companies refraining from embracing the needs and challenges in societies to be included in their business opportunity perception, that otherwise generate costs on governments. [7,8]

"It is an unfortunate issue, as companies fail to grasp the importance of their broader business environment, surrounding their major operations. Without this understanding the opportunity to connect competitive advantage and societal issues is missing. By reconceiving products and markets, redefining productivity in the value chain and enabling local cluster development, companies can reposition themselves to generate shared values on societies for a sustainable business relationship.

To sum it up the business conduct concept to be delivered should offer a prosperity benefit to the societal stakeholders that than yields itself in the form of growth feeding back into the business, as such one could compare it to a product life cycle that instead of having a maturity and decline phase, it would be part of a circular economy with its society forming an ecosystem by creating shared Value." [1]

Measuring of shared value itself, is done on three pillars:

- Reconceiving products and markets, is an orientation level that analyses the outer self of the business, the perception of the business by other external parties. It focuses on the business factors that can be yielded from the shared value creation approach.
- Redefining productivity in the value chain, is an analysis of the internal self of the business conduct, when considering the environmental and social focus areas, exposing opportunities for the internal operating process.
- Enabling local cluster development, is the output level. The shaping of the immediate/local physical area of influence in a mutual beneficial symbiotic relation to benefit the local area with consideration of the environment and social prosperity for the business. [8]

Ports facilitate an economic region's prerequisite to get the goods in and out, in the most efficient way. Thereby this process optimisation is supporting the Port's network maximization.

Exposing plans that have been proven through the test of time, can be identified in history.

These durable alternations, can form the conduct of the future. They require commitment and financing, that in turn can yield its own revenues. These revenues can be outside of the financial realm and support the attractiveness of the waterfront revenue and spatial quality.

A thorough evaluation of the finances is required, from initial financing, operational expenses to maintenance cost. The success of such plans requires the harmonization of activities between institutions such as, local regional and central governments, actors in the private sector and research and educational institutions together with environmental and landscaping societies. [4,5]

Where Patrick Witte's chokepoint on corridors exposes issues in corridors, Watermans and Brouwers Aquapuncture concept offers solutions and actions on these issues with regards to stakeholder's user groups. They do this by exposing the complex interaction of effects a waterfront intertwined balance of entanglement has on its society and environment.

It is remarkable how self-sustaining yet delicate such a balance can be, suspended in the interaction of external forces acting upon it.

4 Equilibrium in institutional relationships

Due to the changing nature of the morphological environment, that is continuously trying to adapt to the new equilibrium of it's surrounding. It is crucial in order to yield the full functional potential of the available space, to apply strategical planning of the functionality requirements.

Even ideas that follow an economic logic, are at risk to be rejected if they do not generate practical value for its participant and stakeholders, within their own surrounding.

One can compare it to a natural ecosystem, that has a new item introduced to interact with. As soon as it does not offer any value to the ecosystem and its inhabitants, the system either rejects it, or generates more efficient alternatives for it.

Nature has thus already set the tone, that for any long term project to be acceptable, a sustainable business model is required. Such a project, is to find a value fit with the societies it is expected to be in a long term interaction with. In turn the accepted models are less invasive when introduced, as these societies are already more receptive to that local configuration to the environmental equilibrium.

Sustainable development projects require long terms investments, that are dependent on sustainable relationships. Even a sustainable project that is not in balance with the institutional interaction, is a possible investment risk. The time constraining transitions of perception and interaction on trade corridor conduct effecting Port investments, require stable relationships.

A significant philosopher dedicated to observing truth was Baruch Spinoza in his works of the Ethics as elaborated by R. Prokhovnik's work Spinoza and Republicanism (2004). Prokhovnik elaborates that Spinoza moved from the notion of balance to be the principle of tolerance between extremes nor did he recognize the golden mean argument. Instead he suggests an equilibrium of the suspension on a moving interaction between actors according to their proportions. As such: "For Spinoza a balance of government or politics represents a group of people, not necessarily in an electoral sense, but as a constituency nevertheless. The critical point for Spinoza is that when a fitting proportion creates balance, things are held in dynamic tension, in balance rather than equal with one another. For Spinoza, this balance is a principle of anti-corruption."

Therefore, one can conclude that balance is the harmonization of dynamic interactive tension between the actors in regard to their values, as Spinoza (1632 Amsterdam-1677 The Hague) concluded so fitting: "On balance I prefer to preserve the etymological connection between affectio and afficere"

Van Tulder and Pfisterer have developed a Taxonomy process on some of these institutional forces acting upon each other. They did this to understand the nature of the current institutional actor's suspension of forces and what does this suspension say on the suitability of possible partnership types. Their article "Creating partnering space" has been motivated by misalignment of action owners, due to the lack of identifying roles to actors in partnership management of sustainable development. This fit on institutional interaction is exposed by the growing knowledge on the conditions, that are needed in sustainable development. Such partnership management is common between the Government, the private sector and civil societies organisations.

Understanding the powerbases of the institutional actors can expose their positioning in the virtual partnering space, which in turn can state something on the delegating role of the Ports partnership program in regard to its inland waterway program, be it for example a public private partnership interacting with a Profit-non-profit partnership.

Misalignment of such relationship often occurs in these bilateral perspectives crowding out one of the parties possibly adding to the institutional divide. A partnership relationship that includes the Government, the private sector and the civil society organisations requires a thorough understanding of the positioning of their roles.

Van Tulder and Pfisterer describe that in order to create an effective partnership with the actors involved in sustainable development, their partnering space can be exposed by asking the following questions:

1. What defines the „space" in which partnerships develop to address sustainable development issues?
2. What kind or relevant roles do parties bring into the partnership?
3. What does this imply for the organizational fit of the parties in partnering space?
4. How does the various degrees of fit contribute to sustainable development?
5. They conclude by questioning whether there is an „optimal fit" at all?

These fundamental questions, drive a collaboration, that is focused on the effects it has on the society it interacts with. An evolution from the traditional models that focus on benefiting the individual partners. These Cross-sector social partnerships (CSSPs) have the potential to drive modern socioeconomic development models. This type of relationship interaction is held in perspective to its feasibility on its societal orientation. For complex intertwined long term relationships, as in sustainable projects, a cross-sectorial social partnership can be considered to be a required management system.

In reflection of this, one can argue that the alignment balance of institutional actors is represented by regional societal values, when confronted with a set of objectives. One can conclude the observation, that a right balance differs from the perception of values. This is dependent on the driving actors to the objectives from each societal region, whether that region is defined by culture, sovereignty, state or other expression of unity in a specific region depending on the corridor. Therefore, each balance perspective is only applicable for a particular corridor and a specific suspension of harmonized tension between actors, the affectio and their conceptualized effects presented by their action items, as the afficare for the observer's consideration, being the right balance in that suspension of state.

5 Application

An approach of mutual interaction, has yielded already many successful results on the international stage in the development of waterfronts. This management style, focusses on facilitating stakeholders to improve their mutual productivity, thereby exposing opportunities to mitigate the challenges of logistical chokepoints.

Such mutual interaction to develop mutual productivity, is based on developing thriving conditions for its stakeholders. These conditions require a base to be developed, fostering complexity and resilience. Such a thriving ecosystem is built on a fundament, that can sustain new complexity adaptations to the interspecies communication with its environment. This communication is then made visible, by the morphological reaction over time to the environment's adaptation. After all an infrastructure or superstructure project is desired to stand the test of time with as little maintenance finance as possible.

These invasions that are introduced by man, create shocks to the environment and climate. The negative effects from these shocks have been observed to be absorbed depending on the diversity of species that is available in the area.

An area rich in diversity has the capability to be attractive for a wider group of users and therefore possible investors. A complexity sustaining area can thus facilitate a cluster of activities that in turn attracts correlated user groups.

This complexity attractiveness has provoked the settlement of people near such waterfronts, who then were facilitated by the clustering of activities to attract other types of modalities to sustain their various volume to time demands of the local distribution of goods and people.

These new business perspectives were driven by social needs, thus creating a shared value between cooperation's and social requirement. That fulfillment of a societal purpose generates opportunities for cooperation's. Allowing the profitability model of a cooperation to include social needs, unlocks the foundation of a new level of business complexity that is build on a purpose to that society, the cooperation is operating in. Such conduct unlocks new profits by harmonizing purpose. It enables a thriving society by feeding it prosperity, which in turn offers new forms of productivity to the private sector, adding to the complexity of available activities and the overall resilience by accepting circular flows of value. This puts in perspective, the value a waterfront offers as a fundament of clustering possibilities not only sustaining life itself, but facilitating the basis for new levels of complexity clusters.

One can thus observe the dependency on the variation of consumer interaction with waterfronts. The size of such clusters such as ports, can attract the interaction with other complexity clusters on the waterfront depending on a healthy water network between them to offer resilience and differentiation on consumer demand. This demand depends on the consumer need of communities, that in turn facilitate business to express that need outside of its local sourcing ability. The business's ability to meet that demand offers in turn favorable conditions for prosperity generation in that community. A community that has facilitated a variety of interaction, has a similar balancing effect to its environmental shocks, as previously described by the balancing effect from species diversity to an environments ability to absorb invasive shocks. A diversity of interaction modes has a favorable potential to address congestion challenges that undermine expressing the community's business needs.

The drive of facilitating an environmental fit of a local area is measurable, both in tangible and intangible effects. Physical adaptations can be measured by health and prosperity of the area's users and even of its commercial entities. An increase in complexity will result in a variety of new knowledge development areas, that in turn develop the intangible capital.

However, such growth requires a livable, attractive area for the effects to be able to manifest themselves.

These clusters of complexity based on a circular value flow, rooted in ecological prosperity, could express their community's business needs by manifesting their own organizational collaboration to absorb challenges on its logistical network nodes. These organizational collaborations have the causal effect of reducing checkpoints on these corridors, facilitating the efficiency of expressing the communities trade interaction.

6 Opportunity

Poland is part of the Paris MOU agreement, its Ports are operating in a region, classified by the International maritime organization as a particular sensitive sea area, referring to the Baltic sea. Therefore, it has an operating expectation on its trade conduct in regards to its port activities. As such it has a good opportunity to reflect on its goals and functional approach, rather than being subjected to a key performance indicator driven perception. Ports ought to include economic entanglement opportunities to work on behalf of their network and reach of activities.

The question a port should ask themselves, in regard to the economic entanglement growth of the knowledge based economy; has the fossil fuel infrastructure matured compared to the sustainable energy efficient infrastructure?

Think of an industry that can stimulate consumerism with a complexity to feed a spectrum of supporting industries in a circular process, feeding and nourishing each other's development. A historical example of such a system is the combustion engine. A car production would require a product lifecycle to match its required fuel supply with an additional spare part after sales industry, to support and maintain this capital investment in consumption commitment. In that the consumer would be committed to purchase fuel for 20 years. This means 20 years guaranteed repeat business. A vessel will do the same on a much larger scale per unit. Imagine than a port and its operations, as such fdi's and production intensive companies consider the energy price of their business conduct, when choosing a location to settle their activities to a new location.

In this perspective, an operational chokepoint such as a Port, has the ability to manifest a consumption unit with a low energy input on its productivity and a wide reach of supporting industries. This consideration delivers today, a competitive advantage. The most energy efficient supply chains are using transportation nodes that have already been provided by nature.

This is well described by the building with nature approach of the Aquapuncture concept by Ronald E. Waterman and Jaap Brouwer 2015a. They connect the historical tendency of civilisations to cluster on the waterfront and utilising waterways to cluster societal activities. These tendencies of slow transit systems joining rivers and lakes were globally reproduced. These core networks clustered a wide range of activities for these societies, from moving cargo and people to delivering live sustaining water for hydration and agriculture to maintaining flood levels. So naturally industry has settled on these time proven cluster enablers.

Due to the rise of faster modes of transferring cargo and people that were introduced to the communities and cities, the waterways have lost their perception of value. This depreciating relevance to waterway systems, came in in par with the raise of the rail and road networks. The waterways spatial structure functionality has been subjected to neglect. This neglect has exposed the sensitive relationship between the rural and urban areas these waterways kept in balance. As such their value is now being recognized with the need to revitalize these water networks.

"Improving the horizontal and vertical competition in the transportation sector should offer additional employment and offer an overall improved attractiveness for foreign direct investors, such as manufacturers to settle in established and upcoming logistical and manufacturing clusters.

A fitting quote by Professor Tadeusz Tillinger describes these interactive opportunities in relationship to waterways: "Waterways are on the one hand a part of the overall economy, and on the other hand, being an element of the transportation system, they are a part of the vascular system of the state body. It is the task of water management to channel the water falling from the sky onto the land to the sea in a way that makes it possible to minimise its harmful effects as an element and to maximise its useful properties as an indispensable matter to the lives of humans, animals and plants, as a material for transportation routes and as mass, i.e. a body with certain energy.

Marciniak (2013) describes that waterways are a part of a hole network consisting of different transport modality types. However, for bulk cargo inland waterways offer the required competitiveness. The Polish legislation states, inland waterways to be surface waters owing to the hydrological conditions and existing water devices, where inland waterway vessels can transport people and cargo under Article 9(1) (18) by the Water Law Act 18th July 2001. In this definition, the authorities declare waterways to be navigable according to local parameters."[1]

7 Research results

The article connected the ports to utilize a range of modal options for specific business needs with its hinterland. It has elaborated the ports functional capacity effected by its infrastructure corridors to address the cumulative effects of congestion and organisational harmonisation. A business perspective on corporate social responsibility introduced the unmet revenue potential of circular value streams, by accepting a social and cooperate interdependency. The inclusion of various user groups and stakeholders in the development process orientation on mutual productivity, offers thriving conditions for complexity. This approach exposes objectives and action owners to work with nature, instead of continually struggling against an ecosystem trying to cope with a new equilibrium, resulting in financial strain to maintain the corridor's infrastructure and superstructure. The challenge of classical methods to mitigate sustainable partnerships have been elaborated with the introduction of creating a partnering space method, to expose the available partnering fit for a durable collaboration on long term sustainability projects. The balancing effects of environmental inclusive methodologies have been introduced to reduce internal costs and develop organisational complexity, creating new competitive capabilities for the elevated complexity availability, thereby unlocking new supporting industry requirements. The mature fossil fuel dependency and the limited growth in competitiveness it can offer on the knowledge based economy by linear value streams has been raised. This maturity has been compared to the opportunity the sustainability field offers to develop new methods of competitiveness in energy efficiency and transferring energy, while developing new skills adding to a knowledge based economy.

This causes a correlated employment potential by connecting the horizontal and vertical competition in the transport sector. In regards of the ports competitive throughput ability of bulk loads, a direct relation has been established to the inland waterway network. That competitive capability has been connected to the polish legislative definition of inland waterways owing to the hydrological cycles and the available water devices.

7.1 Conclusion

As the Sea Ports are part of the hole network consisting from different types of modalities. The ports connectivity to inland waterways in regard to the Port's bulk competitiveness is owing to the hydrological conditions of such surface waters described under Article 9(1) (18). Therefore, the port as a consumption unit is a single driving force of its entanglement of supporting industries. Here competitiveness of this single unit to its alternatives is key. Energy efficiency and entwined industries, create demand from its activities while maintaining a complexity that develops barriers and niche segments on its competitors, is a way to diversify themselves and strengthen resilience. While cultures and traditions are optimally suited to their surroundings opportunities, so are their ultimate manifestations in the form of ports. Thus, it is of critical importance that these entrances as representations to their regional economic arenas are in tune with their surroundings. Prepared for the worst and negotiated for the best, as in natural selection. After all an economic environment is an ecosystem of its own with institutional actors that have their roles and functions."

7.2 Follow up research

It would be interesting to understand, who and if there is a liable party to the loss of revenue for local producers that are dependent on the navigability according to local parameters. Can the local authorities be held accountable for not maintaining local water devices according to the local navigability parameters, or are there even private sector parties that can be held accountable for disrupting the hydrological cycle to uphold these local parameters?

The following article uses parts and sections from the previously unpublished thesis "Balance in the Vistula's Revitalisation" by Bartłomiej Laskowski for the Master of Science in 2017 at the faculty of the Netherlands Maritime University of the STC-Group in Rotterdam the Netherlands

References

1. B. Laskowski, "Balance in the Vistula's Revitalisation", Thesis for the Master of science at the faculty of the Netherlands Maritime University of the STC-Group, Rotterdam, The Netherlands (2017)
2. Ż Marciniak, Acta Energetica, 2/**15** | pp. 152-161 (2013)
3. P. A. Witte a,1, B. W. Wiegmans b,1, F. G. van Oort a,2, T.J.M. Spit a,3 a) Department of Human Geography and Spatial Planning, Faculty of Geosciences, Utrecht University, Heidelberglaan 2, PO Box 80115, 3508 TC Utrecht, The Netherlands b) Department of Transport and Planning, Faculty of Civil Technology and Geosciences, Delft University of Technology, Stevinweg 1, 2628 CN Delft, The Netherlands Research in Transportation Business & Management 5 (2012), Chokepoints in corridors: Perspectives on bottlenecks in the European transport network, pp. 57–66L. T. De Luca, *Propulsion physics* (EDP Sciences, Les Ulis, (2009)

4. R. E. Waterman and Jaap Brouwer, Aquapuncture©: Sustainable Future Of Inland Waterways, Terra et Aqua | Number **140**, pp. 1-4. (2015)
5. R. E. Waterman and Jaap Brouwer 'Aquapuncture' Adaptation & Optimal Use Of Inland Waterways And Their Waterfronts, pp. 1-8. (2008)
6. M.E. Porter and M. R. Kramer, Harvard Business Review January-February, The Big Idea, creating shared value, pp. 63-77 (2011),
7. M. E. Porter and M. R. Kramer, Foundation strategy group June, Measuring shared Value, how to unlock value by linking social and business results, pp. 1-18. (2011)
8. R. van Tulder and S. Pfisterer 'Creating Partnering space' Exploring the Right Fit for Sustainable Development Partnerships, the partnership recourse centre, Published in: Seitanidi, M. M. & Crane, A. (eds). Social Partnerships and Responsible Business. A Research Handbook. Routledge. (2013)

The evaluation of the attractiveness of marine container terminal sector: analysing the polish sector

Robert Marek[1,*]

[1] Gdynia Maritime University, Poland

Abstract. The importance of attractiveness of the marine container terminal operators have significantly increased in the recent years due to them influence on the worldwide economy. This article aim is to evaluate the attractiveness of marine container terminals by using a qualitative analysis. The main point is to define the attractiveness of the Marine Container Terminal Operators (MCTOs) located in Poland. The qualitative analysis has provided arguments to conclude that marine container market in Poland currently is attractive from the point of existing MCTOs in this sector. The developed model helps to estimate the attractiveness of the marine container terminal market (MCTM) in Poland.

1. Introduction

Marine container terminal operators play an important role as a node in many logistic supply chains. These main players operate container transshipment between various transport modalities. The main modalities are deep-sea, short-sea, inland waterway, road and rail (mainly intermodal). Due to the globalization process, marine container flows worldwide are growing very rapidly and it is expected that this grow will be continue during the next decades. Moreover, a new generation of deep-sea cellular container vessels, with a capacity of 18 000-21 000 TEUs, is coming into ship owner's operations. In the future, the vessels may reach cargo capacity 50 000 TEU [1]. These growths and developments urge the main and minor marine container terminal operators to reconsider their equipment and logistics or even to expand. Consequently, currently the attractiveness of the marine container market in Poland has become a particularly relevant, thus influencing not only the market members, but the whole countries' economy as well. The MCTOs constitute rapid growth of their container's transshipment which involves long-term and medium-term investments into operational equipment and support hinterland connections. The attractiveness of the marine container market – the profit potential – is analyzed in accordance with MCTOs. The attractiveness of the marine container market is perceived as the capability to effectively compete in the world while improving most needs of major link in transport chain, the needs of both transport and shippers customers, higher levels of efficiency in land haulage and its infrastructures. The aim of this paper is to

* Corresponding author: r.marek@wpit.am.gdynia.pl

identify, analyze, and evaluate the main factors influencing the attractiveness of marine container terminal operators which are operating in Poland. In order to achieve the aim, the following objectives have been put forward:

- to conceptualize the attractiveness of the marine container market;
- to identify and analyze the main factors influencing the attractiveness of the marine container market;
- to evaluate the main factors influencing the attractiveness of the marine container market in Poland;

In order to implement the objectives, a systematic and comparative analysis of the scientific literature, covering the attractiveness of marine container terminal operators, has been accomplished in the second section of the paper. The third section deals with the methodology of a qualitative research. The fourth section presents and discusses the results of the data qualitative analysis and the evaluation of the main factors influencing the attractiveness of the marine container terminal operators in Poland. The paper is completed with the conclusions based on the generalization of the research results concerning the attractiveness of the marine container terminal operators located in Poland.

2. Literature Review

One of the most inherent developmental features of the present maritime economy is the process of globalization based on the objective nature principle of causality, i.e. how certain activities in certain circumstances always create respective consequences. Globalization processes are seen as a result of the world states' development which opened up their economies and liberalized economic relations among the countries thus motivating the involvement and participation of national companies in the global economic system [2]. Upon joining the European Union in 2004, the Poland container market has become much more accessible to foreign marine container operators; this made it possible to attract internal as well as the foreign marine container terminal operators (MCTOs). The foreign MCTOs thoroughly analyze an unknown Polish market in order to determine whether such investment will be profitable, i.e. they determine the market attractiveness level. When analyzing the general definitions of market attractiveness presented by different authors, one should notice that even though they are quite different, the key factors determining the market attractiveness remain quite similar. The most business and marketing dictionaries define the market attractiveness assessment from the position of a customer, a company, or an investor, i.e. as a profit or benefit opportunity in a particular market or industry. The higher the potential gain or benefit, the more attractive a market is considered to be. The attractiveness of the market in other literary sources is very closely related to market competitiveness. According to the American Marketing Association dictionary, the market attractiveness defines the degree of market opportunities provided by a market segment and a company's ability to meet the segment's needs considering the competitive environment. Meanwhile, the market attractiveness by M. E. Porter [3] is considered to be closely related to the competitive situation in the market, i.e. economy competitiveness of companies and the country. M. E. Porter, based on the five-forces model, argues that an attractive market is the one in which you can get the maximum profit or benefit, in other words, the more relevant market participants getting on for ideal competition, the less attractive it becomes.

Unfortunately, the exiting literature on criteria of MCTOs attractiveness is quite not so large because there are different categorization of terminal operating companies. For example, Bichoou and Bell [4] propose to divide global ports operators on four types: (1) terminal operating shipping lines which are not active in containerized cargo operations; (2) terminal operating shipping lines with terminals operated on a dedicated or a common-user basis; (3) terminal operating port authorities; (4) terminal operating companies. In turn,

Parola and Musso [5] make a distinction among tree groups: pure stevedoring who manage terminals as profit centers; integrated carriers who manage terminals mainly as costs centers, and a group of hybrid terminal operators consisting of shipping lines involved in the stevedoring business and handling both own cargo and third-party traffic to make a profit. Slack and Frémont [6] distinguish two groups of port operators as: transnational terminal operating companies and international shipping lines. Along the same lines, Olivier [7] indicated two groups such as: transnational corporations in the terminal industry and international terminal operators who mainly are ocean carriers, they concentrate on linking terminal operations with ocean carriers business.

When we analyzed the literature we can see that may authors investigated attractiveness of terminal industry from the point of its choice by ocean carriers. Moreover, many authors identify seaport with terminal. Ha [8] has compared and evaluated the service quality factors of major container ports from the viewpoint of ocean carrier and logistics managers. The factors of importance, he has separated into seven group: (1) information viability, (2) port location, (3) port turnaround time, (4) facilities available, (5) port management, (6) port costs, and (7) customer convenience. Similar key determinants of port choice were proposed by Tongzon and Heng [9]. The authors selected the following factors: port (terminal) operation efficiency level, port cargo handling charges, reliability, port selection preferences of carriers and shippers, the depth of navigation channel, adaptability to the changing market environment, landside accessibility and port differentiation. Further studies devoted the port choice by ocean carriers have been run by: Mangan [10], Lirn et al [11], Nirn et al [12], Ahn et al [13], Tongzon and Sawan [14], Tongzon [15], Kim Jin-Young [16], T. Notteboom, J-P. Rodrigue [17].

In summary, factors (determinants) of attractiveness of port choice by ocean carriers are well documented. However, there is a lack of research, which concentrate on marine container terminal sector' attractiveness from the viewpoint of MCTOs. Currently, literature on maritime container terminal market has neither tested such case, nor encountered the challenges of the implementing such attractiveness evaluation.

3. Methodology of the Research

The attractiveness of marine container terminal operators has been assessed by using the weighted score approach to assessment of sectors value created by G. Giereszewska and M. Romanowska [18]. The approach to that analysis has been proved and improved during the meetings with mangers of many firms. During the construction of the analysis's scheme, assuming that, the respondents of the research well-known M.E. Porter's concept "five competitive forces". In the proposed version of the approach, it is necessary to analyses 15 criteria which describe their current attractiveness for the sector's participants and enrol results of the assessment to form which includes the criterion's list with attributed weights. The maximum score which the assessment of the research may receive equally 150 points. The managers has assessed current attractiveness of the sector from its own business unit. All results received from the research has been weighted arithmetic mean.

The list of propose criterions of the MCTOs sector attractiveness has been built on Porter's model and taken into the account the most important factors of the business microenvironment. The general weighs of the criterions, the above-mentioned authors determined weights and criteria on discussion base with different companies' top managers who's filled in a strategic management form.

It should be remember that this method apparently is very easy to run, require a deep knowledge on competition environment and set a lot of traps. One of the most important trap is a subjectivism in assessment, especially in the fields of research, where the research are based on qualitative analysis. This problem has been indicated by the A.A. Thompson and A.J. Strickland [19], they put an attention, that results of the research depend on who are the respondents. For example, the authors indicates that the assessment of the sector will be different by current competitors, potential investors and independent specialists. The current competitors' asses a very high entry barriers very positively but potential entrants the same barriers assess very negatively because they need to invest a huge sum of money and undertake a very high risk. From the other side, the high exit barriers are very crucial for the current competitors which run business in the sector, in turn the potential investors doesn't take into account in the analysis these barriers. Therefore, it should be taken into consideration that we should distinguish real value of the assessment sector from subjective value of the assessment sector indicated by the respondents. The influence of the subjective value assessment on the result of the research is evident when the respondents' asset sector in which run their company.

The second trap which may appear in the research is the time-horizon of the assessment of the sector. So that, we might to assess the current and future perspective of the sector. Because, the research has to assess the current situation in the sector, therefore the respondents had to assess the current attractiveness which exist at the marine container terminal market.

If the know the traps and limitations of the method then we can use it very successfully for assessment of sector attractiveness. Moreover, by using the same criterions and using the same weight allows us to compare researched sector with other related and unrelated sectors.

Target respondents were practitioners in senior positions in marine container terminal operators (see table 1) with sufficient knowledge of their organization's overall activities and operations. Each item offered a five-point scale, anchored by 5 = 'strongly important' and 1 = 'strongly unimportant'.

Table 1. Research respondents.

Marine container terminals	Terminal's abbreviations	Owner
Baltic Container Terminal Gdynia Ltd.	BCT	ICTSI
DB Port Szczecin Ltd.		Deutsche Bahn AG
Deepwater Container Terminal Gdańsk Ltd.	DCT	Macquarie Infrastructure and Real Assets (MIRA)
Gdynia Container Terminal Ltd.	GCT	CK Hutchison Holdings Limited (CK Hutchison)
Terminal OT Port Gdynia Ltd.		OT Logistics Group S.A.
Terminal OT Port Świnoujście S.A.		OT Logistics Group S.A.

Own elaboration.

All responders have given its own ratings on current attractiveness of the marine container terminal industry. The weighted score has been calculated by sum up all indicated ratings and the result for every factor has been divided by amount of the investigated terminal operators. In the research all respondents belonged to top managers of the marine container terminal operators. This study sought answer for the following research questions:

Q1. Is marine container terminal market attractive for MCTOs which are located at the Polish market?
Q2. What are the most important attractiveness criteria for MCTOs?

4. The Factors Attractiveness Analysis and Results

A marine container terminal is a place where container units are transferred by two or more freight transport modes. In this nodal place container units are loaded and unloaded, exchanged, stored and/or distributed. It is noteworthy that, currently in Poland, much effort is aimed at increasing the marine containers flows and the use of new generation terminals which will mean an increase in the use of the container handling operations at marine container terminals. An increase in handling can only be justified by a considerable increase in the new terminal capacity of the current terminals, a decrease in costs, or a combination of the two foregoing developments.

The marine container terminal market is defined as all marine container terminals throughout the Poland where activities are deployed in such a way that facilitate a smooth direct (or indirect) transshipment of marine containers from vessel to container yard stocks and intermodal or truck transport. There is a shape competition among sector players because actors see chances or feel pressure to improve their current positions. The attractiveness of the investigated marine container terminal operators sector depends on the following criteria:

(1) Market size/cargo volume.
(2) Market growth rates.
(3) ROI in the sector.
(4) Degree of company's concentration in the sector.
(5) Shape competitive struggle.
(6) Entry barriers.
(7) Exist barriers.
(8) Substitution threats.
(9) New competitors entrance threats.
(10) Supply assurance.
(11) Technology stability.
(12) Possibilities of service differenation.
(13) Possibility of activity diversification.
(14) Seasonality and cyclicity.
(15) Natural environment pollution threats.

The table 2 shows the results of empirical investigation of evaluation attractiveness of marine container terminal market from the viewpoint of MCTOs.

Table 2. The results of attractiveness assessment of Polish marine container terminal industry in 2018 using the weighed score approach.
Ratings are on a scale of 1 (very unattractive) to 5 (highly attractive)

	Attractiveness criteria	Weight The range from 1 to 3 points	Rating Attractiveness of current situation in the industry (the range from 1 to 5 points)	Weighted Score
1.	Market size	3	4,8	14,4
2.	Market growth rate	3	4,6	13,8
3.	ROI in the sector	3	3,6	10,8

4.	Concentration degree in the sector	1	3,6	3,6
5.	Shape competitive struggle	3	4,4	13,2
6.	Barriers entrance	3	4	12
7.	Exist entrance	1	3,4	3,4
8.	Substitution threats	2	2,6	5,2
9.	New competitors entrance threats	3	3	9
10.	Supply assurance	1	3,6	3,6
11.	Technology stability	2	3,6	7,2
12.	Possibilities of service differentiation	2	3,8	7,6
13.	Possibilities of activity diversification	1	4	4
14.	Seasonality and cyclisity	1	2,6	2,6
15.	Threats of natural environment pollution	1	1	1
16.	Sum the weighted rating scale	30	52,6	111,4

Own elaborated.

The market size/cargo volume factor is present in 1 out of 15 evaluation criteria shown in Table 2. The respondents have indicated that this factor is the most relevant for the decision makers in this investigated sector. The attractiveness of current situation in this sector has been estimated very high (4,8), and the sum of the weighted rating score (14,4). The larger the market, the more attractive it is, because it gives opportunities to quick development of the MCTOs (see figure 1). Additionally, a larger market can indicate the market attractiveness to other companies. Currently, we have a very huge increase of marine container market size measured by loaded and unloaded TEU units in Poland.

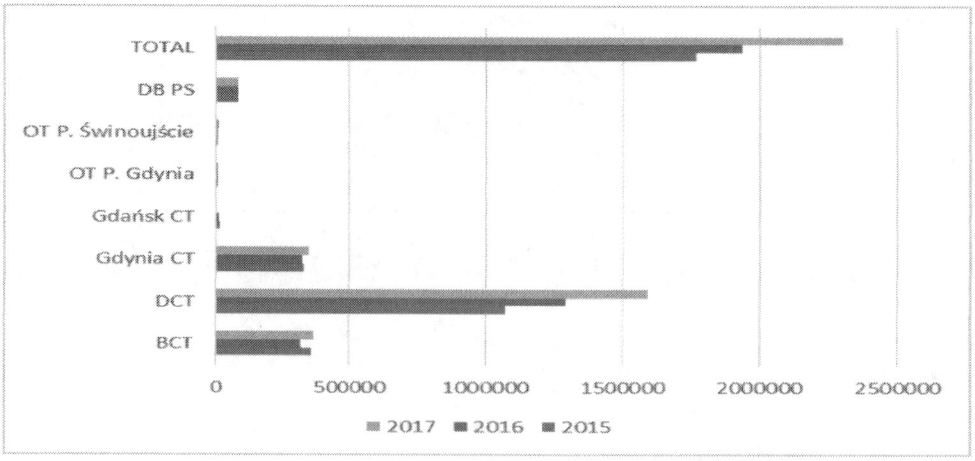

Figure 1. Marine container turnover in Poland 2015-2017 [in TEU]. Own elaborated.

Market growth rate factor indicates the rate at which a marine container terminal market's size is increasing. The market growth rate, as a second of the fundamental factors of marine container terminal market competitiveness, has been present in 2 out of 15 results shown in Table 2. The respondents have indicated that this factor also is the most relevant for the decision makers in marine container terminal sector. The importance of present situation in the sector has been estimated very high by respondents, reaching this results at (4,6), and the sum of the weighted rating score (13,8). This factor is expressed as an annual percentage of container units loaded or unloaded by all MCTOs per annum. Historically high growth rates do not always indicate a high rate of growth looking into the future as sectoral and economic conditions change constantly and are often cyclical (see table 3). Anyway, the market growth rate is gained result 18,74% in 2017 year in Poland. So if, MCTO's customers have high demand, container market growth is quick. If marine container terminal operator develop terminal services assortment or built its service loyalty at a specific price level, than marine container market growth also should increase. The larger the market growth rate factor, the more attractive it is, because it gives opportunities to quick development of current MCTOs.

Table 3. Market container growth rates in Poland during 1994-2017 periods

1994	**1995**	**1996**	**1997**	**1998**	**1999**
6,67%	14,98%	11,12%	13,61%	20,35%	-9,03%
2000	**2001**	**2002**	**2003**	**2004**	**2005**
16,01%	15,43%	11,72%	20,53%	27,86%	12,08%
2006	**2007**	**2008**	**2009**	**2010**	**2011**
13,97%	32,09%	1,29%	-12,94%	57,80%	27,65%
2012	**2013**	**2014**	**2015**	**2016**	**2017**
23,21%	15,31%	7,96%	-14,02%	9,62%	18,74%

Own elaborated.

The next factor which is very important for evaluating attractiveness of the investigated sector is its **return of investment**. This factor is a performance measure used to evaluate the efficiency of an investment in marine container terminal or compare the efficiency of a number of different others investments. Anyway, the higher ROI, the higher attractiveness of the analyzed sector. This factor has been present in 3 out of 15 evaluations criteria shown in Table 2. The respondents have indicated that this factor is relevant for the top decision makers employed in all MCTOs. The importance of this factor for current situation in this sector has been estimated high (3,6), and the sum of the weighted rating score (10,8).

The **concentration degree** of marine container terminal operators in the sector is not relevant for MCTOs decision makers. This factor has been present in 4 out of 15 evaluations criteria shown in Table 2. The respondents have indicated that this factor is not so much important for them. The importance of this criteria for present situation in this sector has been estimated at level (3,6), and the sum of the weighted rating score resulting (3,6).

We need to pay attention that, investigated sector in Poland, has an oligopolistic competition because there are only six main players such as: Baltic Container Terminal Ltd, (located in Gdynia seaport), Gdynia Container Terminal Inc. (located in Gdynia seaport), OT Logistic Terminal Gdynia (located in Gdynia seaport), Deepwater Container Terminal Inc. (located in Gdansk seaport), Gdansk Container Terminal Inc. (located in Gdansk), OT Logistic Terminal (Located in Szczecin-Świnoujście seaport), and finally DB Schenker Terminal Inc. (located in Szczecin-Świnoujście seaport). Unfortunately, Gdansk Container Terminal Inc. is under liquidation proceedings. The above-mentioned marine container terminal operators compete with each other to provide the best services at their terminals for the lowest possible price. The Deepwater Container Terminal Operator gains some advantages because of the terminal's localization (natural deep-water channel at seaport), and receives also additional benefits such as: direct calls ocean carriers at the terminal (for example: 2M Alliance «Hyundai Merchant Marine also uses this loop service», Ocean Alliance), a higher terminal services rates for box then offered by other container terminal operators (higher THC – Terminal Handling Costs), and a faster and higher increase of containers throughputs. So, DCT Gdansk is operating as a hub [20] for others containers terminals located at Baltic Sea. The high degree of competition concentration it means that one marine container terminal operator may become a regional sector maker and impose others MCTOs its own market game rules. Currently, such status of reginal sector maker has DCT Gdansk in Poland. Moreover, the high degree of competition concentration in this sector comes from oligopoly power, because of technologies and economics rules, where terminal services production and its sales must be done at large economic scale to be economic profitable.

The **shape competitive struggle** factor is another relevant factor of sector attractiveness. This factor has been present in 5 out of 15 evaluations criteria shown in Table 2. The respondents have indicated that this factor is relevant for the decision makers in this investigated sector. The importance of this criteria for a present situation in this sector has been estimated very high (4,4), and the sum of the weighted rating score (10,8).

The highest the sector's attractiveness, the lowers a competitive struggle among competitors. The most important information for the potential new comers to the sector is a knowledge on a form a competition. The potential new comers don't want to enter into the sector if there is a shape price war among the competitors. According to all respondents of the investigation there is a strong shape price war in the marine container terminal market in Poland. In the market, we have two important groups of industry competitors. So we can distinguish here, the passive and active marine container terminal operators. The passive operators are Port Authorities and active operators are real terminal services providers. The first group (passive operators-Ports Authorities) consists of container terminal infrastructure owners who are not providing the terminal services by themselves. The only services they provide are some facilitations for the marine container terminal services operators. This group of passive operators are responsible for construction, modernization, development and maintains terminal infrastructure such as: yard stocks, berths, quay, offices space, storages, sheds, warehouses, rail (multimodal) and road infrastructure. These terminal infrastructure requires huge capital investments. Basically, there is, in Poland, only one form of terminal infrastructure ownership: state and municipal-owned terminals. Anyway, whole marine container terminal infrastructure belongs to public and municipal authorities as well. The public container terminal owner (seaports authorities) has to develop terminal infrastructure investments which allow to attract the real marine container terminal operators into the seaport. In this way, the real (active) marine container terminal operator does not lose the time and money for constructing new infrastructure terminal investments and he can mainly concentrate on its own core activity. The main goal of the Polish seaports authorities is to facilitate the central terminal services, or to put it simply by granting concession on container terminal infrastructure per square meters. The price for the square meters depends on the negotiation skills of both parties, on the total square meter volume and the terminal location, and on the market power of both parties. The price is also influenced by the quality of the delivered infrastructure square meters (for example, a new or old rail container terminal infrastructure) [21].

The second group of "sector competitors" consists of the real (active) marine container terminal operators whose provide the terminal service assortment. The marine container terminal operation can be accomplished by international terminal operators (Baltic Container Terminal Ltd. – belonging to ICTSI – International Container Terminal Services Inc. Manila; Gdynia Container Terminal Inc. – belonging to HPH – Hutchinson Port Holding – Hong Kong), rail and logistics operators (DB Schenker Terminal ltd.), capital investors (OT Logistics Terminal Gdynia and Szczecin- Świonujscie – private equity; The Deepwater Container Terminal Gdansk – pension funds). Till now, there is no marine container terminal which is operating by ocean carrier, feeder service carrier, logistics operators or public authorities. However, this situation may change if the capital investors (pension funds or private equity) would like to sell its stocks to above-mentioned potential customer's group. The marine container terminal operator is particularly interested in the costs' optimizing of the terminal services provided to its customers. The costs' optimizing means such looking for costs' saving which are not negatively impact on quality services provided by real MTCO.

Entry barriers factor is that protect "sector competitors" from new entrants to the sector. This factor is present in 6 out of 15 evaluations criteria shown in Table 2. The respondents have indicated that this factor is relevant for the decision makers in this investigated sector. The importance this criteria for a current situation in this sector has been estimated very high (4,0), and the sum of the weighted rating score (12).

Exit barriers factor is that keep a "sector competitor" in the market, even if the return on investment is too low or even negative. This factor is present in 7 out of 15 evaluations criteria shown in Table 2. The respondents have indicated that this factor is relevant for the decision makers in this investigated sector. The importance of this criteria for a present situation in this sector has been estimated very high (3,6), and the sum of the weighted rating score (10,8).

This is especially true of the container terminal market: high investments in different types of vehicles and other terminal facilities (medium or long-term leasing contracts) impose a major exit barrier for the container terminal operator who is already active in the market. The exist barriers may be different for marine container terminal operators which are operate on international and local market. The real international marine container terminal operators may decide to move its own terminal superstructure (yard trucks, AGVs, yard cranes, quay cranes and so on) into the other marine container terminals, which belong to its portfolio if the ROI will be too low or even negative. Contrary to real international container terminal operators will be local terminal operators, which has not possibilities to shift terminal superstructure to another terminal. In that case, exist barrier for local container terminal operator will be higher than for international marine container terminal operator. Exit barriers are relatively low and the entry barriers relatively high this automatically means that the profits in the marine container terminal market can only be high and not very risky. It should be indicated that exist barriers in the investigated sector are very important because they are high; once a marine container terminal is in use, it is relatively difficult to quit the operations because of the huge initial investments which are needed to start operations.

The substitution threats factor may decrease the potential profits of a sector by imposing a barrier on the prices sector competitors. This factor is present in 8 out of 15 evaluations criteria shown in Table 2. The respondents have indicated that this factor is not relevant for the decision makers in this investigated sector. The importance of this criteria for a present situation in this sector has been estimated very low (2,6), and the sum of the weighted rating score (5,2). Each market will have to deal with options that replace current services or products that industry competitors produce. Substitutes for the use of a marine container terminal are surprisingly one of the strongest competitive forces in the market. The most important substitutes in the marine container terminal market are unimodal road and intermodal rail transport. The position of the both transport modes are extremely strong, especially when they carry containers passing over the marine container terminals. This can either be caused by the very competitive solutions offered by unimodal road and intermodal rail transport, by the weak developed link between marine container terminal operator and its hinterland.

New entrants threats factor into the marine container terminal market is newly constructed terminals. This factor is present in 9 out of 15 evaluations criteria shown in Table 2. The respondents have indicated that this factor is a medium relevant for the decision makers in this investigated sector. The importance of this criteria for a current situation in this sector has been estimated medium (3,0), and the sum of the weighted rating score (9,0).The new active marine container terminals will increase capacity as well as competition in the terminal market in a specific terminal service area and thus impact utilization rates of existing terminals. This could result in lower prices and/or higher terminal operation costs and also the transport network could be affected. Even if no extra terminal capacity is added, the take-over of a terminal operator to create a market position also means the entrance of a new industry competitor. This situation takes place when owner of the terminal's infrastructure (passive terminal owner) would like to enter into the terminal operational functions. This game player may not extend a concession agreement with real (active terminal operator) marine container terminal operator and thus take-over resources (mainly human resources whose have the best know-how on terminal operations) created by the previous MCTO. The risk of this situation depends upon ruling political party. The higher risk when ruling Conservative Party, and the lower risk when ruling Liberal Party. Moreover, the potential entrants into the marine container market have quite large barrier because the best places for terminal localization in the seaport area are occupied by the current operating MCTOs in the sector.

Terminal facilities supply assurance factor covers suppliers of technical equipment, mass media, infrastructure and capital resources (money). This factor is present in 10 out of 15 evaluations criteria shown in Table 2. The respondents have indicated that this factor is not relevant for the decision makers in this investigated sector. The importance of this attractiveness criteria for a present situation in this sector has been estimated very low (3,6), and the sum of the weighted rating score (3,6).These suppliers can use their economic power, for example, by raising prices and lowering the quality of their goods and/or services. Another option for the supplier is to threaten integrating of terminal services into their own assortment. The strength of this force further depends on the number of suppliers; if the suppliers of terminal infrastructure are concentrated, in general they have greater economic and bargain power.

Here we will pay attention to **technology stability** factor. This factor is present in 11 out of 15 evaluations criteria shown in Table 2. The respondents have indicated that this factor is relevant for the decision makers in this investigated sector. The importance of this attractiveness criteria for a current situation in this sector has been estimated medium level (3,6), and the sum of the weighted rating score (7,2). High degree of used containers technology also influence on high quality services provided by MCTOs. Every MCTO definite some requirements concerning on container technology use in terminal to aim archiving better competitive advantage over others competitors. The technology influence on amount of capital expenditure used to purchase the technology, technical equipment and on operation organization. The used technology in marine container terminals is quite stable. According to all respondents, marine container terminal technology changes longer than 5 years. Many container terminal operators significant gain in productivity were achieved through advanced terminal layouts, more efficient IT- support ("micro-pcs") and improved logistic control software systems, as well as automated transportation and handling equipment.

Possibilities of service differenation factor. This factor is present in 12 out of 15 evaluations criteria shown in Table 2. The respondents have indicated that this factor is relevant for the decision makers in this investigated sector. The importance of this factor for a present situation in this sector has been estimated low level (3,8), and the sum of the weighted rating score (7,6).The main goal of MCTOs is to provide terminal services that are requested by customers such as: shipping owners, forwarders, logistics operators and so on. In order to provide the best terminal services, the MCTOs should create product which is the terminal service assortment. This assortment differ from terminal to terminal and can be both thick and broad. Moreover, the MCTO may broad its terminal service assortment on: storage, assembling, repacking, consolidating, configuring, cautrylizing, quality controlling, and customizing, packing, finishing, merging, testing, barcode labeling, tracking and tracking and so on. The MCTOs will assure that the terminal service assortment fits into the logistical chain of customers. In this sector, it is very hard to create strong position of the product mark or the company brand. Unfortunately, all respondents of the marine container terminal operators indicated that the customers (especially forwarders) are not loyal to the terminal services providers. The customers (forwarders) change the MCTO very often. The lack of creation of any terminal services mark or company brand in the analyzed market runs to strong increase of shape competitive struggle because there are any possibilities to create some psychologic costs for customers, which theirs should bear, if customer tries to change one service terminal provider to another. The customers transition 'form one marine container terminal operator to another is easy because the container terminals are located very close to each other.

Possibility of activity diversification factor. This factor is present in 13 out of 15 evaluations criteria shown in Table 2. The respondents have indicated that this factor is not relevant for the decision makers in this investigated sector. The importance of the attractiveness criteria for a present situation in this sector has been estimated low (4,0), and the sum of the weighted rating score (4,0). Some MTCOs diversify them activity by entering into the additional terminal services as heavy lift cargo transshipment.

Seasonality and cyclicity factor. This factor is present in 14 out of 15 evaluations criteria shown in Table 2. The respondents have indicated that this factor is completely no relevant for the decision makers in this investigated sector. The importance of the attractiveness criteria for a current situation in this sector has been estimated very low (2,6), and the sum of the weighted rating score (2,6).

In the container activity, we have problem with periodicity which run to terminal production capacity surplus. This situation appeared in the global market during the world crisis in 2007, when containers throughputs has gone down very rapidly. This crisis has driven MCTOs to strong intense competition, because the terminal services providers fight to fill its free terminal production capacity.

The seasonality in marine container terminals has two reasons and is directly connected with the seaborne commodity trades. An important cause of short-term volatility is the seasonality of some commodities carried in marine containers. Nowadays, many agricultural commodities are transported in marine containers and they are subject to seasonal fluctuations caused by harvests, such as: grain, sugar, citrus fruits and others [22]. Moreover seasonality in containerization exist with major holidays such as the Chinese New Year and Christmas, Easter and other local feast days, for example in Poland, All Souls' Day.

The last **Natural environment pollution threats** factor. This factor is present in 15 out of 15 evaluations criteria shown in Table 2. The respondents have indicated that this factor is completely not relevant for the decision makers in this investigated sector. The importance of the attractiveness criteria for a present situation in this sector has been estimated very low (1,0), and the sum of the weighted rating score (1,0). According to the respondents this factor is not relevant for attractive analyzed sector. The answer is quite interesting because there is international policy which concentrate on using low-emission or non-emission cargo handling equipment vehicles.

The evaluation of the attractiveness of the sector is carried out on the base of comparison with other sectors. Unfortunately, we do not have such data and information on attractiveness of other sectors. Therefore, at the end of this analysis we had to create some determine category of the sector. In this aim, we have to sum the weighted rating score and determine category of sector as follows:

1. High attractiveness of the sector => (150;135).
2. Satisfactory attractiveness of the sector => (134;91).
3. Weak attractiveness of the sector => (90;70).
4. Critical attractiveness of the sector => (69;50).

According to the approach above-described we can say that analyzed sector of marine container terminal located in Poland is satisfactory for MTCOs cause total achieved sum equals 111.4. This research, above all, contributes to understanding that attractiveness evaluation of marine container terminal sector depends heavily on such factors as: market size/cargo volume, shape competitive struggle, market growth rate, barriers entrance and return of investment.

5. Conclusions and Proposals

This paper applies evaluation attractiveness approach to assess a marine container terminal attractiveness in Poland through analyzing a survey of six marine container terminals operators. It provides empirical analysis to support the unique and interesting findings that the most important factors influencing marine container terminal operators' investing decisions are: market size/cargo volume, market growth rate, shape competitive struggle, entry barriers and return of investment. It triggers further discussion on marine container terminal operator's competitive factors and development strategies. For instance, it is arguable to ascertain the attractiveness of investigated sector is satisfactory attractive for MCTOs operating in Poland. Market size/ cargo volume, market growth rate become important criteria only when a seaport's infrastructure, draught and its national's political stability satisfy MCTO's basic requirements. The lesson learned from the investigation indicated that market size/ cargo volume, market growth rate, sharp competitive struggle, entry barriers and return of investment, having over 64,2 points importance among the 15 criteria, largely determine the MCTOs attractiveness in Poland.

References

[1] T. Bebbington, 50,000 TEU... the Future or Not?, Maritime Executive, https://maritime-executive.com/ediorials/50000-teu-the-future-or-not. (12.02.2018)

[2] O. Pryhara, Techniques for Analyzing the Attractiveness of International Commodity Markets under Conditions of Economic Globalization, Vol. 14, p. 27 (2006)
[3] M. E. Porter, Competitive Strategy: Techniques for Analyzing Industries and Competitors. New York, N.Y.: Free Press, p. 396 (1980)
[4] K. Bichou, M. Bell, Internationalisation and consolidation of container port industry: Assessment of channel structure and relationships, Maritime Economics and Logistics, 9: 33-51, (2007)
[5] F. Parola, E. Musso, Market structures and competitive strategies: The carrier-stevedore arm-wrestling in northern European ports, Maritime Policy and Management, 34 (3): 259-278 (2007)
[6] B. Slack, A. Frémont, Transformation of port terminal operations: From the local to the global, Transport Review, 25(1): 117-130, (2005)
[7] D. Olivier, Private entry and emerging partnerships in the container terminal industry: Evidance from Asia, Maritime Economics and Logistics, 7(2): 87-115 (2005)
[8] M. Ha, A comparison of service quality at major container ports: Implication for Korean ports, Journal of Transport Geography 11(2):131-137 (2003)
[9] J. Tongzon, W. Heng, Port privatisation, efficiency and competitiveness: some empirical evidence from container ports. Transportation Research Part A 39(5): 405-424 (2005)
[10] J. Mangan, C. Lalwani, B. Gardner, Modelling port/ferry choice in RORO freight transportation, International Journal of Transport Management 1(1): 15-28 (2002)
[11] T. Lirn, H.A. Thnopoulou, A.K. Beresford, Transhipment port selection and decision-making behaviour: Analysing the Taiwanese case. International Journal of Logistics: Research and Applications 6(4): 229-244 (2003)
[12] A. Nir, K. Lin, G. Liang, Port choice behaviour – From the perspective of the shipper, Maritime Policy & Management 30(2): 165-173 (2003)
[12] W. Ahn, S. Lim, S. Ahn, A study on analyzing bottlenecks of logistics in Incheon Port. Focused on container freight. Korean Research of Logistics 14(2): 65-90, (2006)
[14] J. Tongzon, L. Sawant, Port choice in a competitive environment: From the shipping liea's perspective. Applied Economics 39(4): 4777-492 (2007)
[15] J. Tongzon, Port choice and freight forwarders. Transportation Research Port E 45(1): 186-195 (2009)
[16] Kim J-Y., Port user typology and representations of port choice behaviour: A Q-methodological study. Martime Economics & Logistics 16(2): 165-187 (2014)
[17] T. Notteboom, J-P. Rodrigues, The corporate geography of global container terminal operators, Martime Policy & Management, 39(3): 249-279, (2012)
[18] G. Giereszewska and M. Romanowska, Analiza strategiczna przedsiębiorstwa, PWE, 122-126 (2003)
[19] I. Urbanyi, M. Klopott, Container terminals and port city interface – a study of Gdynia and Gdańsk ports, Transportation Research Procedia 16 (2016): 517 – 526 (2016)
[20] A.A. Thompson, A.J. Strickland, Strategic Management. Concepts and Cases, R.D. Irwin, Homewood, Ill, Boston, p. 225, (1996)
[21] B. W. Wiegmans, E. Masurel,P. Nijkamp, Intermodal freight terminals: an analysis of the terminal market, Transportation Planning and Technology, 23: 105-128 (1999)
[22] M. Stopford, Maritime Economics, Routledge, London and N.Y., p. 97 (2009)

The efficiency of a supply chain with the use of maritime transport on the example of the apparel industry

Dariusz Milewski[1,*], *Beata* Milewska[2]

[1]University in Szczecin, Faculty of Management and Economics of Services, Cukrowa 8, 71-004 Szczecin
[2] WSB University in Poznań, Faculty of Finance and Banking

Abstract. The paper presents the problem of the impact of the competitiveness of maritime transport on the efficiency of global supply chains on the example of the apparel industry. With the use of a model developed by the authors, simulations of the profitability of the sourcing in low-costs countries (Bangladesh, China) have been conducted. The model takes into account the specificity of the apparel industry, parameters of cargo, value of goods, and demand characteristics. If freight rates for maritime transport services were to increase significantly, the profitability of production in the Far East compared to local or regional production would be lower particularly for cheaper products. The level of these freight rates is not the main factor of the efficiency of this strategy. Simulations revealed, that changes of costs of production has greater impact on this profitability. In the garment market the possibility of quick response to demand is often more important than costs. The calculations were made on the basis of data obtained from Polish companies, which order production in countries with low production costs and transport garment in containers by sea and from transport and logistics operators.

Introduction

Global supply chains are characterized by geographic dispersal of individual chain links, because producers, their suppliers, sub-suppliers, distributors, and agents are located in various countries and in different parts of the world. There may be various reasons for that, for example, the lack of specific goods in a given country or the higher quality of goods bought abroad. Often, however, the main reason is lower production costs in the so-called low-cost country sourcing (LCCS), e.g. in South-East Asia or South America.

However, when making decisions concerning production or buying in low-cost countries, the trade-off mechanism should be taken into account. On the one hand, thanks to the supply and production in the LCCS countries, production or purchase costs are reduced. On the other hand, the logistics costs can be increased. These costs are: transport costs, the cost of maintaining stocks, costs associated with lower quality of goods, eg with

* Corresponding author: dariusz.milewski@wzieu.pl

deterioration of quality during transport, costs of duties, costs associated with monitoring suppliers and maintaining representative offices, lost sales resulting from the lack of quick response to a demand. Apart from this also other factors should be taken into consideration as the risk of the change of conditions of making business e.g.: increase of labor costs in the LCCS countries and freigth rates in maritime transport (Milewska, 2016, Milewska, Milewski, 2017). Strategy of outsourcing of production to low-cost countries is based on the assumption of low costs production and transport. This strategy is linked with increased lead-times not only because of longer distances but also because of the use of the maritime transport.

The goal of the paper is to present the economic efficiency of global supply chains using maritime transport on the example on the apparel industry. This efficiency has been assessed with the use of a model developed by the authors. The model allows for conducting simulations of the impact of changes of the costs of maritime transport on the profitability of the sourcing in low-costs countries (Bangladesh, China). It allows also to show the impact of the increase of production costs in the LCCS countries on the use of maritime transport. The model takes into account the specificity of the apparel industry, parameters of cargo, value of goods, and demand characteristics. Data for simulations have been collected from the transport and apparel market – among others from the biggest Polish apparel companies.

Literature review

Maritime transport has the biggest share in the global supply chains despite relatively longer time of delivery and bigger risks connected with use of it (unpredictable weather conditions, piracy attacks, terrorist hijacking, and cargo damages, bankruptcies, breakdowns, macroeconomic and political changes (Min, 2012, Manuj, Mentzer, 2008). Although the punctuality of the ocean transport has improved significantly in recent years, the use of maritime transport is linked with higher risks and longer time of reaction to changes of market demand in comparison to other modes of transport.

Containerisation is concidered a key factor of globalisation (Mangan, 2017). The maritime transport competes with other modes of transport with very low freight rates. According to Munim and Schramm (2017) global financial crisis in 2009 have affected the container shipping freight market towards heavily fluctuating freight rate movements. According to Drewry's a maritime research and consulting firm, global freight rates were in the beginning of 2017 37% higher than in 2016 and were forecasted to rise by about 8% (Drewry, 2017).

China have been for many years the main source of supply in the global supply chains. Yet in recent years production in China is becoming less and less profitable because of Chinese currency appreciation, rising labour costs, higher oil prices and reduced value -added tax rebates (Huang, Zhang, Liu, 2013). In the clothing industry the result was the allocation of some part of the production from China to Bangladesh – in this country the biggest brands are present, also the biggest Polish clothing companies like LLP S.A. and Redan.

The sourcing strategies of companies and their supply chains are however diversified and depends largely on the specific of an industry. The study performed by PwC revealed that leaders of supply chains keep the focus on chains that are both fast and efficient (PwC, 2013).

The textiles and clothing markets are characterized by short product lifecycle, high volatility, low predictability, and a high level of impulse purchase (Bruce, Daly, Towers, 2004). The typical problems facing with textile and apparel supply chain are long production lead-time, forecasting errors and minimum batch sizes for production (Lam,

Postle, 2006, Vedel, Ellegaard, 2013). According to Kumar and Samad (2007) outsourcing in this industry is not viable for meeting short term market demands but rather for large seasonal orders.

According to Christopher et. al commercial success or failure in fashion markets is largely determined by the organisation's flexibility and responsiveness (Christopher, Lowson, Peck, 2004). Such views could have been found for years in the scientific literature but business practices seemed to deny it, because more important than flexibility were costs. It seems however, that presently the concept of quick response and agile supply chains begins to be realized.

The specificity of the clothing industry in Poland

Poles' spending on clothes is slowly but steadily growing, and therefore the turnover in the clothing industry is increasing. Data from the Central Statistical Office shows that only in the first ten months of the 2017 the sales of textiles, clothing and footwear increased by 17.5% year-on-year (GUS, 2018).

The industry is characterized by relatively high volatility and low predictability, especially when it comes to „must-have" products, that is, products of the latest fashion trends. Poles make purchases of clothing both in exclusive salons and in discount stores, and more and more often - via the Internet.

The largest Polish clothing companies, such as LPP S.A. (owner of brands: Reserved, Cropp, House, Mohito, Sinsay brands) and Redan Capital Group (brands: Top Secret, Troll and Drywash brands), as well as their foreign competitors operating in Poland (eg H & M or Grupa Inditex), do not have their own production plants. These companies deal with the design, distribution and sale of clothing through their own Logistics Centers and own stores. They outsource production mainly to countries with low production costs (Far East), from where products are transported mainly in containers by sea. These companies order most of their production in Southeast Asia. The production of less complicated products takes place in Bangladesh, where the costs are lower than in China, while the production of more complex products - in China, because for example Chinese sewing factories with which LPP S.A. cooperates, have a modern machine park.

Production of goods in the Far East is much cheaper than in Poland and even after considering the costs of transporting goods, customs duties, but also costs related to monitoring suppliers and maintaining a representative offices. Especially LPP S.A. bear large expenditure on the improvement of working conditions in Bangladesh after tragedy, which took place in Rana Plaza in 2013. Another factor that should also be taken into account are changes in exchange rates, because the costs of clothing production in Asia are settled in dollars, while revenues - in EUR and in Polish currency.

Yet for several years there has been a tendency to increase the share of locally or regionally ordered production of clothes. The main reason is, that in the case of production in countries with low costs, the time from the start of a design to a delivery of goods to a warehouse is 4-6 months. This time is shorter when a new collection is not created, only the "missing" products are sewn on. However, still in case of producing a collecting in Asia, there is no possibility of quick reaction to fashion changes. Therefore, for some products, the biggest Polish clothing companies decide to order production closer - locally or regionally. For example, in Poland, LPP commissions the production of the significant part of the Mohito's brand, while the the share of production of Reserved brand in Poland is small.

Model of economic efficiency of global and local supply chains in the clothing industry

In order to study the possible impact of changes of freight rates in maritime transport and production costs in the Far East on the efficiency of global supply chains - a computer model has been developed by the authors. With the use of this model simulations have been conducted for global and local supply chains in the clothing industry.

General data used in the model are based on financial statements of Polish companies from the garment industry listed on the Warsaw Stock Exchange (table 1). Net operational margins are different – from very low (0,8%) to high, but half of these companies has margin up to 4%. In the case of these companies changes of costs can have significant impact on profits.

Usually costs of production or purchasing in these companies stand for 40 – 50% of sales. The distribution of goods – marketing, sales, physical distribution and logistics costs, which are also on the level of 40 – 50% of sales.

The interesting information can be found in the financial statements of the biggest player on the Polish market – LPP S.A. Since 2013 sales have been growing by 15% - 17% each year. Despite this the net operational margins were falling: from 15% in 2013 to 3,8% in 2016 (in 2017 rose but still was lower than in 2015). What's more the share of the costs distribution and overall costs didn't change much, but the share of costs of purchasing of goods increased from 41,5% in 2013 to 51,3% in 2016. That means that costs of production and purchasing of goods from the suppliers are growing.

Presently the company sells 140 mln items a year. The clothes produced at the request of LPP in low-cost countries are transported by sea in 40' containers to the port of Gdansk.

Table 1 Financial results of listed companies in the clothing industry in 2016 [thous. PLN]

The company's name	Sales revenues	Own cost of sale	Own cost of sale/ Sales revenues	Net margin
PRÓCHNIK S.A.	56 269	24 842	44,1%	0,8%
ESOTIQ & HENDERSON S.A.	124 224	49 869	40,1%	4,0%
Primamoda	35 838	26 343	73,5%	24,9%
Redan	628 487		0,0%	2,0%
Bytom S.A.	152 911	74 453	48,7%	8,3%
Gino Rossi	284 748	149 334	52,4%	4,5%
TXM SA	377 695	219 661	58,2%	4,3%
Silvano Group AS	57 892		0,0%	28,6%
CDRL S.A.	203 675	197	0,1%	9,2%
Monnari Trade SA	231 773	105 450	45,5%	9,3%
Vistula	598 602	286 660	47,9%	8,5%
LPP	6 019 046	3 085 236	51,3%	3,8%

Source: prepared on the basis of financial statements of companies

The rest of data are collected from the transport and logistics market or estimated on the base of available data. For example, in the Asia-Northern Europe connections, the overall Shanghai Containerized Freight Index reached 714 USD / TEU at the end of September 2017 and freight rates on similar level has been assumed as a starting point in the simulations. Part of the data are in the table 2.

Table 2 Parameters of purchased goods

Retail price [PLN/pcs.]	50	100	400
Purchase price (global source) [PLN/pcs.]	17,0	44,9	225,0
Purchase price (local source) [PLN/pcs.]	32,3	75,4	348,5
Difference of purchase prices	90%	68%	55%
No of products in a container [pcs./container]	19 800	15 000	8 000

Source: own study based on market data

The calculations have been conducted for three hypothetical groups of clothes – of low, middle and high value (retail price). It was assumed, that margin of net operational profits in case of global deliveries is on the level of 8% (similar to the operational margin of LPP S.A. In 2017) and in the case of local production – 4%. Basing on this assumptions prices for product sourced globally and in Poland have been determined. It is interesting, that the differences between these prices decrease when the value of the commodity increases – for the cheapest goods (50 PLN/item), the purchase price in Poland resulting from simulation was 90% higher than in the low cost country. For the price of 100 PLN/item the difference was lower – 68%, and for the most expensive (400 PLN/item) – 55%. That could mean, that for more expensive garments the cost advantage of global suppliers is relatively lower and thus probably greater opportunity for domestic producers to compete with global suppliers is in case of more expensive products, more complex and of higher quality.

Next important assumption is that a local supplier can be more flexible than a global one. So in the model it was also taken into consideration, that in case when the demand is higher than it was forecasted, a global supply chain won't be able to react quickly and the costs of the lost sales will rise. In case of lower demand inventories will be higher.

Simulations of the impact of changes of transport costs and changes production costs for three levels of prices of products have been conducted. The results are shown in table 3 and on graphs 1 and 2.

Table 3 Impact of changes of maritime freight rates and production costs on profitability of supply strategies

Supply strategies	**Increase of freight rates**				**Increase of production costs**			
	0%	**180%**	**220%**	**500%**	**0%**	**8,0 %**	**10,0 %**	**13,0 %**
Production in Poland	4%	4%	4%	4%	4%	4%	4%	4%
Global supply -	9%	4%	3%	-5%	9%	6%	5%	4%

50 PLN/item								
Global supply - 100 PLN/item	9%	5%	5%	-1%	9%	5%	4%	3%
Global supply - 400 PLN/item	9%	7%	7%	4%	9%	4%	3%	1%

Source: own study based on market data

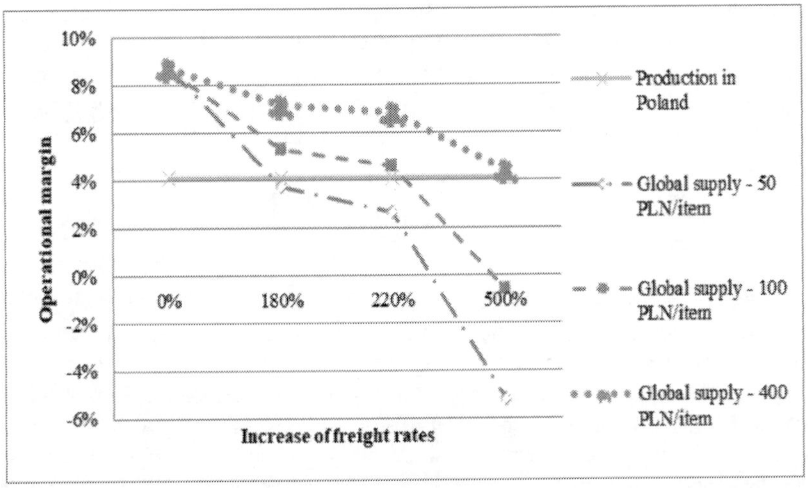

Figure 1 Impact of changes of maritime freight rates on profitability of supply strategies

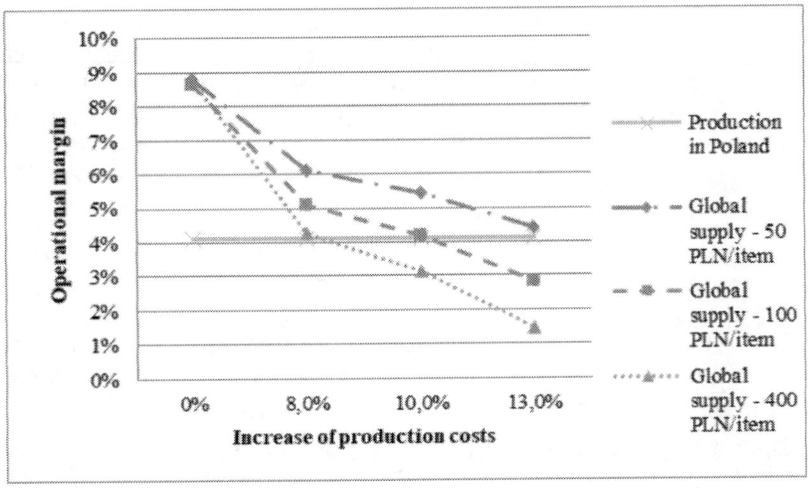

Figure 2 Impact of changes of production costs on profitability of supply strategies

It results from the simulations, that if the freight rates for containers rose mostly the cheapest goods would be affected, what can be explained by the biggest share of transport costs in the value of goods (graph 1). This rise would have to be very big if it were to be a threat for profitability of global supply chains – even in the case of the cheapest goods this rise would have to be more than 180%, what doesn't seem possible in the nearest future. Such small impact of the growth of the freight rates on this profitability is the result of the small share of transport costs in the ocean transport in the value of goods. These results are similar to the OECD data concerning this problem (OECD, 2005)

Second simulation concerned changes of production costs in the low-costs countries. Here – quite contrary than in the first simulation – profitability of the most expensive products have been mostly affected (graph 2). The increase of only 8% of production costs of global suppliers will make production in Poland so profitable as in the low costs countries. For the cheapest goods this is - 13%. In all three cases however these increases of prices are still very much lower when compared with the increases of the freight rates. Although then it is very unlikely that changes of prices in the transportation market could affect this market, it very possible, that rising costs of production will have impact on the profitability of global supply chains and indirectly on the volume of goods carried out by sea.

Conclusions

The conducted research confirm, that the phenomena in the global supply chains in the clothing industry described in the scientific and professional literature indeed take place. Simulations based on the obtained data can be helpful in explaning the nature of these phenomena, that means reasons for changes taking place in the global supply chains. These changes can relate localization of production but also changes in global transport market. Even not very big increases of costs production of garment in Far East could decrease economic efficiency of the global outsourcing, what could have impact on the market of maritime transport.

However, it should be emphasized, that simulations have been conducted taking into account greater flexibility of local sources and higher costs of the lost sales in the global supply chains. Without this assumption the profitability of the local sourcing would much lower, what proves, that flexibility in the supply chains is of very great importance. The model and simulations conducted with the use of this model confirm the rightness of opinions expressed by the practicioners and researchers (e.g. M. Christopher) concerning the importance both of costs of production and possibility of quick response in the supply chains. While it is difficult to find justification for the views on the large role of changes in freight rates on the location of production in these chains.

References

1. Bruce, M., Daly, L., Towers, N., (2004) "Lean or agile: A solution for supply chain management in the textiles and clothing industry?", International Journal of Operations & Production Management, Vol. 24 Issue: 2, pp.151-170
2. Christopher, M., Lowson, R. H., Peck, (2004) Creating agile supply chains in the fashion industry. International Journal of Retail and Distribution Management, Vol. 32, Issue 8, pp. 367 – 376.
3. Global Container Terminal Operators Annual Review and Forecast 2017, Drewry
4. Huang, G. Q., Zhang, A., Liu, X., (2013) "A supply chain configuration model for reassessing global manufacturing in China", Journal of Manufacturing

Technology Management, Vol. 24 Issue: 5, pp.669-687, Kumar, S., Samad Arbi, A., (2007) "Outsourcing strategies for apparel manufacture: a case study", Journal of Manufacturing Technology Management, Vol. 19 Issue: 1, pp.73-91

5. Lam, J. K.C., Postle, R., (2006) "Textile and apparel supply chain management in Hong Kong", International Journal of Clothing Science and Technology, Vol. 18 Issue: 4, pp.265-277
6. Mangan D. J. (2017), Foresight – Future of the Sea Evidence Review Foresight, Government Office for Science
7. Manuj, Ila, Mentzer, John T., (2008) "Global supply chain risk management strategies",International Journal of Physical Distribution & Logistics Management, Vol. 38 Issue: 3, pp.192-223,
8. Milewska B. (2016), Wyzwania w zarządzaniu międzynarodowymi łańcuchami dostaw z uwzględnieniem uwarunkowań kulturowych, Studia Zarządzania i Finansów, Wydawnictwo WSB 11/2016
9. Milewska, B. Milewski, D. Efektywność strategii zakupów, Eurologistics 3/2017, s. 18-21.
10. Min, H., (2012), Maritime Logistics and Supply Chain Security, in Dong-Wook Song , Photis M. Panayides (ed.) Maritime Logistics, pp.91 – 116
11. Munim, ZH, Schramm, HJ (2017). Forecasting container shipping freight rates for the Far East - Northern Europe trade lane. Maritime Economics & Logistics 19 (1), p 106-125
12. OECD (2005) DSTI/DOT/MTC(2005)5/REV1 (https://people.hofstra.edu/geotrans/eng/ch3en/conc3en/table_containershippingcosts.html)
13. Production of major industrial products November 2017, GUS, 2018.
14. PwC. (2013) Global supply chain survey 2013: Next generation supply chains efficient, fast and taylored.
15. Vedel, M., Ellegaard, C., (2013) "Supply risk management functions of sourcing intermediaries: an investigation of the clothing industry", Supply Chain Management: An International Journal, Vol. 18 Issue: 5, pp.509-522

SHS Web of Conferences **58**, 01020 (2018) https://doi.org/10.1051/shsconf/20185801020
GLOBMAR 2018

Analysis of the seaport value chain as a method for assessing its strategic potential

Andrzej Montwiłł[1,*]

[1] Faculty of Engineering and Economics of Transport, Maritime University of Szczecin, 71-051 Szczecin, Wały Chrobrego 1-2, Poland

Abstract. The aim of the paper is to describe the analysis of the seaport value chain as a method for assessing its strategic potential. This potential is understood as the port's ability to develop and improve its competitive position in sectors that are defined as part of the assessment of this potential. In the paper, key aspects of the seaport value chain analysis were described in the form of a summary of previous research on this subject. The seaport is understood as a system in which a number of interpenetrating processes for passengers, freight, means of transport and enterprises are carried out. To better illustrate key aspects of the value chain analysis, examples of selected processes implemented at seaports were used. Next, the paper presents the results of research interpreting the impact of various internal factors on the efficiency of the port value chain that translates into its strategic potential, where the issue was repeated using specific cases. In the final part of the paper, based on previous research, the significance of the analysis of the seaport value chain to assess the strategic potential of the seaport was pointed out. Moreover, in this part of the paper the chosen case was used to indicate the level of applicability of the research carried out.

1. Introduction

The modern system of the global economy forces enterprises to operate in a specific way, where the key issue is positioning the company's products on a given market in such a way that, in a given segment, the enterprise, or rather its product, obtains a competitive advantage. It can be a global market, a regional market or a local market. In the subject literature related to business management and building a competitive advantage, two issues are usually highlighted: profit and competitive advantage in the market in terms of the entire enterprise [1]. And while in the case of enterprises, or rather global capital groups, this approach may seem right, the author believes that this approach is too schematic.

It seems that the modern global economy should rather address the problem of competitiveness to specific products or groups of products or specific markets. This can be perfectly illustrated with seaports, where the largest of them are currently the most complex socio-economic structures operating in a cluster system. However, seaports are also spatially and structurally small organizations that are perfectly competing, thanks to a selected market offer, on selected markets, building a competitive advantage on them. From here it can be concluded that the approach to the competitiveness of economic entities or

* Corresponding author: a.montwill@am.szczecin.pl

complex multi-stakeholder socio-economic structures should be an individualized approach considered not only in the global system but in regional or local systems.

The aim of the paper is to describe the analysis of the seaport value chain as a method for assessing its strategic potential. This potential is understood as the port's ability to develop and improve its competitive position in sectors that are defined as part of the assessment of this potential The conducted research confirmed the need for an individual approach to the problems of seaport competitiveness, indicating that the size of the port, in the context of cargo traffic, does not have to be a key factor in improving its competitiveness on a given market of port services.

2. Methodology

The research process, the purpose of which was to assess the usefulness of the seaport value chain value analysis to assess its strategic potential, was based on a methodology consisting of three main stages [2]. The first stage was aimed at analyzing the functioning of modern seaports where the key issues of seaport organization, the functions and related areas and areas of activity in each of them were identified in the cognitive process. The area of analysis also included issues that qualify ports for a particular generation reflecting the cargo prevailing in port traffic, the level of cooperation of seaport users and the significance of the seaport in the network arrangement of supply chains. This stage of research included, on the one hand, the author's many years of experience related to work in the port sector, and on the other, research methods such as document examination, analysis and criticism of the literature, case analysis and observation.

The second stage of the research process focused on the processing of collected material, grouping facts about the evolution of seaports and the genesis of this evolution in the last decades, defining the connections between individual port processes as well as institutions and enterprises creating a joint organizational and functional structure. The basic method used at this stage was the method of analysis and logical construction in the part concerning the analysis, which allowed to organize the collected material regarding the functioning of seaports.

The third stage of research process was focused on modeling, based on previously collected and ordered materials, the value chain of the seaport as the basic method of the overall analysis of the strategic potential of the seaport. Using the research method, which is the analysis and logical construction (also called the "mind experiment" method) through the analysis and synthesis process, a research thesis has been proved. This thesis provided that the classically used analysis of the enterprise value chain to determine its strategic potential [3] can be used for seaports, where the sequence of port processes is usually carried out by a number of companies, often unrelated capital, whose proper cooperation creates added value in the value chain of the seaport. In the third stage of the research process, analysis of key success factors based on the 80/20 rule [4] was also used as an auxiliary method for determining the strategic potential of the seaport.

3. Results

3.1. The generation of the seaport is a key factor in the complexity of its value chain

Contemporary seaports are multifunctional socio-economic structures composed of the functional segment and the spatial segment. The functional segment includes the administrative and management sphere and the operational and service or operation-service-production sphere (in the situation when the production function is implemented in the port space). The administrative and management sphere is composed of: the managing entity (port management), responsible for the development of the sea port and state institutions responsible for the legal aspects of port processes implemented for: means of sea transport and means of land transport (railway, road, inland waterway, transmission) and cargo in export and import. The operational and service sphere (operational, service and production) is created by enterprises implementing services for means of transport (mentioned above) and cargo on a commercial basis as well as enterprises involved in production (in the case of a developed production function). The aforementioned areas of the seaport create a series of connections necessary for the efficient implementation of economic processes taking place in the port area. A significant part of these connections is the effect of applicable legal provisions imposing the obligation to carry out economic processes on the state administration, managing entities, port users. The rest result from the organization of the port and civil-legal relations between the managing entity and its users.

The spatial segment of the sea port consists of: waters and areas with the necessary infrastructure. These are elements of technical port utilities that determine its potential handling, distribution, logistics and industrial capabilities. These are also, what is very important, elements of the transport infrastructure of the state, which are subject to the "rigors" of state interests in the transport policy segment. Hence, the turnover of port infrastructure and land real estate within the boundaries of seaports is subject to many legal regulations and is prohibited in many countries. The above is due to the fact that coastal areas are treated by the states as strategic areas for their development and security.

There are many definitions of seaports in the literature. According to Grzelakowski A. and Matczak, from the economic point of view, the seaport is: "a spatially large economic area located at the interface of the land with the sea with a complex structure of activity, mainly of a transport, commercial and industrial character, prepared in terms of technical and technological, organizational and administrative, and legal and economic aspects for servicing sea and land means of transport as well as performing other transport, logistics, forwarding and commercial activities, entailing the necessity of their comprehensive service in the transport chain and the supply chain." [5]. In turn, according to the author of the article, *the seaport is a socio-economic space, with multifaceted impact on the surrounding environment, that combining in transport processes, thanks technical and technological equipment, sea with land, where are realized interpenetrating, interdependent and related, objective and spatial functions related to with trade and movement of people.* [6]

The seaport definitions presented above indicate its multifunctional character, where the key to its development is the effective implementation of processes for the means of transport, cargo and passengers that are part of the network of supply chains. Processes carried out both in the services sector and in the production sector. What is also important is that the greater the range of services, complemented by production, is implemented in the space of a seaport and the higher the level of cooperation of its users, the greater its strategic potential. The level of generation of the seaport is an expression of evolution in the development of ports that translates into their competitive position. Currently functioning seaports belong to different generations (from 1st to 3rd, where part of the third generation ports are transformed into seaports of the 4th generation) [7][8]. Table 1 presents key elements decisive for including the port in a given generation.

Table 1. The key elements determining the competitive position of a seaport

	1st generation	**2nd** generation	**3rd** generation
Dominant factors	Labour and capital	Capital	Capital, technology and knowledge
Dominant cargo	Dry bulk, General cargo	Dry bulk, Liquid bulk General cargo	General cargo Containers Ro-Ro Mobile
Main functions	Transport	Transport, industrial, commercial	Transport, logistics and distribution
Strategy	Conservative	Expansive	Market
The basic principle of the strategy	Port is waiting for the cargo	Port stimulates the development of the port industry, and thus acquires cargo	Port co-creates supply chains
The scope of operation	Handling and storage services for cargo. Simple administrative, handling and control services for cargo.	Handling and storage services for cargo. Processing of cargo Complex administrative, handling and forwarding services	Handling and storage services for cargo. Distribution of cargo Logistics services for cargo Information flow Organization of supply chains
VAL + VAF[1]	**Low**	**Higher**	**High**
The operating principles of the port (internal environment of the port)	Atomization Informal connections with port users.	No connections between port industries Close links with port users Ad hoc connections with the city.	Unity of action within the created organizational structures Organizational or capital integration of the port with its users Close links with the port city and the region

1. VAL – Value Added Logistics, VAF – Value Added Facilities

Table 1 presents the key elements determining the generation of a seaport, and thus the form of operation of its functional segment. Segment determining the manner and scope of operation of the seaport, functions performed in its space, and thus the range of services and the importance of the port in the network supply chain. It can be considered that third generation seaports act as logistics centers [9] or logistic platforms, as illustrated in Figure 1. Ports of this generation, where we distinguish three main areas (fields) of activity, presented in the referenced figure, are characterized by the implementation of a wide range of processes as part of the objective functions: transport, industrial, commercial and logistic-distribution.

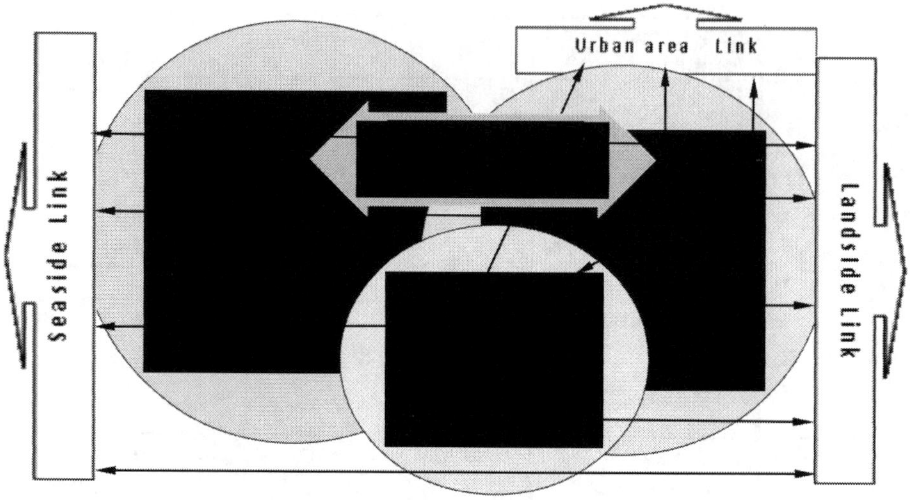

Fig. 1. Seaport as a logistic platform

In seaports of the 3rd generation, being logistic platforms, where a number of processes are carried out simultaneously in the framework of a network of many supply chains, we distinguish three key areas of activity. These areas can be assigned to individual objective functions and thus a specific range of processes (services and production) within which added value (VAL and/or VAF) is created. In Table 2, o functions are assigned to key areas of seaport activity.

Table 2. Correlation of key areas of seaport activity and objective functions

Objective functions: / Key fields of activity	Transport function	Industrial function	Commercial function	Logistics and distribution function
Maritime logistics centres (**MLC**)	**basic**		auxiliary	
Industrial park (**IP**)	auxiliary	**basic**	auxiliary	
Inland logistics centres (**ILC**)	**basic**		auxiliary	**basic**
Logistics platform (MLC + IP + ILC = **LP**)	**basic**	**basic**	auxiliary	**basic**

As shown in the research, the generation of a seaport is a key factor affecting the complexity of its value chain, which is the result of the level of development of objective functions, and the range of processes (services and production) in each function.

3.2. The value chain as a tool to analyze the strategic potential of the seaport.

The problem of the value chain as a tool for the general analysis of activities (essential and auxiliary) important for the enterprise's effectiveness in the context of building their competitive advantage on the market was taken by, among others, M. Porter. He pointed out that the value chain is a strictly defined sequence of various company activities that are implemented as part of its operation, creating a common system of values [10]. Importantly, M. Porter recognized the company's value chain as one of the important factors of the company's strategic approach to creating added value in the course of ongoing processes, combining both the company's values and the resources it owns. As J. Borowski points out, M. Porter, as the creator of the new theory of strategic management, connected positional theory and resource theory, combining their achievements and considering the value-enhancing activities and its value chain as crucial for the company's competitiveness [11].

The value chain is widely taken up by many authors in the context of the scientific approach. It should be pointed out, however, that the majority of scientific considerations and assessments concern the value chains of manufacturing enterprises. This issue is much less often addressed in the context of service enterprises and very rarely in the context of the value chain of complex multi-entity economic structures such as seaports, integrated logistics centers or network supply chains. Hence, in the research process aimed at assessing the usefulness of the value chain analysis to assess the strategic potential of the seaport, it was necessary to create a specific value chain for seaports. Classic chains were used for this purpose, starting from the one formulated by M. Porter. The value chains of enterprises in the tourism sector or the TSL sector were also analyzed. Based on, among others, the definition of the value chain by A. Stabryly [12], the author of the article attempted to define the value chain of the seaport, considering that it is *a sequence of horizontally and vertically related executive processes (for passengers, cargo, means of transport) and management processes (organizational, technical and legal) implemented by connected with each other organizational and functional areas of the seaport, where during these processes, the value added for the customer determining the competitive position of the seaport is created, depending on the degree of development of the objective functions.*
Based on the above definition, which is consistent with the definitions of the enterprise value chain previously formulated by many authors, a model of the seaport value chain was created. This model takes into account the postulate of M. Porter on the strategic character of this tool [10] in creating a development strategy, in this case, a seaport. And the same is not limited to internal processes, but includes aspects of external environment analysis and strategic positioning of a seaport on a given market of port services.

Figure 2 presents the scheme of the value chain of the seaport, where we can distinguish the stage related to the positioning of the seaport on the market of port services and adjusting its potential in the area of management and performance to the needs of selected markets located on the foreland or hinterland of the seaport. These activities are related to the organizational, technical and legal adaptation of the port to the handling of cargo, passengers and means of transport moved in land-sea transport chains between selected markets on the foreland and hinterland of the seaport.

In the process of adapting the port to the needs of selected markets in the area of cargo and passenger shipment, strategic decisions are made regarding the scope of planned processes and, consequently, decisions on the development of specific objective functions and port generation. Hence, in the value chain of the seaport (Figure 2), four variants of its functionality are indicated, depending on what functions are developed, which is the result of previous strategic decisions about the range of executive processes available in a given port for customers from previously selected markets (strategic segmentation).

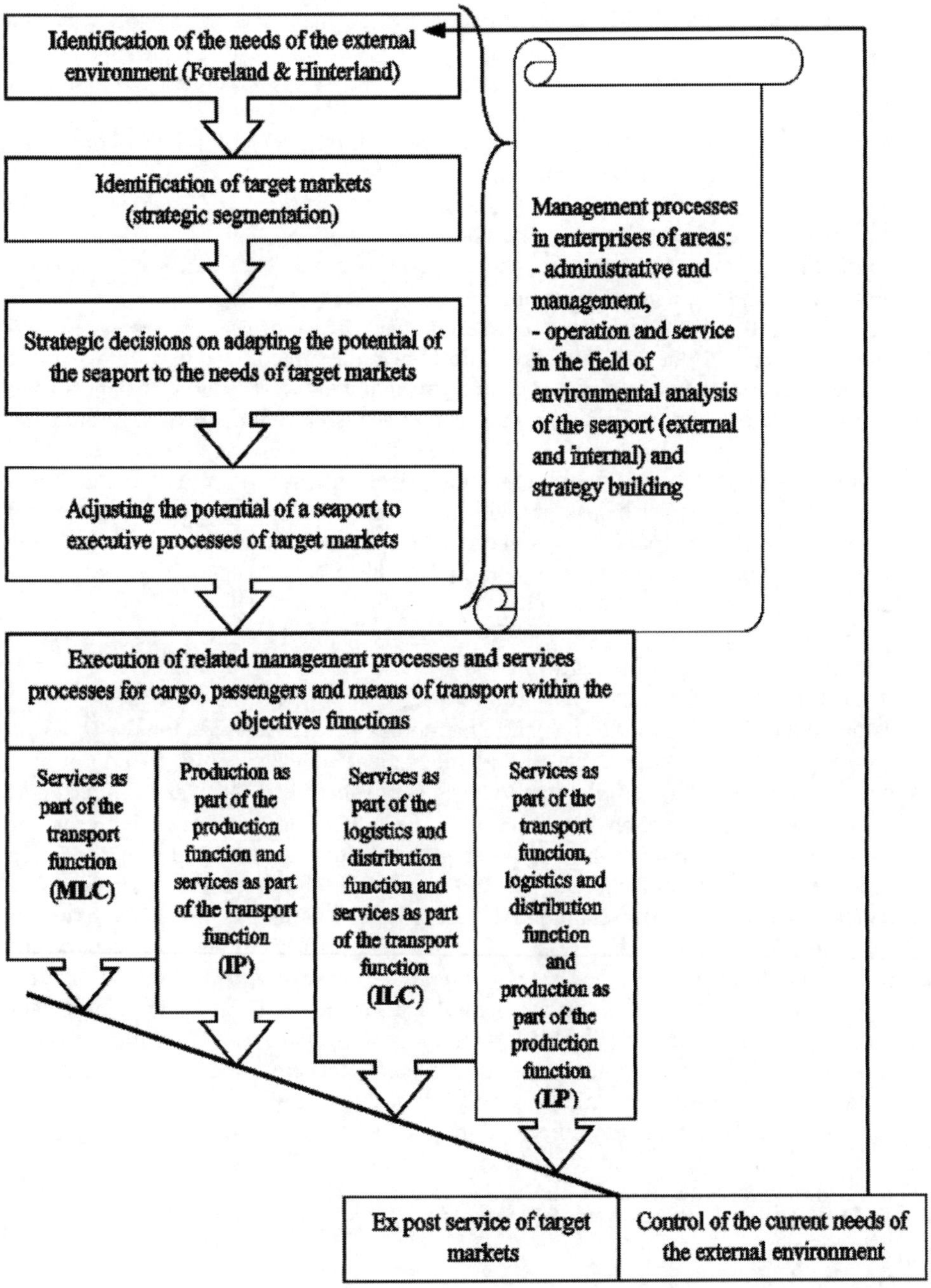

Fig. 2. The value chain of the seaport

As indicated in Figure 2, the shape of the seaport value chain in the part concerning the implementation of processes for cargo, passengers and means of transport depends on the generation of the seaport. The generation of the seaport determines the form of its operation, which in turn determines the range of processes carried out in its space, each of which creates a specific added value. Thus, depending on the level of development, the seaport may function as:

1. Maritime logistics centres (**MLC**) where the basic function is the transport function.
2. Maritime logistics centres (**MLC**) + Industry Park (**IP**), where the basic functions are: transport and production functions.
3. Maritime logistics centres (**MLC**) + Inland logistics centres (**ILC**) where the basic functions are: transport, logistics and distribution functions.
4. Logistics Platform (**LP**) where the basic functions are: transport, logistics, distribution and production functions.

The evaluation of the correlation between the number of various port processes realized within particular objective functions and the value of created value added carried out in the research process unambiguously indicates a proportional relationship between these two numbers. It is not a directly proportional relationship because the number of variables that this proportionality depends on is significant. Examples of variables are: load structure [13] or the efficiency of operations of individual enterprises operating within the functional sector of the seaport [14]. Nevertheless, this dependence is a fact. The greater the number and variety of port processes for cargo, passengers and means of transport, the greater the added value created in the space of the seaport. In the case of seaports, there can be distinguished three types of added values [13] [14]:

– value-added logistics (**VAL**) which is generally created as part of basic services for cargo; here we can distinguish cargo port services (**MLC**) and cargo logistics services (**ILC**),
– value-added facilities (**VAF**) which is created as part of services for sea and land transport and as additional cargo and intermodal transport units (ITU) services which are an essential complement to basic cargo services (**MLC** and **ILC**),
– value-added in production processes (**VA**) which is the difference between the value of production and the value of all expenditures incurred in production processes by manufacturing enterprises operating within the seaport.

The model of the seaport value chain presented in Figure 2, taking into account various forms of its operation (MLC / MLC + IP / MLC + ILC / LP), allows to determine the strategic potential of a given port. The strategic potential is measured by the size of tangible resources (technical infrastructure, suprastructure and other technical equipment of the seaport) and intangible resources (employees, know-how, organization of cooperation of port enterprises, principles of cooperation of seaport areas) that limit the seaport's ability to carry out a range of processes for cargo, passengers and means of transport as well as the effectiveness of these processes, measured by the value of the added value arising in its space. Strategic potential, as indicated in the model (Figure 2) also allows for continuous adaptation of the functional area of the seaport to changing needs on selected target markets. Within the research it was found that the key factors determining the strategic potential of a seaport are:

1. Quantitative and qualitative handling capacity (transshipment, storage and other services and activities in the land and sea operation) of selected target markets, cargo (strategic segmentation).
2. Dominant cargo categories in port operation with their structure affecting the possible range of forwarding and logistics services and the potential size of VAL / VAF.

3. Quantitative and qualitative ability to provide forwarding and logistics services for land and sea and land-land cargo (logistic services for cargo as part of supply logistics or distribution logistics).
4. The range of offered cargo forwarding and logistics services, including the capacity of port operators to manage stock in the supply and distribution processes that are part of the supply chains.
5. Range of services (quantitative and qualitative criterion) for transport and intermodal transport units and other maintenance services for port users.
6. Quantitative and qualitative ability to produce and process manufacturing enterprises operating within the seaport.
7. Principles of cooperation of the port (functional segment) in the context of efficiency of port processes (e.g. speed of handling of means of transport, where the speed of loading/unloading of vessels or customs and phytosanitary procedures determining the time of cargo flow, especially containerized and rolling cargo, by the port is crucial) .
8. The speed of reaction of the port (functional segment) to changes in the port environment, in particular in the target markets, which aims to adapt the port's potential to new expectations of target markets.
9. The total value of the possible value added stream (VAL + VAF + VA) created in the seaport, indicating the functional segment's ability to invest in the development of the quantitative and qualitative capacity of the port to service more sophisticated target markets [13].

Presented key factors determining the strategic potential of the seaport can be described by means of a value chain, the model of which was proposed in the research process. It is obvious that the proposed model is of a general nature. The evaluation of the strategic potential of a given port requires the specification of the value chain and giving it characteristics of specificity for the analyzed case. Such simulations of an individualized value chain were used to analyze selected seaports in Europe. The aim was to confirm the usefulness of this method of studying seaport resources to assess its strategic potential. This part of the research also uses (auxiliary) analysis of key success factors based on the 80/20 rule [4], which allowed to choose in the analysis of selected European seaports the factors that decided on their competitive advantage in European port services sector markets.

4. Main conclusions

The research process, the key results of which are presented in this article, was aimed at assessing the usefulness of the value chain analysis for determining the strategic potential of the seaport. As indicated in the methodology, the research process was divided into three stages, in which, among others, a number of analyzes and assessments concerning modern seaports were made and a model of their value chain was developed. The seaport seen as a multi-functional socio-economic space, where processes for cargo, passengers and means of transport have a network character both in horizontal and vertical layout. These processes are carried out by a number of enterprises associated with organizational and legal rules, acting in the common interest which is building a competitive advantage in serving selected markets located on the foreland and hinterland of the seaport. Moreover, these processes require to create a specific strategic potential, the definition of which requires specific analytical tools. The research process proved the usefulness the analysis of the value chain to assess the strategic potential of seaports widely used in the case of individual enterprises.

References

1. T. L. Whleelen, J. D. Hunger, *Strategic Management and Business Policy Toward Global Sustainability International* Edition (Thirteenth Edition), Pearson (2012).
2. J. Apanowicz, *Metodologia ogólna*, W.D.P. „BERNARDINUM", (2002), https://wsaib.pl/images/files/E-Publikacje/MO.pdf.
3. G. Gierszewska, M. Romanowska, *Analiza strategiczna przedsiębiorstwa*, PWE (2017).
4. R. Koch, *The 80/20 Principle The Secret of Achieving More with Less*, NICHOLAS BREALEY PUBLISHING, (1998), http://leadershipcoachingblog.com/wp-content/uploads/2012/03/the-80-20-principle-to-achieve-more-with-less-effort1.pdf.
5. A. Grzelakowski, M. Matczak, *Współczesne porty morskie. Funkcjonowanie i rozwój*. Wyd. Akademii Morskiej w Gdyni (2012).
6. A. Montwiłł: *The Role of Seaports as Logistics Centers in the Modelling of the Sustainable System for Distribution of Goods in Urban Areas*, Procedia Social and Behavioral Sciences 151 (2014), pp. 257-265.
7. *Port marketing and the challenge of the third generation port*, UNCTAD TD/B/C.4/AC.7/14, Geneva (1991)
8. S. Szwankowski, *Funkcjonowanie i rozwój portów morskich*, Wyd. Uniwersytetu Gdańskiego, (2000),
9. Assessment of a seaport land interface: an analytical framework. Report by the UNCTAD secretariat. UNCTAD/SDTE/TLB/MISC/2004/3
10. M. E. Porter, *Przewaga konkurencyjna. Osiąganie i utrzymywanie lepszych wyników*, Helion, (2006).
11. J. Borowski, Łańcuch wartości jako nowa teoria zarządzania strategicznego, Optimum. Studia ekonomiczne nr 2 (62) (2013), s. 11-25, http://repozytorium.uwb.edu.pl/jspui/bitstream/11320/637/1/02_Jerzy%20BOROWSKI.pdf,
12. A. Stabryła, , *Zarządzanie strategiczne w teorii i praktyce firmy*, PWN, (2000).
13. Port reform Toolkit, Second Edition, Module 3, *Alternative Port Management Structures and Ownership Models*, The World Bank (2007).
14. D. Bernacki, Usługi wartości dodanej jako element koncepcji logistycznej portu morskiego, Logistyka 4/2012.

SHS Web of Conferences **58**, 01021 (2018) https://doi.org/10.1051/shsconf/20185801021
GLOBMAR 2018

Sea port as an element of a city sustainability based on city-port of Gdańsk

Aneta Oniszczuk-Jastrząbek[1*] , *Barbara* Pawłowska[2*], and *Ernest* Czermański[3]

[1] Institute of Maritime Transport and Seaborne Trade, Department of Global Economy, Faculty of Economics, University of Gdańsk, Poland
[2] Chair of Transport Economics, Faculty of Economics, University of Gdańsk, Poland
[3] Institute of Maritime Transport and Seaborne Trade, University of Gdańsk, Poland

Abstract. Cities and ports are elements of the socio-economic space of coastal regions and they are integrally interrelated with each other while their development is mutually conditioned. Cooperation of the port and the city usually would bring development of both these elements in the historical perspective. The seaport has always grown and developed along with the city and the region. On the one hand, a port-city performs functions related to maritime economy in the broad sense of the word, i.e. maritime functions, and on the other hand - land functions that result from the city's location in relation to its land base. When the maritime function is overwhelming, the port has a powerful impact on the city, its spatial and economic layout. The very nature of the port has a significant impact on shaping the face of the city and the region. Contemporary cities are currently facing many various problems resulting from their rapid development: debt, commercialization of the public space, degradation of the infrastructure, deteriorating quality of public services, or excessive and inefficient consumption. Since the time concept of sustainable development came into being, it has been adopted as the basis for all activities at various levels of territorial organization, including at the city level. Nonetheless, each tier of development planning has its own specificity, similarly to individual territorial units within a specific tier. This is related but not limited to cities, particularly port-cities, in respect of which sustainable development should be understood a little differently than in other cities, due to the complexity of their relationships and the special role that they play in the economic system. The aim of this article is to indicate such a way of understanding the sustainable development that refers to the specificity of a port-city. Ports and cities interact across many dimensions, however, a more detailed insight how port-cities integrate the port and urban functions is still lacking. The survey conducted among city representatives, businesses and port authorities is to help identify the directions of activities for sustainable development and indicate the common grounds of these activities, where both parties could support each other.

*Corresponding author: ekoao@ug.edu.pl

Introduction

The 21st century marks a globally progressing urbanization process. According to UN estimates, only 30 per cent of people in the world lived in the cities in 1950, while by 2014, well over half of the population, i.e. 54 per cent, were inhabiting urban areas. By 2050, the size of urban populations is forecast to have grown by another 2.5 billion. The percentage of urban dwellers will rise to 66 per cent [31]. Economists and politicians have proved that cities are the driving power essential to the sound operation of the economy. It suffices to point out that in 2017, the world's largest six hundred agglomerations generated over half of global gross domestic product (GDP), and the figure has probably increased since then [26]. According to the European Commission, around 85 per cent of the EU's GDP is generated by urban areas [12]. Therefore, there is a good reason to believe that urban development will determine the EU's future economic, social and territorial growth. This is particularly true because cities are home to the majority of the workforce, business enterprises and centres of higher education, and their operation is decisive for accomplishing social cohesion.

There is no doubt that cities are centres of change driven by innovation, a spirit of enterprise and economic growth [13]. Modern-day cities are also facing a wide variety of problems brought about by their rapid expansion, such as debt, commercialization of public space, infrastructural decay, deteriorating quality of public services as well as excessive and inefficient consumption. Ever since the idea of sustainable development came into existence, it has been adopted at various levels of territorial organization, including at the city level, as a basis for all activities.

Port cities are a special type of city owing to the complexity of relations existing within them and the unique role they play in the economic system. In economic terms, mutual interactions between cities and ports favour synergy. Cities supply sea ports with human resources, transport networks and a demand for commodities, while ports reciprocate by stimulating urban growth through creating workplaces and generating the need for various services (e.g. banking, insurance, etc.). These two entities have been shown to have a positive impact on the economic sphere, but the areas in which they interact with each other are much more numerous. The purpose of this article is to establish an understanding of sustainable development, which is consistent with the specific nature of a port city. The survey carried out among respondents representing the urban population, businesses and port management authorities was designed to help identify the directions of action for sustainable development and to indicate areas for joint action, in which these entities could support each other.

The idea of sustainable development in the city

Sustainable development is one of the major trends which emerged at the turn of the 20th and 21st centuries. This idea of socio-economic development arose in response to a social, economic and environmental crisis which had taken on global proportions in the second half of the 20th century [19]. The idea and implementation of sustainable development calls for a different perspective on the problem of governance. The Brundtland Report of 1987 defines sustainable development as meeting the needs of the present without compromising the ability of future generations to meet their own needs [18]. The report ties in economic and environmental goals with social ones and, by extension, with a fair distribution of natural resources or, to use more cautious language, with fair opportunities for access to these resources. Sustainable development is the only idea to provide comprehensive coverage of the problem of the modern economy's long-term ability to develop in conformity with the criterion of intergenerational justice. The economy, environment and society are understood here as an integrated developmental complex requiring an interdisciplinary approach. FigFig. **1** groups the goals of sustainable development into three areas: (1) social solidarity, (2) economic effectiveness and (3) environmental responsibility. As these areas are inextricably connected together and mutually interactive, it is important for the implementation of this concept to integrate activities and approaches.

To define the mutual relations between the individual areas, it is necessary to apply a number of principles, including the principles of integrity of dimensions, subsidiarity and sustainable participation, including intersectoral partnership. While the first principle emphasizes equal treatment of the three dimensions of development, the second is founded upon the assumption that entities or groups of people who can single-handedly perform tasks / activities should not relinquish this responsibility to administrative authorities.

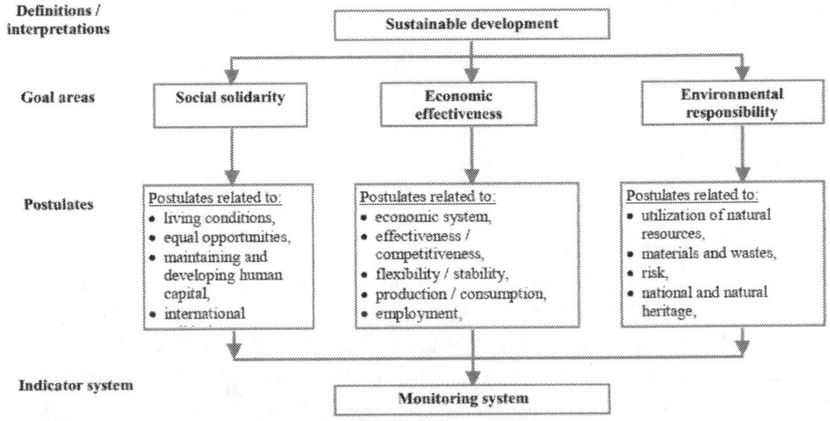

Fig. 1. Goals and postulates of sustainable development
Source: [23].

Tadeusz Borys presents sustainable development as a harmonization of the economic, social and environmental dimensions [3], with its successful implementation being conditional upon taking into account the spatial, civilizational, institutional and political aspects in addition to the already considered three dimensions.

The spatial dimension relates to spatial policy and spatial planning, the civilizational dimension refers to human capital and assures progress in the domain of knowledge and social communication, while the institutional and political aspects are bound up with politics, governance and management at various tiers of social organization [7].

This type of sustainable development can be pursued successfully where the following four policies are interlinked: environmental protection, economic, social and spatial protection, as well as a management and governance system for a specific area, country, region or community [4].

Ever since the concept of sustainable development has been disseminated around the world, it has been adopted in many countries at various tiers of territorial organization as a basis for all activities. Priority is given to activities undertaken at the local level, as this particular tier of organization is assumed to be the most suitable for implementing the concept [8; 2].

The city is an entity consisting of a large number of constituent components interconnected through various types of relations and operating in a specific area, in a specific environment; as such, it should be regarded as a dynamic, functional whole, or – in other words – as a system, specifically, a territorial social system. At the core of a social territorial system, regarded as an especially useful concept for the city, lies the effective control of territory by the population that inhabits it, that is to say, social participation.

According to Chojnacki, this is a necessary requirement for a successful implementation of sustainable development [5]. Within it, we may distinguish the social, natural and economic urban subsystems, with the system's overall structure constituted by the relations existing between these subsystems and their parts, on one hand, and those with the surroundings on the other [15].

The complexity of social, economic, environmental and spatial problems characteristic of cities and urban areas, together with a rapid rate of change witnessed there, force the municipal authorities to seek for more efficient and cost-effective methods of managing their affairs. Their main goal should be to improve the living standards within the urban areas, with the overall effect of greater social and economic well-being [23].

A picture of mutual relations between port cities and ports

The lines, along which port cities developed historically ran closely side by side with those of the evolution of maritime transport technology, including methods of cargo handling. This has had an enormous impact on the rise, expansion and – eventually – decay of urban waterfronts[2], with their former and present harbour facilities [9]. Driven in the last 50 years by the ever more intense process of globalization, the maritime transport industry has undergone a technological breakthrough, with containerization and other important modifications to cargo-handling machinery.

These changes have brought about two different but interconnected phenomena. These are, on one hand, the growth of ports, with terminals being relocated to more peripheral sites in order to comply with the present-day standards related to vessel size and land access [1], on the other, the restoration of urban coastal areas, which had become obsolete and disused, through installing modern cargo handling equipment [16] and a demand for the revitalization of the natural environment [11].

At the outset of the discussion of the subject, we should note that the literature lacks any commonly accepted definition of a port city. It should be kept in mind that a port city, on one hand, performs maritime functions, or ones related to the maritime economy, and on the other – land functions, or ones tied with its location relative to the hinterland. The history of mutual relations between sea ports and cities has followed a winding course and continues to do so until today. B.S. Hoyle has named six periods in the evolution of sea port-to-city relations (Table 1).

Table 1. Stages of mutual relations between sea port and port city

Historical stages		Period	Description
I	A primitive type of relations between a sea port and a city,	prevalent until the 19th century	A spatial and functional unity of port and city
II	Developments in the relations of ports and cities	19th century / early 20th century	A rapid commercial and industrial development shifts the expanding ports outside the city limits
III	A modern city and industrial port	mid-20th century.	Expansion of the industrial function (especially where this relates to processing kerosene oil); the onset of container and ro-ro technologies increases demand for space, leading to the utilization of new areas
IV	Recession of ports	1960s and 1980s	The technological change in maritime transport lead to the

[2] Waterfront – in urban planning, this term refers to areas located alongside a natural body of water, such as a lake, river or sea. In practice, however, the word relates to the contact areas lying between the urban and port structures. Waterfronts are the site of conflict and cooperation at the same time between port and city authorities in an administrative, economic and spatial sense. The space is not marked out once and for all; on the contrary, it follows a twisted historical course of growth and shrinkage (according to Wikipedia [27]).

	from the coastline		port industrial operations being receded from areas in the direct vicinity of the coastline
V	Continued coastline development	1970s and 1980s	A large, modern port occupies a sizeable area of water and land; re-opening of old docklands
VI	Port relations re-established	1980s until 2000	Globalization and intermodal transport transform the role of ports; their relations with cities are re-established; urban revitalization of old port areas.

Source: [10].

Mutual relations of sea ports and port cities change as time passes by. Until the 19th century, in the so-called primitive stage, the city and port were arranged into a single spatial and functional unit. In the 19th and early 20th centuries, commercial and industrial growth moved the expansion of ports outside the city limits. The growth of the industrial function of sea ports in the middle of the 20th century led to increased demands for space and utilizing new areas located far away from port cities. The tendency persisted in the 1960s and 1970s when technological developments in maritime transport made it necessary for the industrial operations of ports to recede from areas in the direct vicinity of the coastline. The coming back to life of the old docklands became visible again in the 1980s and 1990s. The coastal ports also saw a rebirth at that time.

The research carried out on the relations of cities and ports focused on sea ports acting as centres of activity for the local community thanks to their direct and indirect impact. Especially noteworthy is the appeal of cities to new residents who arrive to take advantage of the large number of businesses operating in the ports and in their vicinity. For this reason, port cities are often contemplated in terms of port clusters [21]. The socio-political aspect of relations between city/region and port drives a job market of its own and generates workplaces offered by business entities cooperating with the port. This has the effect of reducing unemployment in the region, while salary levels contribute to the personal wealth of employees and their families, as well as stimulating demand in the port city / region. The multiplier effect is present here, in that we can see direct impact (port's core operations), indirect impact (entities cooperating with the port) and induced impact (resulting from demand related to the salaries earned by port employees) [19].

Another crucial aspect addressed by the research is the question of advantages enjoyed by various groups of stakeholders (businesses, local authorities, society, etc.) because of a port's proximity. Due to the size of these groups and the complex relations between them, it is difficult to reconcile their various interests. The beneficial impact of port operations on the local community is also related to financial assets flowing in from local self-government budgets. However, taking into account the complexity and specific nature of business activities related to the operation of a port and numerous companies run on its premises, the complexity of these transfers should be stressed (real estate tax, poll tax, income tax). In the case of European and US ports, the financial aspect of their impact on the nearby communities is seen in the added value generated by the sea port [25].

In exploring the relations between city ports, attention should be paid to cost economy for entities located in the vicinity of ports and to economies of scale for specialized service providers located close to port cities.

The literature provides extensive coverage of the negative effects of port operations, caused by external costs generated by their activities. In the past, ports were mostly concerned with their own activity and economic efficiency, rather than the impact of port operation on the surrounding environment [30]. In line with this, port areas used to be regularly expanded to accommodate the increasing volume of cargo handled. However, as noted by Wiegmans and Louw [29], it has become more difficult than ever before for port areas to expand, due to a growing environmental awareness and the resulting community opposition. Therefore, it seems that the era of easy and free-of-controversy port expansion has come to an end [6]. The idea of sustainable development and internalization of external costs has been included in the sectoral development policy and strategies of individual sea ports. It is necessary to step up efforts to incorporate the principles of sustainable development in the strategies of sea ports and port cities. The city of Gdańsk and the port of Gdańsk have been used as examples in the research.

Purpose and methodology of research

The main purpose of the research was to determine the level of awareness among the members of the City Hall of Gdańsk (UM) and the Port of Gdańsk Management Authority (ZMPG) with regard to the principles of sustainable development and to identify the mutual relations of city and port.

The structural organization of the City Hall of Gdańsk (UM) and the Port of Gdańsk Management Authority (ZMP) was used as a basis for a database of directors and managers of selected sections of the City Hall and departments of the Port of Gdańsk Management Authority. Next, all the respondents appearing on the list were contacted for an interview. Some of them refused to respond due to a lack of time, lack of interest in the interview or lack of knowledge of the subject-matter. As a result, it was decided to use relevance-based selection, which yielded a list of 12 employees of the City Hall of Gdańsk and 6 representatives of the Port of Gdańsk Management Authority, whose job profiles included issues related to sustainable development.

The research project was in progress from 28 March to 9 April 2018. Due to the short time-limit for completing the research and the relatively small number of interviews, the survey should be seen as a poll for reference purposes.

The interview questionnaire consisted of ten questions (for the City Hall) and nine questions (for the Port of Gdańsk Management Authority). The third question used a modification of Likert's scale, i.e. a complex attitude scale. All items on Likert's scale were assigned degrees of intensity representing intervals of the measurement scale [Wieczorkowska i in, 2003; Sagan, 2003; Oniszczuk-Jastrząbek, 2016].

The steps of the scale were described in verbal form. In that question, the respondents rated the areas of sustainable development as "very important," "important," "I don't know," "not very important," and "unimportant."

Survey results

The survey found that 100 per cent of the respondents, both from the City Hall and the Port Management Authority, were familiar with sustainable development. This means that management-level employees working for both of these institutions demonstrated a high level awareness of the concept of sustainable development. 80 per cent of respondents gave positive answers to the question concerning the actions related to sustainable development, implemented by the City Hall and the Port of Gdańsk. The answer *I don't know, it's hard to say* was returned by 20 per cent of respondents. The implementation of these actions by the city and the port was evaluated as *very good* or *good* by 67 per cent of respondents from the City Hall and 50 per cent of those from the Port of Gdańsk. The remaining respondents said *I don't know*. The implementing instruments used for sustainable development which were the most frequently named in the survey were the city / port's development strategy and internal regulations.

The next question asked the respondents to select what they believe were the most important areas of sustainable development. Fig. 2 presents what the employees of the City Hall of Gdańsk thought were the most important areas of sustainable

development, while

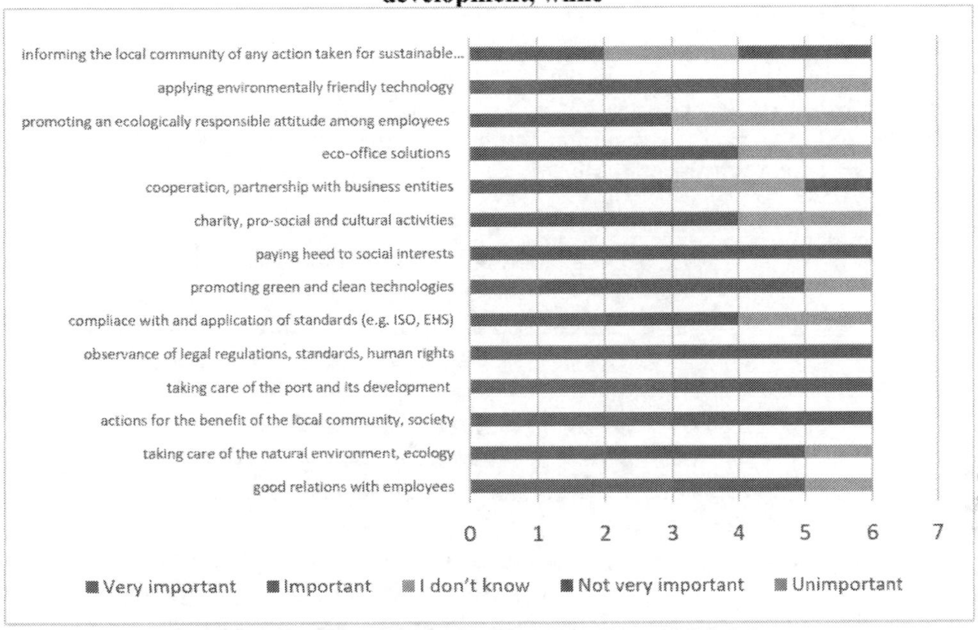

Fig. **3** presents responses from the employees of the Port of Gdańsk.

Fig. 2 demonstrates that the City Hall of Gdańsk's employees on average assigned the highest ratings in their responses to the following areas: "taking care of a high quality of urban life" (4.83), "compliance with legal regulations, standards and human rights" (4.75) and "actions for the benefit of the local community" (4.5). Further down the scale were "consideration for social interests in urban development planning" (4.42), "charity, pro-social and cultural activity"; "compliance with and application of standards (e.g. ISO, EHS)"; "taking care of the natural environment, ecology" (4.25). The following ranked further down the scale in the ranking: "good relations with employees"; "active information policy to raise ecological awareness" (4.17), "cooperation, partnership with business entities; inclusion of citizens in decision-making on the direction of action in the city" (4.00); "applying environmentally-friendly technology to limit water and energy consumption" (3.92); "promoting renewable sources of energy" (3.83).

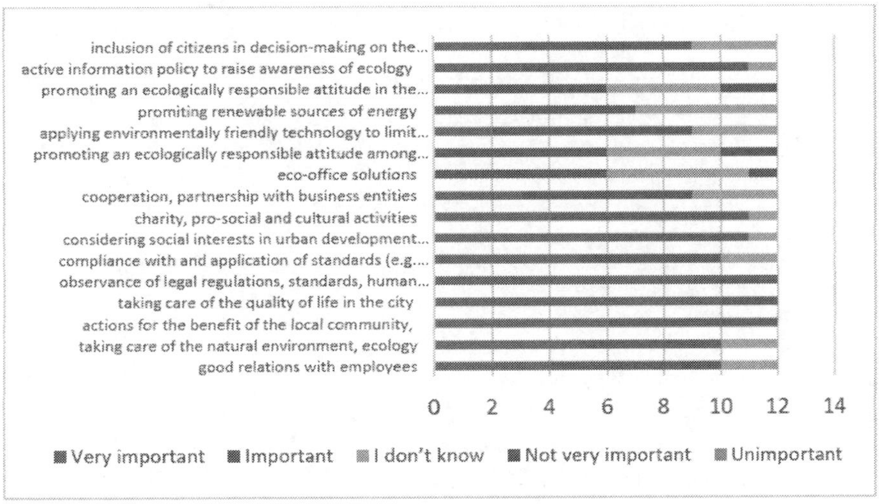

Fig. 2. Importance assessment of sustainable development areas according to the employees of the City Hall of Gdańsk
Source: The author's own elaboration.

The respondent ratings were the lowest for the following areas: "eco-office solutions"; "promoting an ecologically responsible attitude in the local community" (3.42) and "promoting an ecologically responsible attitude among employees" (3.33).

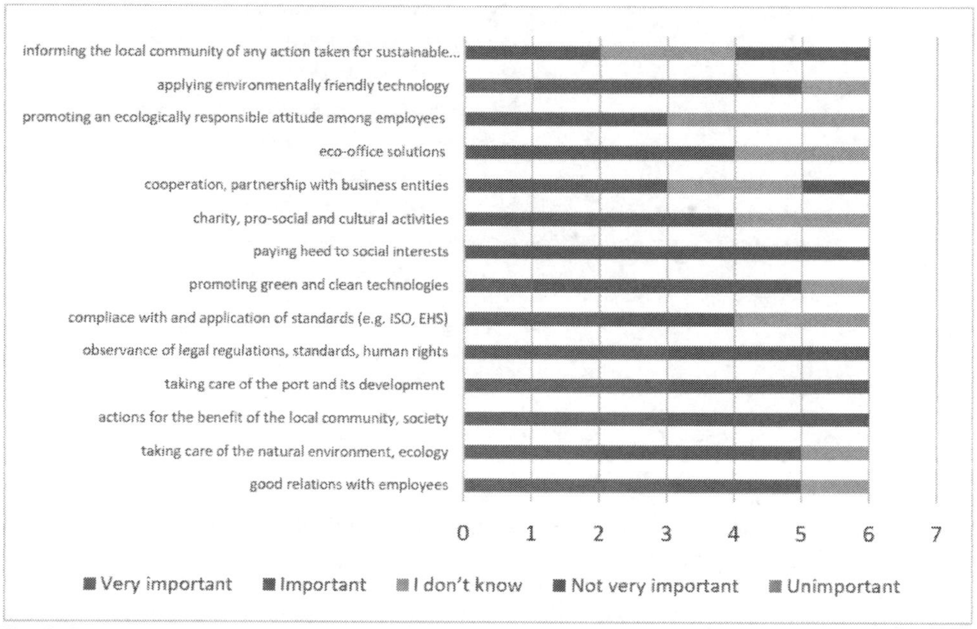

Fig. 3. Importance assessment of sustainable development areas according the Port of Gdańsk Management Authority's employees
Source: The author's own elaboration

Similarly, the Port of Gdańsk Management Authority's employees (

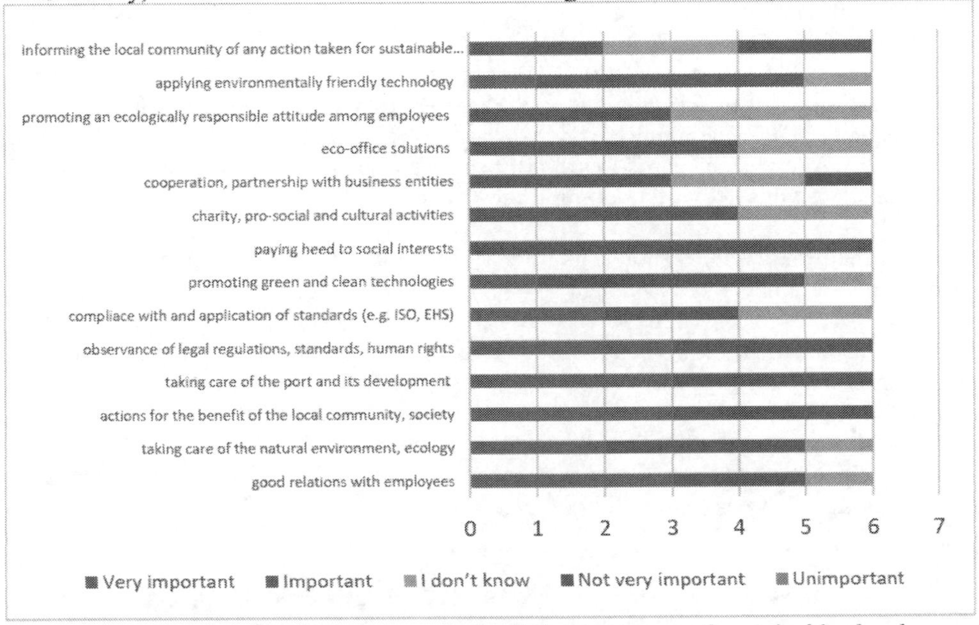

Fig. **3**) assigned the highest ratings to the following areas of sustainable development: "action for the benefit of the local community"; "taking care of the port and its development"; "compliance with legal regulations, standards and human rights" (4.50), "good relations with employees" (4.33); "taking care of the natural environment, ecology (4.17); "compliance with and application of standards (e.g. ISO, EHS)"; "promoting pro-ecological solutions and clean technology solutions"; "consideration of social interests"; "using environmentally-friendly technology" (4.00). Lower ratings were assigned to such areas as "charity, pro-social and cultural activity"; "eco-office solutions" (3.67); "promoting an ecologically responsible attitude among employees" (3.50)"; "cooperation, partnership with business entities" (3.33)"; "informing the local community of actions taken in respect of sustainable development" (3.00).

The next question asked the respondents of the City Hall of Gdańsk to identify what they thought were the benefits and problems related to the application of principles of sustainable developments. The responses are presented in Fig 4 and Fig. **5**.

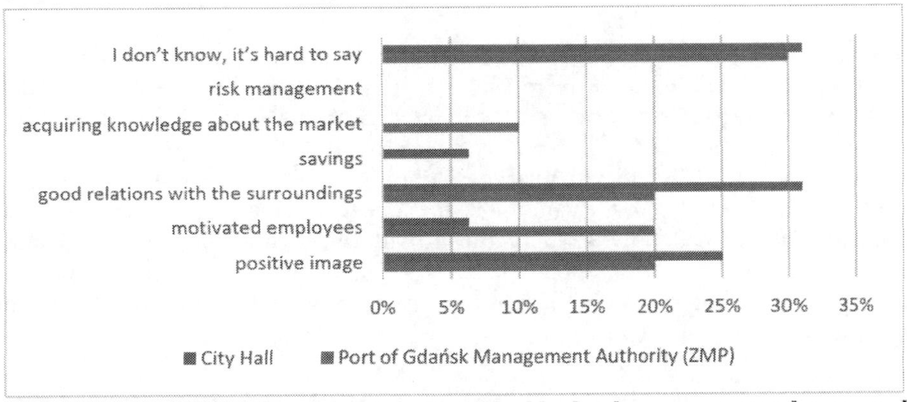

Fig. 4 **Benefits of applying the principles of sustainable development as seen by respondents from the City Hall of Gdańsk and the Port of Gdańsk Management Authority**
Source: The author's own work.

Fig. **4** shows respondents identifying the most important benefits of applying the principles of sustainable development as "good relations with the surroundings" (31 per cent – City Hall of Gdańsk; 20 per cent – Port of Gdańsk Management Authority) and "positive image" (25 per cent – City Hall of Gdańsk; 20 per cent – Port of Gdańsk Management Authority). The employees of the Port of Gdańsk Management Authority assigned equally high value ratings to "actions to motivate employees" (20 per cent), while "acquiring new knowledge" came fourth (10 per cent). The employees of the City Hall of Gdańsk rated "actions to motivate employees" and "savings" (6 per cent) as third in importance.

Besides the benefits, the respondents were asked about problems related to applying the principles of sustainable development (Fig. 5). The most frequent responses included "complexity of the matter" (26 per cent – City Hall of Gdańsk; 25 per cent – Port of Gdańsk Management Authority). Next in the ranking by the City Hall of Gdańsk came "a lack of impact on everyday practice" (15.9 per cent) and "need for financial outlays"; "lack of time"; "lack of knowledge or skills" (10.5 per cent), while the respondents from the Port of Gdańsk Management Authority ranked "need for financial outlays"; "lack of knowledge or skills" (17 per cent) as second in importance and "lack time"; "lack of commitment from the city authorities" (8 per cent) as third.

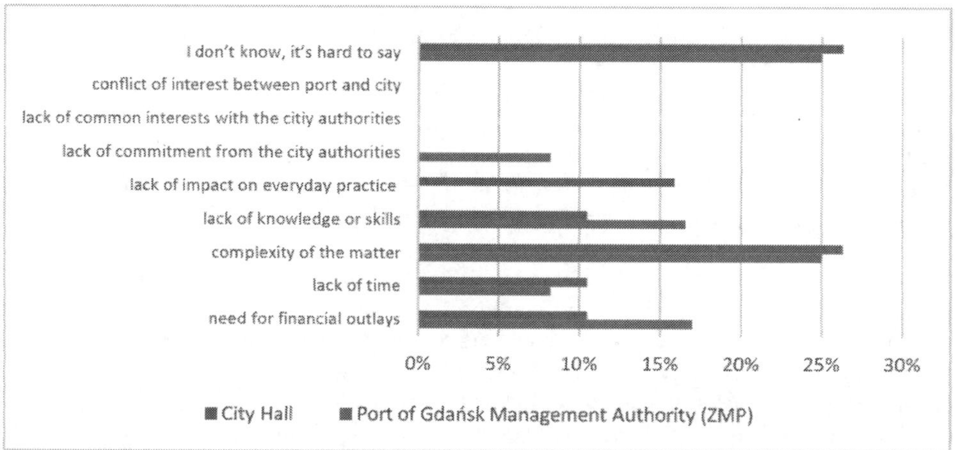

Fig. 5. Problems related to applying the principles of sustainable development as seen by respondents from the City Hall of Gdańsk and Port of Gdańsk Management Authority
Source: The author's own work.

Another important problem addressed by the survey was mutual relations between city and port. The respondents were asked if they thought that the port or the city gauges the external effects generated by port operations which have an impact on both port operations and the local community. 25 per cent of the City Hall respondents answered in the affirmative, while 8 per cent answered in the negative, with 66 per cent having no opinion on the matter. Among the Port of Gdańsk respondents 50 per cent answered in the affirmative, while the other half had no opinion. As a result, the respondents were asked to specify the principles of sustainable development which they believe were reflected in the mutual relations between the city and the port (

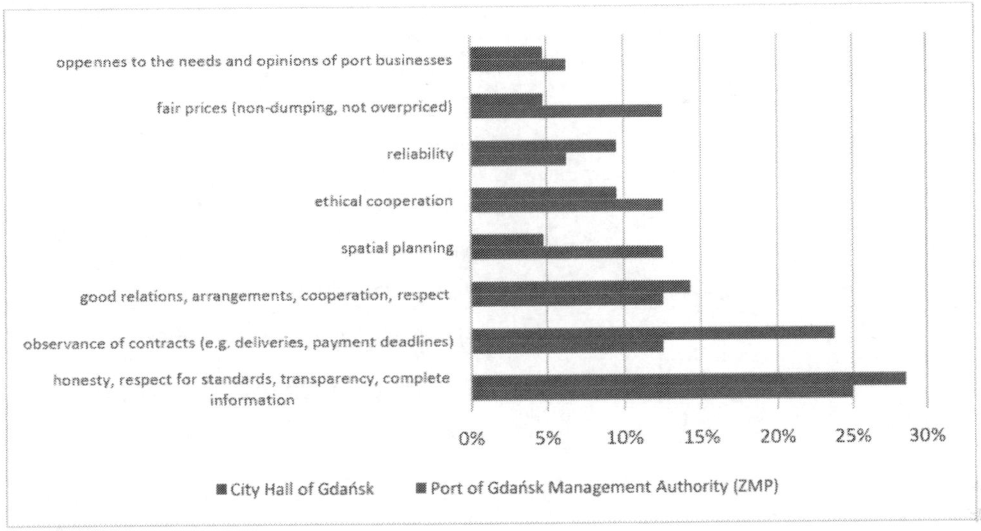

Fig. **6**).

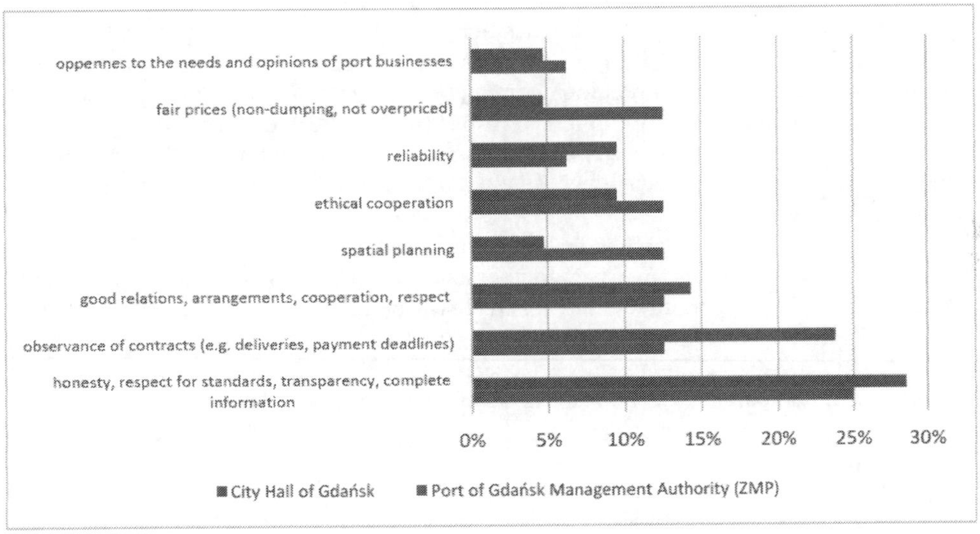

Fig. 6 **The principles of sustainable development in city - port relations**
Source: The author's own work.

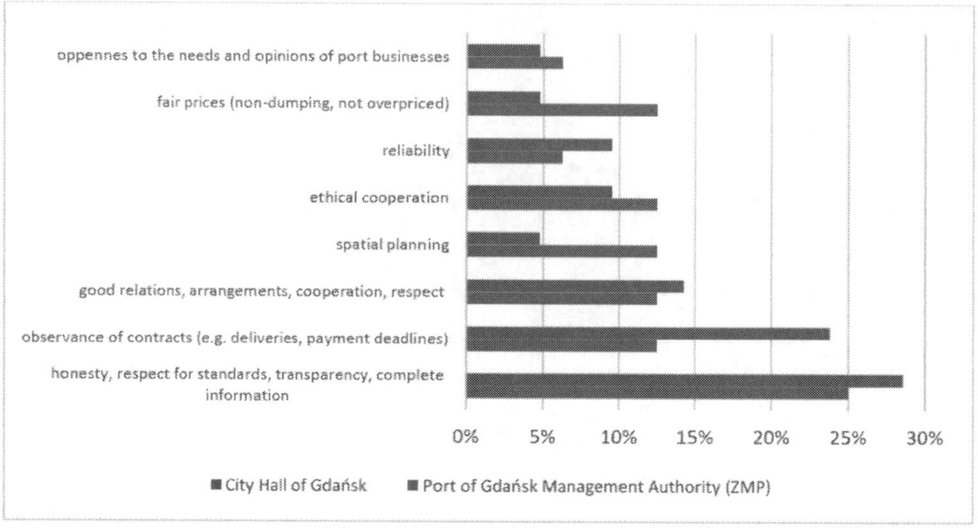

Fig. **6** shows that both respondent groups were of the opinion that the entities under analysis were driven in their actions mainly by "honesty, respect for standards, transparency and reporting complete information" (29% - City Hall; 25 per cent – Port of Gdańsk). Ranking lower in the City Hall respondents' assessment were "observance of contracts" (24 per cent): "good relations, arrangements, cooperation, respect" (14 per cent) and "ethical cooperation and truthfulness" (10 per cent). In the Port of Gdańsk respondent group's ranking, second place was held by "observance of contracts"; "good relations, arrangements, cooperation, and respect"; "spatial planning"; "fair prices" and "ethical cooperation" (13 per cent).

In addition, the City Hall respondents were asked to order the areas, in which the city can support port development, by assigning points to each of the areas on a scale of 6 (most important) to 1 (least important). The respondents assigned the highest ratings (5 and 6) to "legal principles", "investment in the infrastructure", "port accessibility", "spatial planning" and "transparent cooperation between city and port", meaning that these items were the most important in their view for good mutual relations.

Conclusions

Cities and ports feature prominently in the social and economic landscapes of coastal regions, they are integrally interconnected and rely on each other for continued development. Historically, they have evolved side by side, cooperating with each other for mutual benefits. The sea port has always grown and developed in conjunction with the city and the region. The development of port areas is a test of strategic and operational maturity for cities, as it requires a comprehensive and integrated approach and commitment from a large number of stakeholders in implementing the principles of sustainable development. In this way, it is the product of a compromise reached by different stakeholder groups in order to integrate their various goals, responsibilities and limitations. As a result, a socially acceptable city's strategic maturity is reflected in the sustainable creation, implementation and control of activities expected to fulfil a considerable number of requirements. Consequently, the city must be able to act as a broker for the conflicted parties who come forward with incompatible expectations.

The development strategies of the city of Gdańsk and the port of Gdańsk contemplate the mutual relations of city and port as one of the most essential for the development of both. Stemming from a compromise reached between various community members, such as social and business partners, the city of Gdańsk's development strategy pointedly emphasizes that the city would not have been able to thrive as it has since its beginnings if it had not been for transport. Located at the intersection of waterways and overland routes, the city is a major transit junction between Scandinavia and south-eastern Europe. A flourishing sea port is among the pieces that come together to form a positive picture of a metropoly now and in the future. It is also a good starting point for the development of other services and resources to fuel a modern economy. Today, the greatest assets for a city intent on building up a strong competitive edge are its seaside location, economy and transport systems [32].

Research has shown that, when it comes to mutual relations between the city and the port, respondents see eye to eye in reporting honesty, transparency, reporting complete information, good relations and cooperation as the most important areas of sustainable development, all of which are reflected in the Port of Gdańsk's Development Strategy till 2027. The document stresses that the port can develop only through cooperation with the Community of Gdańsk and with active support from the central government. The city of Gdańsk and the neighbouring sea port have developed side by side for centuries and they must continue on that path in the future. An example of a project which may result in a success due to combined efforts of the port and the city is to build an industrial zone around the port [33]. It should be stressed that cooperation based on trust and respect provides a foundation of safety and social order to strengthen the bonds across populations, institutions, businesses and governmental bodies in the metropolis, leading to a stronger sense of individuality and shared responsibility.

It should be kept in mind that high-quality space allowing effective, economic and ecological management will be attractive not only for the residents, but will also appeal to tourists and businesses willing to invest their money. Spatial planning appeared frequently in the survey as one of the principles of sustainable development which needs to be considered by the city and the port working together for improved relations. The importance of spatial planning is confirmed by the city of Gdańsk's strategy to the extent that it relates to the laying down of clear and open principles of sustainable management of space which considers the preservation of land reserves necessary for housing, recreational, service and manufacturing purposes.

It should also be mentioned that a report by The Milken Institute, an American think-tank, has ranked Pomorskie province as Europe's fourth best and most efficiently managed region. The authors of the report agree that one of the factors that have led to this is the key role of the Port of Gdańsk for transport and international trade. The development of the Port of Gdańsk means growth not only for the city or the region, but for the entire country. The port in Gdańsk is the largest Polish harbour and the only one on the Baltic Sea to operate a deep water container terminal. The port contributes considerable revenues for the state budget in the form of taxes, duties and excise. In 2017, the port set the record by reaching an annual cargo-handling throughput of 40,000,000 tons. Ports have always been of key importance for the wealth of the entire country, being the powerhouses of the national economy, which is why it is so important for cities and ports to cooperate.

References

[1] Asteris, M., Collins, A., *Developing Britain's port infrastructure: markets, location and policy.* "Environment and Planning" A, **39(9)** (2007).

[2] Banachowicz B., Danielewicz J., *Współrządzenie jako czynnik zrównoważonego rozwoju miast – przykład Łodzi*, [in:] *Polityka zrównoważonego rozwoju oraz instrumenty zarządzania miastem*, ed. by J. Sołdczyk, D. Rajchel, Wyd. Uniwersytetu Opolskiego, Opole (2006).

[3] Borys T., *Teoretyczne aspekty konstruowania wskaźników ekorozwoju*, [in:] *Sterowanie ekorozwojem. teoretyczne aspekty ekorozwoju*, ed. by B. Poskrobko, Wyd. Politechniki Białostockiej, Białystok (1998).
[4] Borys T., *Wskaźniki ekorozwoju*, Wyd. Ekonomia i Środowisko, Białystok (1999).
[5] Chojnicki Z., *Koncepcja terytorialnego systemu społecznego*, Przegląd Geograficzny **60 (3)**, (1989).
[6] del Saz-Salazar, S., García-Menéndez L., Merk O., *The Port and its Environment: Methodological Approach for Economic Appraisal*, OECD Regional Development Working Papers, **24**, (2013).
[7] Dembińska-Niemiec A., *Zrównoważony rozwój a funkcje miast. Badanie związków między zrównoważonym rozwojem średnich miast w Polsce a ewolucją ich struktury funkcjonalnej*, Uniwersytet Opolski „Studia i monografie" nr. **540**, (2017).
[8] Gajdzik B., *Urbanizacja w zrównoważonym rozwoju województwa śląskiego*, JecolHealth **14** (2010).
[9] Hall P., *Waterfronts: A new Urban Frontier*, [in:] *Waterfronts. A New Frontier for Citiess on Water*, ed. by R. Bruttomesso, Citta d'Acqua, Venice (1993).
[10] Hoyle B.S., *The redevelopment of derelict port areas* [in:] The Dock & Harbour Authority, no **887**, (1998).
[11] Hoyle, B.S., *The port-city interface: trends, problems, and examples*, Geoforum, **20(4)** (1989).
[12] Komunikat Komisji do Parlamentu Europejskiego, Rady, Europejskiego Komitetu Ekonomiczno-Społecznego i Komitetu Regionów, *Wspólne dążenie do osiągnięcia konkurencyjnej i zasobooszczędnej mobilności w miastach*, KOM(2013) 913.
[13] Komunikat Komisji, *Europa 2020 Strategia na rzecz inteligentnego i zrównoważonego rozwoju sprzyjającego włączeniu społecznemu*, KOM(2010) 2020.
[14] Kourtit K., Nijkamp P., *The New Urban World – the challenges of cities in decline*, Romanian Journal of Regional Science **7** (2013).
[15] Mierzejewska L., *Zrównoważony rozwój miasta – wybrane sposoby pojmowania, koncepcje i modele*, Problemy Rozwoju Miast" Kwartalnik Naukowy Instytutu Rozwoju Miast Rok XII, **Zeszyt II**/(2015).
[16] Olivier, D., Slack, B., *Rethinking the port*, Environment and Planning **A, 38(8)** (2006).
[17] Oniszczuk-Jastrząbek A., *Przedsiębiorczość w budowaniu zdolności konkurencyjnej przedsiębiorstwa. Analiza danych ilościowych, część 1* [w:] Współczesna Gospodarka. Czasopismo Elektroniczne, Vol. **7**, Issue **1**, (2016), http://www.wspolczesnagospodarka.pl.
[18] *Our Common Future*, World Commission on Environment and Development, Oxford University Press, Oxford 1987.
[19] Pawłowska B., *Zrównoważony rozwój transportu na tle współczesnych procesów społeczno-gospodarczych*, Wyd. UG, Gdańsk (2013).
[20] Pluciński M., *Polskie porty morskie w zmieniającym się otoczeniu zewnętrznym*, CeDeWu.pl, Warszawa (2013).
[21] Pluciński M., *Współistnienie miast portowych i porów morskich - wybrane aspekty teoretyczne i praktyczne*, [w:] *Porty morskie w perspektywie przestrzennej, ekonomicznej, transportowej i społecznej*, ed. by H. Klimek, Wyd. UG, Gdańsk (2016).
[22] Sagan A., *Analiza satysfakcji i lojalności klientów*, [w:] Zastosowania statystyki i data mining, StatSoft Polska, Warszawa-Kraków (2003).
[23] Stawasz D., Sikora-Fernandez D., *Koncepcja smart–city na tle procesów i uwarunkowań rozwoju współczesnych miast*, Wyd. Uniwersytetu Łódzkiego, Łódź (2016).
[24] Sustainable development in Switzerland. Indicators and comments. The Swiss Federal Statistical Office, Nauchatel (2004).
[25] Suykens F., *The city and its port—an economic appraisal*, Geoforum Vol. **20**, Iss. **4**, (1989).
[26] *Urban world: Mapping the economic power of cities*, McKinsey Global Institute, Toronto (2011).
[27] Waterfront (front wodny) - https://pl.wikipedia.org/wiki/Waterfront.
[28] Wieczorkowska G., Kochański P., Eljaszuk M., *Statystyka. Wprowadzenie do analizy danych sondażowych i eksperymentalnych*, Warszawa (2003).
[29] Wiegmans, B.W, Louw, E., *Changing port–city relations at Amsterdam: a new phase at the interface?*, Journal of Transport Geography **19**, (2011).

[30] Wooldridge, C., Stojanovic, T., *Integrated environmental management of ports and harbours. The European experience – from policy to practice*, [in:] *Shipping and ports in the twenty-first century. Globalization, technological change and the environment*, eds. by Pinder D., Slack B., Routledge, New York (2004).
[31] *World Urbanization Prospects. The 2014 revision*, *The highlights*, United Nations, Department of Economic and Social Affairs, Nowy Jork (2014).
[32] Gdańsk 2030 Plus Strategia Rozwoju Miasta, City Council Resolution LVII/1327/14 of 25 September 2014.
[33] Strategia Rozwoju Portu Gdańsk do 2027 roku; http://www.portgdansk.pl/zmpg-sa/strategia-rozwoju
[34] Jackson J., Ratnatunga M., Devol R., Best performing cities. Europe 2017, Milken Institute (2017).

Sustainability issues in ports: content analysis and review of the literature (1987- 2017)

Nergis Özispa[1] and *Gamze* Arabelen[2*]

[1]Dokuz Eylül University, Maritime Faculty, Department of Maritime Business Administration, İzmir, Turkey

[2] Dokuz Eylül University, Maritime Faculty, Department of Maritime Business Administration, İzmir, Turkey

Abstract. Sustainability, which has environmental, social and economic dimensions, is a phenomenon that has been used extensively in recent times. As decided in Brundtland Conference in 1987, it basically aims to meet the needs of the present without compromising the ability of future generations to meet their own needs. When it is taken into consideration in terms of sustainability, it is known that ports are one of the most polluting sectors due to their nature. However, in recent years many ports have started to different sustainability programs and invested in large amounts over the years to reduce emissions and to gain a competitive advantage. When examined the past studies, it is determined that the sustainability concept is a relatively recent approach in the maritime literature and still a gap exist in this field. It is thought that the analysis of these studies in a systematic way will contribute to the port sustainability literature. Therefore, in this study, it is aimed to review port sustainability concept with the help of existing literature. For the methodology, studies obtained from different databases were evaluated systematically by using content analysis. The study has been concluded with the findings for the related literature, recommendations for the sector and guidelines for future researches.

1 Introduction

The sustainability concept is based on forestry, fisheries and soil science and associated with the concept of productivity in the fields of renewable sources like agriculture ecology and fishery [1] and it has been described as "*ability to maintain productivity despite obstacles*" [2]. The concept of sustainability has been defined in diverse fields like management and social science, environmental science or technical science [3], but the most adopted definition of sustainability was made in Brundtland Report for the World Commission on Environment and Development in 1987 as; "*Meeting the needs of the present without compromising the ability of future generations to meet their own needs*" [4]. Sustainability concept consists of three dimensions which are; environmental, economic and social, and these three dimension

* Corresponding author: gamze_arabelen@hotmail.com

also referred as the triple bottom line of sustainability [5]. These three value driven dimension are visualized and help monitor the port sustainability performance [6].

Ports are very complicated because of their nature, various services and a wide range of environmental issues such as water releases (limber water), harmful emissions, waste and noise production and pollution [6, 7]. They are composed of different companies, dealing with different activities and offering a wide range of services. Seaports are notorious to be serious polluting industry, however, they have many opportunities for reducing emissions and pollutions [8]. Sustainability in the port industry is of growing source of fear for port authorities, policy makers, port users and local communities [9]. Therefore, sustainability is one of the important concepts for port industry and it was thought that more detailed examination would be useful. Accordingly, the aim of the study is determined as to contribute to the literature by systematically analyzing port sustainability studies published in various databases. To address this aim, the rest of this paper is organized as follows. Port sustainability literature in different databases is reviewed in the following section. Subsequently, the data collection, the method used to analyze studies and the results are described. Finally, we discuss the results in terms of their implications for practical and scientific, and their limitations and directions for further researches.

2 Literature review

2.1 Sustainable development

When analyzed current literature in port sustainability concept, it is seen that most of the studies is written on sustainable development subject. Starting with Daamen's (2007) study that investigate port sustainable development in two different geographies which USA and Europe aiming to find a new typology and new interpretation of the contemporary port-city interface [10]. Same year Abood made a case study at New York port and demonstrate several examples to show how to fit more advanced programs into a sustainability framework. To limit sustainability subject, Abood used eight port operational topics which are; dredging, ballast water, habitat restoration, air quality, water conservation, energy conservation, material conservation and waste handling, and used them as sustainability perspective. At the end of the study Abood proposed a more quantitative sustainability rating system specifically designed to deal with port activities [11]. In (2012) Lam and Van de Voorde made a case study with 4 ports that the port of Los Angeles, Long Beach, Antwerp and Hong Kong to fill the gap in the concept of green and sustainable port strategy by formulating a new framework. Hence an original framework and propositions have been developed in green and sustainable port strategy as findings of the study [12]. Jonathan et al. (2014) made a literature review to propose a sustainable port project planning model while adding specific components to the basics of sustainability. Their model collected all individual components of sustainability concept into one roof for port project planning (PPP) [13]. Hosseini et al. conducted survey to 381 people at Torkaman city to acquire relatively exact description of condition of urban sustainable development from citizens' rights point of view in 2014. At the end of the study they determined most and least effected identifiers on sustainable development of Torkaman port city which is respectively civil rights and political identifiers [14]. Kutkaitis et al. (2014) conducted interviews with specialists and consultants of port and logistics sector practitioners to identify the criteria of port sustainable development. As a result, they designed seven steps that determine sustainability of Port of Klaipeda's logistics activity. Hence they prepared a set of indicators that determine different levels of sustainability of ports [15]. In addition, in 2014 Martino created an alternative managerial perspective in the decision making process of port authority using the literature review method [16]. In the year of 2015 Wang et al. investigated energy efficiency of port operations in their case study, aiming to improve energy efficiency with financial success, to report the effectiveness of sustainable development policy and to highlight the workable plans for ports in China. Finally, they reported a framework of assessing energy efficiency of port operations in China based on the data from public domain by using the combination of DEA (Data Envelopment Analysis) and PDE (Panel Data Estimation) methods [17]. Case study method was also conducted by Boulos in 2015 in Port Said to propose a model framework for sustainable city-port connectivity. In order to develop that model framework, they investigated the development of port-city relationship. As a result of study Boulos emphasized the importance of the balance of both port and city growths. While port expand and modernize, its impacts of the origin settlements (port cities) should be studied for each unique region as well [18]. In 2016 two different studies were published in the subject of port sustainable development. First of them belongs to Laxe et al. With the aim of using the Synthetic Indexes of sustainability in the analysis of port sustainability they examined 16 Spanish port authorities' datum with literature review method. Thus they demonstrated how the ports that have higher Synthetic Indexes on Sustainability move more goods at their facilities than the rest. According to their findings ports that have higher Synthetic Indexes pay more attention to a global and integrated management while being very attentive to both demands of users and permanently adapt themselves to new conditions of international competition [19]. In (2016) Roh et al. conducted in-depth interviews with Asian port authorities to determine the main factors that shaping sustainable port development. As a result, they identified key elements of sustainable port development from port authorities' perspective [20].

2.2 Sustainability performance

Sustainability performance is the second important subject in port sustainability literature according to current literature review. Nine study published in port sustainability performance subject in the last 30 years. However, it should also be noted that, even the sustainability concept was born in 1987, the papers about sustainability performance of ports written between the years of 2008 and 2017. First, Hakam and Solvang used case study method on Narvik container port in 2009 to link flexibility and sustainable advantages together in the context of container ports. According to Hakam and Solvang to increase the container port's flexibility one should increase the port's economic and environmental sustainability [21]. Hakam and Solvang also conducted a literature review study about container ports sustainability in 2013. They examined 334 papers to look at port-related initiatives and gain a better insight on market trends about container ports sustainability. Their results demonstrated that sustainability performance and port sustainability has domination on literature between 1985 and 2012. With this study they revealed that the environmental dimension of sustainability seen as an opportunity to improve performance not as an expense anymore [22]. In 2013 Lirn et al. used multi-criteria decisions making methods to measure a port's green performance and choose three major container ports from China as sample. As a finding of study they determined three most critical sustainability indicators chosen by respondents in two of three ports investigated, chosen indicators are respectively; avoiding pollutants during cargo handling and port maintenance, noise control, and sewage treatment [23]. Elzarka and Elgazzar (2014) used three methods which are multi-criteria decisions making, survey and literature review with the purpose to propose a performance measurement system to evaluate and improve green performance of ports. As a result of the study they designed a green port performance index using a weighted averaging aggregation method. With this single index a port can compare its overall performance with other competitive ports [24]. In 2015 Hakam published a study about Nordic container ports sustainability performance. The aim of that study is to propose a conceptual sustainability performance management framework for container ports. Hence as a result of the study Hakam developed a prototype based on hypothetical data and suggested a list of key port sustainability performance indicators list [6]. Asgari et al. (2015) also used multi-criteria decisions making methods to measure sustainability performance of UK major ports. The main aim of that study is to investigate 5 largest ports in UK sustainability performance according to environmental and economic aspects of sustainability [25]. In 2016 Lu et al. aimed to examine the effects of sustainable supply chain management on sustainability performance in the port context. To achieve this purpose, they conducted survey with 132 respondents who is supervisor or above in three major international port authorities in Taiwan. Studies results demonstrated that external sustainable collaboration is positively associated with internal sustainable management, and internal sustainable management positively influences sustainability performance [26]. Çağlar (2016) used three methods which are interview, survey and literature review in Turkey and USA, aiming to demonstrate sustainable advancements and challenges in efficient and sustainable container terminal operations. As a result, Çağlar emphasized the importance of connectivity of seaports from railroads and gate technologies to approach channel depth and even customs regulations. Study also mentioned advance technology in container terminal equipments as a crucial point of value creation in terms of sustainability of ports [27].

2.3 Sustainable management

Port sustainability issues are also studied in management subjects. In this study eight study found written in sustainable management subject. First, Denktaş Şakar and Karataş Çetin integrated the organizational relationships between supply chain and port stakeholders with the aim of providing a conceptual framework based on the socially responsible marketing in 2012. Managerial implications and port authorities' roles in different dimensions of sustainability and framework that demonstrate supply chain stakeholder's influences on port sustainability are obtained as two important outcomes of study [28]. Broesterhuisen et al. (2012) aimed to make a procurement process at ports as sustainable as possible. To reach that aim and to develop a tool, method or model that used by all stakeholder easily interview method was conducted. As a result of study, even they found consensus about some topic, they also found disagreement so many topics. For instance, while air quality is the number one topic in the several sustainable initiatives of international ports, most respondents did not mention air quality as an important criterion for sustainability. Hence they suggested that more research needs to be done about this subject to obtain some reliable and valid results [8]. In 2014 Kim and Chang conducted interviews in Busan Port to guide ports to implant sustainability practices into operations of ports. As a result of study, they identified four important work processes that would improve sustainability performance of ports if conducted better. These processes are listed as; container traffic growth, low environmental impacts and corporate responsible image making, operational efficiency, efficiency use of the port area. They also proposed the simultaneous implementation of those business processes that improved sustainability [29]. Fedai and Madran conducted both case study and interview methods to two different marinas in Turkey in Antalya in 2015. Sustainability issues of ports are examined and sustainable management strategies are mentioned in this comprehensive study. At the end of the study Fedai and Madran found lack of sustainable management perspective in ports and marinas as a mainstream management approach in studied marinas in Turkey, Antalya [30]. Dynamics in sustainable port and hinterland operations was examined by Hou and Geerlings in 2016 with the combination of literature survey, interview and case study methods. They aimed to present a conceptual framework and simulation of sustainability measures and its effectiveness. In that study they considered sustainable port related hinterland transport as an organizational issue and they conclude that the greatest challenge for a port to reach sustainable management standards is not to develop new technology but to develop new management regulations [31]. In 2016 Sislian et al. aimed to review the maritime literature of two interrelated concepts which are Port Sustainability and Ocean's Carrier Network Problem (OCNP), they first reviewed the port sustainability concept and then relate it to the OCNP, to integrate port sustainability indicators in the OCNP as a conceptual framework. Findings of this study revealed that the most common sustainable development measures focus on environmental measures hence they suggested that the future researches should focus triple bottom line of sustainability also [5]. In 2017 Wang et al. aimed to develop a conceptual model for the Chinese oil ports in the sustainability policy framework. At the end of the study they contribute current literature with two outcomes. First they developed specific measures to assess oil port sustainability and a holistic oil port sustainability framework [32].

2.4 Port construction

Sustainable port construction is also an important and mostly studied subject in port sustainability literature. Between the years of 2003 and 2016 several papers published about sustainable port construction. First, Dooms and Macharis conducted in-depth interviews and multi criteria decision making methods in 2003 to present a conceptual framework for inland port planning which takes into account stakeholder preferences and objectives, in order to realize sustainable port development. As a result, they proposed a framework to develop a strategic Master plan with the port authority of the Port of Brussels [33]. Then, in 2010 Girard aimed to discuss under what conditions port areas can become an opportunity for the whole city development and how could it be possible. Girard used and suggested strategic implementation and management planning in order to combine tangible (hard) and intangible (soft) values in this study [34]. Wahab et al. (2011) used multi criteria decision making, survey and case study methods to ensure a set of standards and technical criteria during construction, design and operation processes of maritime and harbor structures. With developing an integrated sustainable port construction model Wahab introduced a new methodology to assess constructability and operability of port building designs [35]. In (2014) Megahead aimed to outline an approach for exploring the architectural styles of Port Said and proposed the challenges and opportunities for sustainable development in that study. Interviews with 63 residents, 18 planners and architects were conducted in the scope of the study. As a conclusion of the study the importance of building heritage together with environmental, economic and social aspects of sustainability as a main concern on any sustainable development emphasized [36]. A case study conducted on Port of Koper in (2014) by Pavlic et al. and they aimed to provide a methodological approach to solve the problems of future development challenges of seaports like implementing the green port concept in current infrastructures of ports. As a finding of the study they demonstrated the importance of state-of-the-art technologies, developing pilot initiatives based on modern energy solutions, improve efficiency in fuel consumption and emissions reduction in rubber tired gantry cranes and improve the overall energy efficiency improvement based on testing based on the case studied port [37]. Awad-Nunez et al. published two interrelated study in 2016. Multi criteria decision analysis (MCDA) and case study methods conducted both studies samples that consist of 10 different dry ports. They aimed to research different variables that influence sustainability of dry port location. Consequently, they found the most significant variables effect sustainability which is related to environmental protection [38].

2.5 Environmental sustainability

Even though environmental sustainability is just one feature of triple bottom line concept of sustainability, it is the most studied one in the current literature. In this study 5 publishing in the concept of environmental sustainability, are examined according to their aims, methods and findings chronologically. In (1999) Hoyle has examined some projects designed to measure urban coastal change processes and models in Canadian port cities, the influence of process and modals on the community groups. Hoyle conducted interviews with some community groups in selected Canadian port cities in the scope of this study. Study revealed that there is a clear distinction between geographical, area-based or neighborhood groups [39]. Vujicic et al. worked on zero emission cargo handling equipments in 2012. Presenting and evaluating the up-to-date trends in the cargo handling equipment (CHE) industry was the main aim of their study. Reducing the environmental impact of container terminal operations and contributing to the sustainability of ports was another aim of this study. As a result of the study they concluded that electrification system of cargo handling equipments is a step forward to reduce the environmental impacts of ports and it is found as feasible and sustainable solution by the authors of the study. In addition, they also found that zero emission objective in cargo handling equipments are impossible to achieve yet [40]. In (2013) interviews with 5 ports and 2 local government-based harbor authorities are conducted by Dinwoodie et al. to analyze the smaller port authorities' ways to manage environmental sustainability in Southwest England. Dealing with smaller ports became their motivation unlike the current literature and as an outcome they provided a different perspective on environmental management in ports. They revealed that larger ports offer much more revenues compared to smaller ports as a matter of course for that reason it is impossible to transfer application of current environmental management systems that prepared to large ports to smaller ports. At the end of the study necessity of preparing new environmental management systems for smaller ports are emphasized [41]. Borriello et al. (2013). conducted a case study on Valencia port as a successful sample on environmental port management issues and aimed to adapt its management strategies in other port cities and port areas especially in the Southern ports of the Mediterranean. As a result of the study necessity of new integrated evaluation tools for managing the conflict between the port areas and the port cities emphasized [42]. Lastly, a case study conducted on Port of Koper in (2015) by Beskovnik and Bajec with the aimed to identify environmental initiatives and measures that affect sustainability of ports. Study introduced a new six pillar environmental sustainability model that could be used environmental measuring in the region. The outcomes of the study intend to help port management for faster implementation of environmental and social objectives and measures in order to follow sustainable development [43].

2.6 Sustainability indicators

Port sustainability indicators used for measuring port sustainability and sustainable development of an industry, in addition it is important to mention that "Indicators" and "metrics" are interchangeably words used for referring to measurement of sustainability, yet somehow term of "indicators" used more broadly in current literature (44). Hence term of "indicators" used in this study instead of "metrics" and 4 studies in the concept of sustainability indicators, are examined according to their aims, methods and findings chronologically. Peris-Mora et al. (2005) published first article in this subject and they developed system of indicators for sustainable port management with the combined use of multi criteria decision making and case study methods. Port of Valencia was chosen for case study, other Spanish and European ports are chosen for Multi Criteria Analysis (MCA) in the scope of the study. Aim of the study was to propose a system of sustainable environmental management indicators that can be used by any port authorities finally they proposed 17 system indicators as an outcome of the study [45]. In 2014 Puig et al. aimed to develop a new methodology which is user-friendly to encourage port authorities to adopt the selected indicators within their environmental program. To reach this aim they conducted focus groups, survey, literature review methods, conferences and workshops. As an outcome to assist port authorities in reporting and calculating the proposed indicators, a user friendly tool has been developed. List of indicators grouped in two categories. First qualitative indicators category consists of nine components of environmental management that measure the Port Authority's ability to deliver compliance and effective environmental protection, second quantitative indicators category designed to measure of the actual impact related to specific aspects, namely Carbon Footprint, Waste Management and Water Consumption [46]. Bergmans et al. (2014) conducted a case study method to Port of Antwerp and aimed to examine the port of Antwerp's sustainability reporting initiatives, sustainability reporting performance of the harbour community and process. According to the study that kind of initiatives helps to create new discursive spaces, new practices and new reporting standards but also they contribute to the creation of community identity. Hence it is found that discussing and negotiating sustainability indicators is the way of reaching desired futures of an ideal harbour community [47]. The last study of sustainability indicators subject is published by Shiau and Chuang in 2017. The main aim of that study was to develop an indicator generator to generate initial port sustainability indicators (PSIs). After conducting case study and interview methods in Keelung Port it is suggested that 34 expert based PSIs for the scientific, operational, public relations and non-professional frame in the scope of this study [48].

2.7 Sustainability policy and coastal development

Sustainable policy and coastal development subjects are also studied in port sustainability concept. First Gilman studied in port sustainable policy concept in 2003. The main aim of the study was to decide in which Government's strategic objectives could be achieved in the scope of a private port system for the infrastructure, economy and environment. At the end of the study as the workability of the development approval process, the impact of the new approaches, the principle of clear need and particular problems of integration with in inland modes four main issues have emerged [49]. The other study in the concept of port sustainable policy conducted as multi case study with European and American port authorities by Kotowska in 2016. The article aimed to demonstrate port authorities' roles in creating sustainable growth. As an outcome of the study the actions taken by ports to improve the port environment and promote the modal shift have been grouped by author [50]. In 1998 Turner et al. studied the concept of sustainable coastal development for the first time, to decide main forces that influence the development of coastal areas. Consequently, they revealed the need for more open, flexible, integrated and participatory approach to coastal science and management to reduce the inevitable uncertainties [51]. The other study in the concept of sustainable coastal development was conducted by Yap and Liam in 2013. Research aimed to examine ports growth paths and sustainability issues in coastal development. Hence the study contributed to current literature by addressing the need for a balanced approach in sustainability for port and coastal development both political and research subjects [52].

2.8 Economic and environmental sustainability, ecological footprint, social sustainability, sustainability reporting and measurement

Economic and environmental sustainability, ecological footprint, social sustainability, sustainability reporting and sustainability measurement subjects also studied in the current literature, there has been found just one study for each. In 1999 Tindall published an article in social sustainability subject, using the the survey method aimed to examine the attitudes, opinions, and perceptions of Port Alberni residents and Anti-Environmentalism Movement organization members. After several analyses it is found that the general public is more likely to acknowledge ecological limits, and to be more open to change than members of Anti-Environmentalism Movement organization [53]. Soler et al. studied about port sustainability reporting subject in 2009. Study aimed to develop sustainability reporting guide standards for the Spanish port sector to satisfy the requirements of stakeholder groups. Study also considered indicators that defined in the Spanish Port System Balanced Scorecard and defined a set of information including ecological, financial and social variables as non-financial indicators and financial indicators [54]. A case study conducted on a port in Northern China (2015) by Guo et al. in ecological footprint subject with the aimed to build a port ecological footprint model, considering biological resource, energy and construction land ecological footprint while transferring resources and energy consumed in port production and operation to four ecological productive lands. Consequently, the model could demonstrate port ecological carrying capacity and provide a basis for evaluating the ecologically sustainable development of port but also because of nature of ports waters are accepted as the main sources of port ecological carrying capacity, hence study suggested strengthen the protection of waters to ensure sustainable development of ports [55]. Laxe et al. studied on economic and environmental dimensions of sustainability in 2015. Statistical analyses conducted on 16 Spanish ports sustainability reports to enable the existence of links between ports economic and environmental indicators. After analyzing data, it is found that not all Port Authorities fulfill the requirement specified in the Port Law, just four ports sustainability reports having a series of indicators of sustainability. According to study results to carry out the establishment of penalties to those entities not fulfilling the requirement or showing a negative evolution in these indicators have suggested [56]. Finally, a case study conducted on Singapore Port in (2017) by Xiao and Lam in sustainability measurement subject aiming to narrow the gap between adopting the systems theory and to develop a systems framework for sustainable development using three aspects of sustainability. According to analysis results, study demonstrates that there is a positive relationship between the port and the city, hence study arrived the result of Singapore port could keep on developing by reducing the negative relationships environmentally while increasing the positive relationships economically and socially [57].

3 Content analysis

Content analysis is a widely used qualitative research method that uses a set of procedures to make valid inferences about the senders or audience of the message or message itself (58, 59). It is used for, to analyze written, verbal or visual communication messages (60). According to Mayring (2000), qualitative content analysis is "an approach of empirical, methodological controlled analysis of texts within their context of communication, following content analytic rules and step-by-step models, without rash quantification" [61]. Content analysis was first used in the 19th century as a method for analyzing newspapers, magazine articles, advertisements and political speeches [62]. According to Hseih and Shannon (2005), there are three different approaches of content analysis; conventional, directed, or summative. They differed from each other features like; timing of defining codes or keywords, source of codes or keywords and the main motivation to start study. Summative content analysis will be conducted in this study. In summative content analysis, study starts with keywords, these keywords are identified before or during data analysis and they derived from interest of researchers or review of literature [63].

In this study, content analysis will be carried out by examining the studies on sustainability in the ports. Hakam and Solvang (2013) define the aim of content analysis as to summarize existing researches by identifying patterns and issues [22]. To reach that specific aim, first studies are collected regarding their main subjects. To obtain an overview of the increase of writings on the topic, an online search was made in the following electronic databases: Google Scholar, Ebscohost, Proquest, Science Direct and Scopus. The search terms "port sustainability", "sustainable port", "environmental management in ports", "green port", "sustainable supply chain management", "corporate social responsibility in ports" and "port sustainable development" are used on all databases in English and Turkish from the period of 1987 to 2017 April. These seven keywords are shown in below Table 1.

Table 1: The Search Keywords

Number	Corresponding Keyword(s)
1	Sustainable Port (*Sürdürülebilir Liman*)
2	Port Sustainability (*Limanlarda Sürdürülebilirlik*)
3	Environmental Management in Ports (*Limanlarda Çevresel Yönetim Uygulamaları*)
4	Green Port (*Yeşil Liman*)
5	Sustainable Supply Chain Management (*Sürdürülebilir Tedarik Zinciri Yönetimi*)
6	Corporate Social Responsibility in Ports (*Limanlarda Kurumsal Sosyal Sorumluluk*)
7	Port Sustainable Development (*Limanlarda Sürdürülebilir Gelişme*)

While conducting the scan, search was limited with some criteria. As a source, articles and conference proceedings are included. There also has a time limit in the search criteria. As mentioned before, sustainability concept was born in 1987 Brundtland Report [64], so the literature between 1987 and 2017 was screened. 5 databases were used that free Access provided by Dokuz Eylül University. All scans were carried out in the title of paper, with "containing all of the words" condition. Each keywords and their relevant paper count are given in Table 2 for each database.

Table 2: Number of Studies Available in Port Sustainability Related Publications from Different Electronic Databases

Keyword(s)/Databases	Google Scholar	Ebscohost	Proquest	Science Direct	Scopus
Sustainable Port	58	8	7	21	15
Port Sustainability	74	11	5	28	44
Environmental Management in Ports	17	6	3	11	0
Green Port	125	5	6	20	20
Sustainable Supply Chain Management	1.245	362	167	216	321
Corporate Social Responsibility in Ports	4	0	0	1	0
Port Sustainable Development	77	5	7	7	1

As shown in Table 2, the number of studies on port sustainability issues is quite limited compared to the number of studies on issues like "sustainable supply chain management" or "green port". Considering this gap in the literature, in current study a detailed analysis was carried out with the keywords "sustainable port" and "port sustainability" in databases shown in Table 2. Each study group under the "port sustainability" and "sustainable port" headings in Table 2 has been examined separately and the studies that scanned in more than one database have been separated. In this instance studies which are not completely suitable for the scope of this study according to their subjects have been excluded to ensure the reliability and internal validity of the research. 53 papers selected at the end of the scanning period, which are appropriate for the criteria determined, and they were examined in terms of "research subject, type of research and data analysis methods. Detailed information about these 53 studies are given in Appendix 1.

The detailed scanning of the databases includes articles and conference proceedings which include the whole words of "sustainable port" and "port sustainability" in the title of the study and published between 1987 and 2017, the year in which sustainability came into being.

3.1 Distribution of publications to subjects and years

The distribution of studies that included in the content analysis is shown in Table 3, in terms of their subjects and years. According to Table 4, although the concept of sustainability emerged in 1987, no studies were conducted in the scope of sustainability in the ports between 1987 and 1997. The first study on port sustainability issues was published in 1998 and after that eight more studies published between the years of 1998 and 2007 which were examined in the scope of this study. After 2008, the importance given to the sustainability issue has increased, and a number of studies have been conducted on a variety of topics, primarily sustainable development, sustainability performance, and sustainable management concepts. 45 of these studies have been found directly related to the subject and examined within the scope of the study.

Table 3: Distribution of Research Subjects by Years

Subject	**1987 to 1997**	**1998 to 2007**	**2008 to 2017**	**Frequency**	**Percent**
Sustainable Development	-	2	9	11	20,75
Sustainability Performance	-	-	9	9	16,98
Sustainable Management	-	-	8	8	15,09
Port Construction	-	1	6	7	13,21
Environmental Sustainability	-	1	4	5	9,43
Sustainability Indicators	-	1	3	4	7,55
Sustainability Policy	-	1	1	2	3,77
Sustainable Coastal Development	-	1	1	2	3,77
Economic and Environmental Sustainability	-	-	1	1	1,89
Ecological Footprint	-	-	1	1	1,89
Social Sustainability	-	1	-	1	1,89
Sustainability Reporting	-	-	1	1	1,89
Sustainability Measurement	-	-	1	1	1,89
Total	-	**8**	**45**	**53**	**100**

According to 53 studies' content analysis regarding to port sustainability subjects, the most common five subjects are sustainable development, sustainability performance, sustainable management, sustainable port construction and environmental sustainability respectively. The most common factor which is related to the sustainable development is stated in 11 studies. Details of other studies and subjects are given above Table 3. The distributions of the examined papers in terms of publishing years are shown in below Figure 1.

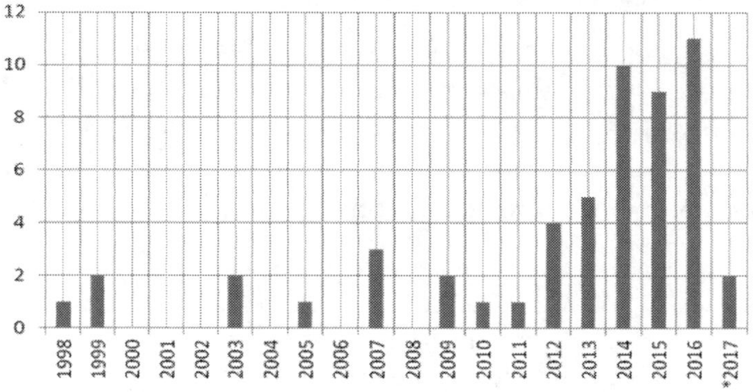

Figure 1: Number of Articles and Conference Proceedings Published per year between 1998 and 2017
* Values for 2017 include publications issued until April.

It should also be noted that, in the situation of that the examined study involves more than one subject listed in Table 3 the keyword rankings specified by the authors of the studies were taken into account. Hence studies included the subject with respect to their keywords.

3.2 Distributions of publications to journals

Table 4 demonstrates the lists and the distributions of the journals in which the studies published in the scope of content analysis. According to Table 4, 22,64% of the studies consist of various conference proceedings. In accordance 77,36% of total studies consist of articles that published various journals. 4 articles that have 7,55% of total studies published in Marine Pollution Bulletin. Its followed by Transportation Research Procedia, Maritime Policy and Management, Research in Transportation Business & Management, The Asian Journal of Shipping and Logistics, Marine Policy and Sustainability journals respectively. 20 articles' diverse journal names also listed in Table 4.

Table 4: Distribution of Studies to Journals

Name of Journal	Frequency	Percent
Conference Proceedings	12	22,64%
Marine Pollution Bulletin	4	7,55%
Transportation Research Procedia	3	5,66%
Maritime Policy & Management	3	5,66%
Research in Transportation Business & Management	2	3,77%
The Asian Journal of Shipping and Logistics	2	3,77%
Marine Policy	2	3,77%
Sustainability	2	3,77%
Thermal Science	1	1,89%
Open Journal of Social Sciences	1	1,89%
Problems of Sustainable Development	1	1,89%
Advanced Engineering Forum	1	1,89%

European Online Journal of Natural and Social Sciences	1	1,89%
Procedia Social and Behavioral Sciences	1	1,89%
Ocean & Coastal Management	1	1,89%
The Geographical Journal	1	1,89%
Journal of Korea Port Economic Association	1	1,89%
Journal of Transport Geography	1	1,89%
Journal of Service Science and Management	1	1,89%
Journal of Cleaner Production	1	1,89%
International Journal of Architectural Research	1	1,89%
Journal for Communication Studies	1	1,89%
Journal of Mechanical Engineering	1	1,89%
Journal of Black Sea Studies	1	1,89%
International Journal of Sustainable Development	1	1,89%
Transportation Research Part E	1	1,89%
International Journal of Physical Distribution & Logistics Management	1	1,89%
Scientific Papers of the University of Pardubice Series D	1	1,89%
Others (No info)	3	5,66%
Total	53	100,00%

3.3 Distributions of publications to research methods

Studies analyzed according to their research methods in the scope of the study. According to Table 5, 50,94% of studies used quantitative research methods, 37,74% of them used qualitative research methods as a research method. In addition, 11,32% of total paper used both qualitative and quantitative research methods to reach their aims.

Table 5: Distribution of Studies to Research Methods

Research Methods of Study	Frequency	Percent
Quantitative Study	27	50,94%
Qualitative Study	20	37,74%
Qualitative and Quantitative	6	11,32%
Total	53	100,00%

3.4 Distributions of publications according to methods used for analysis

Data analysis methods of the papers were also analyzed in the scope of the study. Results are given in below Table 6. Case Study is the most frequent data analysis method in examined port sustainability studies with the ratio of %29,85. Literature review method was also used quite a lot, generally with the help of sustainability reports and annual reports of ports and current literature. Interview, MCDM (Multi Criteria Decision Making), Survey and Statistical Methods like SPSS, DEA, PDE and PSIR followed them respectively. In addition, just one paper found in literature that used Focus Group and Workshop methods.

Table 6: Distribution of Studies According to Methods Used for Analysis

Methods Used for Analysis	Frequency	Percent
Case Study	20	29,85%
Literature Review	14	20,90%
Interview	12	17,91%
MCDM (Multi Criteria Decision Making)	9	13,43%
Survey	7	10,45%

Statistical Methods (SPSS, DEA, PDE, PSIR)	4	5,97%
Focus Group and Workshop	1	1,49%
Total	67	100,00%

In addition, it is also identified that, 41 papers used just one method listed in Table 6, 9 studies used two different methods and 3 studies used three methods to reach their findings.

4 Conclusion

The importance given in sustainability issues increased significantly in recent decades. While the concept of sustainability emerged in 1987, no study was conducted in the scope of sustainability in the ports between 1987 and 1997. After 2008, the importance given to the sustainability issue has increased, and a number of studies have been conducted on a variety of topics. Although port sustainability issues come on the agenda in the last two decades, it is seen as one of the most important subjects in the sustainability concept due to the nature of the industry. In this direction, this study aimed to investigate current port sustainability literature in a systematic way.

First; the number of studies on port sustainability issues is quite limited compared to the number of studies on issues like "sustainable supply chain management" or "green port". In this direction, the results of the study demonstrated that there is still a significant gap in port sustainability issues and the emphasis should be given on port sustainability studies. When examined the 53 studies about ports sustainability subject in the current literature, it is determined that there exist very limited number of studies written about quite various fields. The most common five subjects are sustainable development, sustainability performance, sustainable management, sustainable port construction and environmental sustainability respectively. Addition to these subjects it is seen that there are neglected subjects like; sustainability policy, sustainable coastal development, economic and environmental sustainability, ecological footprint, social sustainability, sustainability reporting and sustainability measurement. It is thought that these neglected topics are quite important to the future studies. Especially sustainability reporting and sustainability measurement subjects should be studied in-depth and should also be developed to get more clear scores on sustainable performance indexes.

When the studies examined in terms of the methods they used, it was observed that they mainly focused on 6 methods, mainly case study, literature review and interview methods. The intense use of methods such as case study, literature review and interview, which are generally used to identify the current situation, is an important indicator of the fact that the concept is still in the process of development and it should continue to be studied. It is also seen that, MCDM methods used intensively in studies that examined sustainability measurement and sustainability indicators.

The study contains a list of major journals that publish ports sustainability studies. This list was created to provide a roadmap not only to authors who wish to obtain these studies but also to publish their port sustainability related studies to contribute to current literature.

The most important limitation of this study is the very limited number of publishing on port sustainability literature. Hence, it is suggested to the authors to carry out studies on issues that are found to be neglected like sustainability reporting and sustainability measurement in future studies.

References

[1] R. Bozlağan. Sosyal Siyaset Konferansları Dergisi, **50**, 1011-1028 (2005).
[2] Ç. Aksoy. Marmara University Institute of Social Sciences (2013).

[3] M. Taşçıoğlu. Georgia Southern University (2014).
[4] G. H. Brundtland. United Nations (1987).
[5] L. Sislian, A. Jaegler, and P. Cariou. Research in Transportation Business & Management, **19**, 19-26. (2016).
[6] M. H. Hakam. Journal of Service science and Management, **8**(01), 14 (2015). https://file.scirp.org/pdf/JSSM_2015012211245289.pdf
[7] R. M. DARBRA. Marine pollution bulletin, **50.8:** 866-874 (2005). https://www.sciencedirect.com/science/article/pii/S0025326X05001918
[8] E. F. M. Broesterhuizen, T. Vellinga, L. Docters van Leeuwen, J. W. Zwakhals, P. Taneja and M. Nijdam. CESUN2012, (2012). https://www.researchgate.net/profile/Tiedo_Vellinga/publication/266051258
[9] M. Acciaro, H. Ghiara, and M. I. Cusano. Energy Policy, **71**, 4-12 (2014). https://www.sciencedirect.com/science/article/pii/S0301421514002262
[10] T. Daamen. In ENHR-conference, 25-28 (2007).
[11] K. A. Abood. In Ports 2007: 30 Years of Sharing Ideas: 1977-2007 (2007).
[12] J. S. L. Lam and E. Van de Voorde. International Forum on Shipping, Ports and Airports (IFSPA), 27-30 (2012).
[13] Y. C. E. Jonathan, Kader, A. S. A., bin Ahmad, Z., Mokhtar, I. https://www.researchgate.net/profile/Jonathan_Yong_Chung_Ee/publication/269221509
[14] S. A. Hosseini, H. Amozad and M.K. Mafi. European Online Journal of Natural and Social Sciences, **3(3),** 810 (2014).
[15] A. Kutkaitis, L. Šimanskienė and D. Burgis. Scientific Papers of the University of Pardubice. Series D, Faculty of Economics & Administration, **21**(30) (2014). http://citeseerx.ist.psu.edu/viewdoc/download?doi=10.1.1.463.3768&rep=rep1&type=pdf#page=66
[16] M. De Martino. In Advanced Engineering Forum Trans Tech Publications. **11**, 87-95 (2014). https://www.scientific.net/AEF.11.87
[17] H. Wang, D. Huo and J. Ortiz. Open Journal of Social Sciences, **3**(05), 28 (2015). https://file.scirp.org/pdf/JSS_2015052016462023.pdf
[18] J. Boulos. Procedia-Social and Behavioral Sciences, **216**, 974-985 (2016). https://www.sciencedirect.com/science/article/pii/S1877042815062746
[19] F.G. Laxe, F.M. Bermúdez, F.M. Palmero and I. Novo-Corti. Marine pollution bulletin, **113**(1-2), 232-239 (2016). https://www.sciencedirect.com/science/article/pii/S0025326X16307615
[20] S. Roh, V.V. Thai and Y.D. Wong. The Asian Journal of Shipping and Logistics, **32**(2), 107-118 (2016). https://www.sciencedirect.com/science/article/pii/S2092521216300281
[21] M.H. Hakam and W.D. Solvang. SOLI'09. IEEE/INFORMS International Conference on IEEE 417-422 (2009). http://ieeexplore.ieee.org/abstract/document/5203969/
[22] M.H. Hakam and W.D. Solvang. Container ports sustainability-a literature review. In Cognitive Infocommunications (CogInfoCom), 2013 IEEE 4th International Conference on IEEE 803-810 (2013). http://ieeexplore.ieee.org/abstract/document/6719209/
[23] T.C. Lirn, Y.C. Jim Wu and Y.J. Chen. International Journal of Physical Distribution & Logistics Management, 43(5/6), 427-451 (2013). https://www.emeraldinsight.com/doi/full/10.1108/IJPDLM-04-2012-0134
[24] S. Elzarka and S. Elgazzar. In International Forum on Shipping, Ports and Airports (IFSPA) (2014). https://www.researchgate.net/profile/Sara_Elzarka2/publication/280385412
[25] N. Asgari, A. Hassani, D. Jones and H.H. Nguye. Transportation research part E: logistics and transportation review, **78**, 19-39 (2015). https://www.sciencedirect.com/science/article/pii/S1366554515000150

[26] C.S. Lu, K.C. Shang and C.C. Lin. Maritime Policy & Management, **43**(8), 909-927 (2016). https://www.tandfonline.com/doi/abs/10.1080/03088839.2016.1199918
[27] V. Çaglar. Karadeniz Arastirmalari, (**49**), 141 (2016). https://search.proquest.com/docview/1788513671?pq-origsite=gscholar
[28] G. Denktas-Sakar and Ç. Karatas-Cetin. The Asian Journal of Shipping and Logistics, 28(3), 301-319 (2012). https://www.sciencedirect.com/science/article/pii/S2092521213000035
[29] S. Kim and B. Chiang. Sustainability Practices to achieve Sustainability in International Port Operations. **30**(3), 15-37 (2014).
[30] A. Fedai and C. Madran. II. Ulusal Liman Kongresi, (2015).
[31] L. Hou and H. Geerlings. Journal of Cleaner Production, 135, 449-456 (2016). https://www.sciencedirect.com/science/article/pii/S0959652616308113
[32] X. Wang. A conceptual model for oil port sustainability policy research–A case of Chinese ports. (2017).
[33] M. Dooms and C. Macharis. 43rd Congress of the European Regional Science Association. (2003). https://www.econstor.eu/handle/10419/116041
[34] L.F. Girard. International Journal of Sustainable Development, 13(1-2), 161-184 (2010). https://www.inderscienceonline.com/doi/abs/10.1504/IJSD.2010.035106
[35] R. Wahab, T. Zayed, T. Goh and C.M. Wang. Proceedings of MTEC2011, 13, 15 (2011). https://www.researchgate.net/profile/Roshan_Wahab/publication/257418895_
[36] N. Megahed. International Journal of Architectural Research: ArchNet-IJAR, 8(1), 94-107 (2014). http://198.58.80.116/index.php/IJAR/article/view/299
[37] B. Pavlic, F. Cepak, B. Sucic, M. Peckaj and B. Kandus. Thermal Science, 18(3), 935-948 (2014).
[38] S. Awad-Núñez, N. González-Cancelas, F. Soler-Flores and A. Camarero-Orive. Transportation Research Procedia, 13, 124-133 (2016).
[39] B. Hoyle. Journal of Transport Geography, 7(1), 65-78 (1999). https://www.sciencedirect.com/science/article/pii/S0966692398000301
[40] A.M. Vujicic, N. Zrnić and B. Jerman. Strojniški vestnik-Journal of Mechanical Engineering, 59(9), 547-555 (2013). http://ojs.sv-jme.eu/index.php/sv-jme/article/view/sv-jme.2012.933
[41] J. Dinwoodie, A. Kuznetsov, D. Gibbs, M. Sansom and H. Knowles. Marine Policy 54, 59-68 (2013).
[42] F. Borriello. Sustainability, 5(10), 4288-4311 (2013). http://www.mdpi.com/2071-1050/5/10/4288/htm
[43] B. Beškovnik and P. Bajec. Problems of Sustainable Development.10 (1), 99-106 (2015). https://papers.ssrn.com/sol3/papers.cfm?abstract_id=2556857
[44] D. Tanzil and B.R. Beloff. Environmental Quality Management, 15(4), 41-56 (2006).
[45] E. Peris-Mora, J.D. Orejas, A. Subirats, S. Ibáñez and P. Alvarez. Marine Pollution Bulletin, 50(12), 1649-1660 (2005). https://www.sciencedirect.com/science/article/pii/S0025326X05003048
[46] M. Puig, C. Wooldridge and R.M. Darbra. Marine pollution bulletin, 81(1), 124-130 (2014). https://www.sciencedirect.com/science/article/pii/S0025326X14000873
[47] A. Bergmans, F. Vandermoere and I. Loots.ESSACHESS-Journal for Communication Studies, 7(1), 107-123 (2014). https://www.ssoar.info/ssoar/handle/document/40725
[48] T.A. Shiau and C.C. Chuang. Maritime Policy & Management, 42(1), 26-42 (2015). https://www.tandfonline.com/doi/abs/10.1080/03088839.2013.863436
[49] S. Gilman. Maritime Policy & Management, 30(4), 275-291 (2003). https://www.tandfonline.com/doi/abs/10.1080/0308883032000145591
[50] I. Kotowska. Transportation Research Procedia, 16, 236-243 (2016). https://www.sciencedirect.com/science/article/pii/S2352146516306378

[51] R.K. Turner, I. Lorenzoni, N. Beaumont, I.J. Bateman, I.H. Langford and A.L. McDonald. Geographical Journal, 269-281 (1998). http://www.jstor.org/stable/3060616?seq=1#page_scan_tab_contents
[52] W.Y. Yap and J.S.L. Lam. Ocean & coastal management, 71, 13-25 (2013). https://www.sciencedirect.com/science/article/pii/S096456911200289X
[53] Tindall. Anti-Environmentalist Thinking About Change and Sustainability in Port Alberni. (1999).
[54] C.C. Soler, J.M.D. Orejas, A.G. Fillol and V.R. Feliu. Revue Sciences de Gestion, (70) (2009). http://eds.b.ebscohost.com/eds/detail/detail?vid=0&sid=2c6c6a90-10db-4ce5-920b-48f8b7e0af12%40sessionmgr102&bdata=Jmxhbmc9dHImc2l0ZT1lZHMtbGl2ZSZzY29wZT1zaXRl#AN=47027958&db=bth
[55] Z. J. Guo, Q. Zhang, W.Y. Wang and C. Wang. In Proceedings of the Eastern Asia Society for Transportation Studies (10) (2015). https://pdfs.semanticscholar.org/9b3e/b6bd66b96af0fec998a8781da2267b21575e.pdf
[56] F. G. Laxe, F.M. Bermúdez, F.M. Palmero and I. Novo-Corti. Marine pollution bulletin, 113(1-2), 232-239 (2016). https://www.sciencedirect.com/science/article/pii/S0025326X16307615
[57] Z. Xiao and J.S.L. Research in Transportation Business & Management, 22, 255-262 (2017). https://www.sciencedirect.com/science/article/pii/S2210539516300694
[58] K. E. Rosengren. SAGE Publications, Incorporated. 9 (1981).
[59] R.P. Weber. Sage. (1990).
[60] F. L. Cole. Clinical Nurse Specialist 2(1), 53–57 (1988).
[61] Mayring, P. VS Verlag für Sozialwissenschaften. 601-603 (2010). https://link.springer.com/chapter/10.1007/978-3-531-92052-8_42
[62] T.G. Harwood and T. Garry. The Marketing Review 3, 479–498 (2003)
[63] H. F. Hsieh and S.E. Shannon. Qualitative health research, 15(9), 1277-1288 (2005). http://journals.sagepub.com/doi/abs/10.1177/1049732305276687
[64] T. Kuhlman and J. Farrington. Sustainability, 2(11), 3436-3448 (2010).

Appendix: Details of the Reviewed Studies

No	Name of Study	Author	Year	Research Method	Method	Subject
1	Coastal Management for Sustainable Development: Analyzing Environmental and Socio-Economic Changes on the UK Coast	Turner et al.	1998	Quantitative	Statistical Methods	Sustainable Coastal Development
2	Anti-Environmentalist Thinking About Change and Sustainability in Port Alberni	Tindall	1999	Quantitative	Survey	Social Sustainability
3	Scale And Sustainability: The Role of Community Groups in Canadian Port-City Waterfront Change	Hoyle	1999	Qualitative	Interview	Environmental Sustainability
4	Sustainability and National Policy in UK Port Development	Gilman	2003	Qualitative	Literature Review	Sustainability Policy
5	A Framework for Sustainable Port Planning in Inland Ports: A Multi Stakeholder Approach	Dooms and Macharis	2003	Quantitative and Qualitative	MCDM/ Interview	Port Construction
6	Development of a System of Indicators for Sustainable Port Management	Peris-Mora et al.	2005	Quantitative	MCDM/ Case Study	Sustainability Indicators
7	Sustainable and Green Ports: Application of Sustainability Principles to Port Development and Operation	Abood	2007	Quantitative	Case Study	Sustainable Development
8	Sustainable Development of The European Port-City Interface	Daamen	2007	Qualitative	Literature Review	Sustainable Development
9	Sustainability Reports as a Key Tool for Communication Improvement in the Spanish Port System	Soler et al.	2009	Qualitative	Literature Review	Sustainability Reporting
10	On Flexibility and Sustainability in Container Ports	Hakam and Solvang	2009	Quantitative	Case Study	Sustainability Performance
11	Sustainability, Creativity, Resilience: Toward New Development Strategies of Port	Girard	2010	Quantitative	MCDM	Port Construction

	Areas Through Evaluation Processes					
12	Sustainability and Maintainability of Port Infrastructure Using BIM/4D CAD Simulation Model	Wahab et al.	2011	Quantitative	MCDM/ Survey/ Case Study	Port Construction
13	Port Sustainability and Stakeholder Management in Supply Chains: A Framework on Resource Dependence Theory	Denktaş-Şakar and Karataş-Çetin	2012	Qualitative	Literature Review	Sustainable Management
14	Sustainability as a Procurement Criterion for Port Investments	Broesterhuisen et al.	2012	Qualitative	Interview	Sustainable Management
15	Ports Sustainability: A Life Cycle Assessment of Zero Emission Cargo Handling Equipment	Vujicic et al.	2012	Quantitative	Case Study	Environmental Sustainability
16	Green Port Strategy for Sustainable Growth and Development	Lam and Van de Voorde	2012	Quantitative	Case Study	Sustainable Development
17	Towards a Port Sustainability Management System for Smaller Ports	Dinwoodie et al.	2013	Qualitative	Interview	Environmental Sustainability
18	Green Performance Criteria for Sustainable Ports in Asia	Lirn et al.	2013	Quantitative	MCDM	Sustainability Performance
19	Container Ports Sustainability: A Literature Review	Hakam and Solvang	2013	Qualitative	Literature Review	Sustainability Performance
20	The Sustainability of Mediterranean Port Areas: Environmental Management for Local Regeneration in Valencia	Borriello et al.	2013	Quantitative	Case Study	Environmental Sustainability
21	80 Million-Twenty-Foot-Equivalent-Unit Container Port? Sustainability Issues in Port and Coastal Development	Yap and Lam	2013	Qualitative	Literature Review	Sustainable Coastal Development
22	Sustainability Practices to Achieve Sustainability in International Port Operations	Kim and Chiang	2014	Qualitative	Interview	Sustainable Management

23	Sustainable Port Infrastructure, Practical Implementation of the Green Port Concept	Pavlic et al.	2014	Quantitative	Case Study	Port Construction
24	Green Port Performance Index For Sustainable Ports in Egypt: A Fuzzy AHP Approach	Elzarka and Elgazzar	2014	Quantitative	MDCM/ Survey	Sustainability Performance
25	Identification and Selection of Environmental Performance Indicators for Sustainable Port Development	Puig et al.	2014	Quantitative and Qualitative	Survey/ Focus Group/ Workshop	Sustainability Indicators
26	Heritage-Based Sustainability in Port Said: Classification of Styles and Future Development	Megahead	2014	Qualitative	Interview	Port Construction
27	Co-Producing Sustainability Indicators for the Port of Antwerp: How Sustainability Reporting Creates New Discursive Spaces for Concern and Mobilization	Bergmans et al.	2014	Quantitative	Case Study	Sustainability Indicators
28	Assessment of Sustainability of Logistic Activity of Port of Klaipeda	Kutkaitis et al.	2014	Qualitative	Interview	Sustainable Development
29	Green and Sustainable Port Project Planning: Paradigm Shift for an Integrated Container Handling System Expansion	Jonathan et al.	2014	Qualitative	Literature Review	Sustainable Development
30	Planning of Sustainable Development Based on Civil Rights in Torkaman Port	Lam and Van de Voorde	2014	Quantitative	Survey	Sustainable Development
31	Sustainable Development Strategies of the Port Authority: The Network Approach	Martino	2014	Qualitative	Literature Review	Sustainable Development
32	Nordic Container Port Sustainability Performance: A Conceptual Intelligent Framework	Hakam	2015	Qualitative	Literature Review	Sustainability Performance
33	Sürdürülebilir Liman Yönetimi ve Antalya'da İki Yat Limaninda Vaka İncelemesi	Fedai and Madran	2015	Quantitative	Case Study	Sustainable Management

34	Sustainable Development of Coastal Cities-Proposal of a Modelling Framework to Achieve Sustainable City-Port Connectivity	Boulos	2015	Quantitative	Case Study	Sustainable Development
35	Assessing Energy Efficiency of Port Operations in China: A Case Study on Sustainable Development of Green Ports	Wang et al.	2015	Quantitative	Statistical Methods	Sustainable Development
36	Application of Environmental and Social Sustainable Measures by Port of Koper: The Basis for the Regional Approach	Beskovnik and Bajec	2015	Quantitative	Case Study	Environmental Sustainability
37	Using Ecological Footprint Model for Port Ecosystem Sustainability Assessment	Guo et al.	2015	Qualitative	Case Study	Ecological Footprint
38	Towards a Sustainability Management System for Smaller Ports	Kuznetsov et al.	2015	Qualitative and Quantitative	Interview/ Case Study	Sustainable Management
39	Sustainability Ranking of the UK Major Ports: Methodology and Case Study	Asgari et al.	2015	Quantitative	MCDM	Sustainability Performance
40	Social Construction of Port Sustainability Indicators: A Case Study of Keelung Port	Shiau and Chuang	2015	Quantitative and Qualitative	Interview/ Case Study	Sustainability Indicators
41	Sustainability and the Spanish Port System. Analysis of the Relationship Between Economic and Environmental Indicators	Laxe et al.	2016	Quantitative	Statistical Methods	Economic and Environmental Sustainability
42	Dynamics in Sustainable Port and Hinterland Operations: A Conceptual Framework and Simulation of Sustainability Measures and Their Effectiveness, Based on an Application to the Port of Shanghai	Hou and Geerlings	2016	Quantitative and Qualitative	Literature Review/ Interview/ Case Study	Sustainable Management
43	A Literature Review on Port Sustainability and Ocean's Carrier Network Problem	Sislian et al.	2016	Qualitative	Literature Review	Sustainable Management

44	Examining Sustainability Performance at Ports: Port Managers' Perspectives on Developing Sustainable Supply Chains	Lu et al.	2016	Quantitative	Survey	Sustainability Performance
45	Sustainable Container Terminal Operations: Challenges and Enhancements	Çağlar	2016	Quantitative and Qualitative	Interview/ Survey	Sustainability Performance
46	Assessment of Port Sustainability Through Synthetic Indexes: An Application to the Spanish Case	Laxe et al.	2016	Quantitative	Statistical Methods	Sustainable Development
47	A Methodology for Measuring Sustainability of Dry Ports Location Based on Bayesian Networks and Multi-Criteria Decision Analysis	Awad-Nunez et al.	2016	Quantitative	MCDM/ Case Study	Port Construction
48	How Should the Sustainability of the Location of Dry Ports Be Measured?	Awad-Nunez et al.	2016	Quantitative	MCDM/ Case Study	Port Construction
49	Functional Differentiation and Sustainability: A New Stage of Development in the Chinese Container Port System	Wang et al.	2016	Qualitative	Literature Review	Sustainability Performance
50	Towards Sustainable ASEAN Port Development: Challenges and Opportunities for Vietnamese Ports	Roh et al.	2016	Qualitative	Interview/ Literature Review	Sustainable Development
51	Policies Applied by Seaport Authorities to Create Sustainable Development in Port Cities	Kotowska	2016	Quantitative	Case Study	Sustainability Policy
52	A Conceptual Model for Oil Port Sustainability Policy Research: A Case Study of Chinese Ports	Wang et al.	2017	Qualitative	Literature Review	Sustainable Management
53	A Systems Framework for the Sustainable Development of a Port City: A Case Study of Singapore's Policies	Xiao and Lam	2017	Quantitative	Case Study	Sustainability Measurement

A new relationship between government, industry and knowledge institutes, the case of the maritime sector

Chris Peeters[1], and *Ruben* Pilon[2]

[1]Department of Transport and Regional Economics, University of Antwerp, Belgium
[2]Policy Research Corporation, The Netherlands

Abstract. The paper titled 'A new relationship between government, industry and knowledge institutes, the case of the maritime sector' intends to provide a set of principles for the interaction of governments with industrial players. In order to furnish industry (and knowledge institutes) with sufficient flexibility to meet the challenges of the maritime world – with the goal of enhancing its sustainability – unfettered interaction between government, industry and knowledge institutes is key. Each of these players can extend the scope of their activities and responsibilities to enhance cooperation, so that the conglomerate of these parties can achieve the goal of increased sustainability in an increasingly dynamic maritime world. The goal of this paper is to prove the necessity of such cooperation, hence the research question: "Is a closer cooperation between government, knowledge-institutes and industry necessary to cope with modern day challenges within the maritime sector?" The question shall be answered by constructing a theoretical, descriptive argument, expanding on the challenges faced in the maritime sector, showing that current decision making processes are unfit to deal with these challenges, introducing novel principles guiding decision making processes based on commonality between government, industry and knowledge institutes and finally expanding on the role these principles may play in realising the goal of sustainability in the maritime sector.

Keywords. Public policy, triple helix, maritime industry, ecological sustainability, cybersecurity, overcapacity, boardroom innovation

1 Introduction

In the 1990s, the concept of the triple helix experienced its first significant breakthrough because of papers by Etzkowitz [1] and Etzkowitz and Leydesdorff [2]. The concept of the triple helix concerns the joint relationships between three institutions: government, knowledge-institutes and industry. Note that the triple helix does not come in one form. Although the basis is cooperation between government, knowledge-institutes and industry, the triple helix model can take many configurations. In recent decades, a clear influence of the triple helix model can be identified on the structuring of institutional sources and the way of conceptualizing innovation [3].

This paper will use the concept of the triple helix in exploring the added value of closer cooperation between government, knowledge-institute and industry within the maritime sector. Although applicable to the entire maritime sector, the focus of this paper will mainly be on shipping and ports. The maritime sector has not been immune to a changing economic environment the last decades. Although the maritime sector has never been autarkic, due to the globalisation of the recent decades, the maritime sector is one of the most international sectors around the world. This international nature brings new opportunities, especially for maritime transport, but, as is often the case, new challenges the maritime sector never had to deal with as well. This paper aims at showing the necessity of closer cooperation in combating those (relatively) new structural challenges of the maritime sector, by pointing out the shortcomings of the current level of cooperation. The main difficulty in describing the 'current level of cooperation' between the three parties of the triple helix is that there is not *one* current level of cooperation. Logically, the level of cooperation between government, industry and knowledge-institutes differs between countries. Note that a *level* of cooperation is discussed, this way cooperation is not treated as an opposition (no cooperation/full cooperation), but as a degree (more cooperation/less cooperation). For instance, a closer cooperation between the three parties can be identified in The Netherlands [4][5] and Japan [6] compared to Germany [7]. In general, no maritime sector around the world has the level of cooperation between the three parties that this paper is exploring. Granted, certain maritime niches have the intended level of cooperation (f.e. naval construction in The Netherlands), but the intended level of cooperation within an entire maritime sector cannot be identified.

After the relation between a lack of cooperation and the current structural challenges of the maritime sector is discussed, this paper will propose a closer cooperation between government, knowledge-institutes and industry. Lastly, a conclusion and recommendations are drawn up.

2. Challenges

The maritime sectors around the world do not fully utilize the potential of triple helix cooperation. Closer cooperation between government, industry and knowledge-institutes will be decisive in combating four structural challenges facing the maritime sector: volatility of demand, safety & security, ecological sustainability and an outdated decision making process. Working towards solving these four challenges asks for closer cooperation. More specifically, these challenges ask for a triple helix cooperation. When looking at it the other way around, these structural challenges show the shortcomings of the current level of cooperation. Closer cooperation requires an incentive to cooperate, both for cooperation between industrial players as for cooperation between the three parties of the triple helix. The current lack of incentive leads to a lower level of knowledge and less financial power when compared to a triple helix model.

One of the most structural concerns within the maritime transport sector is the **divergence between supply and demand** within the shipping market. To illustrate what fluctuations can be identified, the 2000 - 2008 period was characterized by an undersupply of capacity, while shipping companies currently have to deal with oversupply of capacity. The last five years in a row, the growth rates of the size of the world fleet have been decreasing, however, supply still outgrew demand. BCG [8] investigated whether the decelerating demand is cyclical or the new standard. Their analysis concluded the latter. Due to the new standard of decelerated demand, the BCG report predicts that demand will trail supply in the coming years as well, thereby increasing the already existing overcapacity. This seems to be the general prediction from leading consultancies [9]. This oversupply is threatening the profitability of shipping which leads to decreasing freight rates in all segments [10] (liner, dry bulk, tanker). Vice versa, the period of undersupply of capacity was a period in which shipping companies made significant profits. This divergence of supply and demand within the maritime transport sector is the result of a strong derived demand that characterizes maritime transport. The demand for maritime transport is at the whim of macro-economic forces, it is induced by the labour, transport, capital and, foremost, commodity markets. This results in a rather fluctuating demand, which in turn leads to a rather fluctuating divergence of over- or undersupply of capacity. Even on a yearly basis, the differences in total tonnage over- or undersupply of capacity vary significantly [11]. Shipping companies struggle structurally with managing the divergence between supply and demand. The result of this volatility of demand is not to be taken lightly, it is the cause of many bankruptcies in the maritime transport sector during a time of oversupply of capacity.

The solution for this challenge is rather complicated. The root of the problem, fluctuating demand combined with rather inflexible supply, is a given fact due to the derived demand of maritime transport sector. If derived demand is a given fact of the maritime transport sector, what about combating the symptoms of derived demand? Combating the symptoms, over- and undersupply of capacity, might be as complex. Logically, combating undersupply of capacity is counterintuitive. Undersupply of capacity brings significant profits to the maritime transport sector and will enable the maritime transport sector to grow until an undersupply of capacity no longer exists. When looking at countering overcapacity of supply, the most obvious countermeasure is slowing down the procurement of new ships. Some argue that, with the current scrapping rate and without the placement of new orders, overcapacity will no longer be a problem in shipping in 2022[1]. However, it is questionable whether this is the case. Ships keep being ordered, even in a situation of overcapacity, due to a fourfold of reasons. Firstly, multiple countries are trying to revive or improve the market position of their shipping industry. Examples are South Korea [12] and Taiwan [13]. Secondly, due to fierce competition in the construction of ships, the price of building a ship is relatively low. Thirdly, in order to keep the cost per slot low. Recently built ships are bigger and more efficient, therefore result in lower cost per slot. Fourthly, and foremost, due to irrationality in shipping asset management [14]. This shows the complex nature of coping with the diversion between demand and supply, a structural challenge that is not simply solved. This does not mean that all investments in times of overcapacity are a form of irrationality. Skilled entrepreneurship leading to acyclical investments is vastly different than irrationality.

Further professionalization of the maritime transport sector, in particular the shipping industry, could prove helpful in combating the divergence between supply and demand. The term professionalization is, in this sense, meant to indicate the increase of the influence of external parties in the maritime transport sector, thereby raising the level of knowledge within the maritime transport sector. Further professionalization will rationalize the debate and increase the level of knowledge within a rather fragmented and conservative maritime transport industry[2]. In managing the volatility of demand, shipping companies have a tendency to base decision making chiefly on their entrepreneurship. Skilled entrepreneurship by shipping companies

[1] F.e. Maersk Group CEO Soren Skou in his presentation on Capital Markets Day 2016

[2] The view of the maritime industry having a rather conservative nature is a view shared by multiple scholars, amongst others: [24][25][26]

is not to be underestimated. Most shipping companies have a long history in the shipping industry, which results generally in a good perception of the shipping business. However, when aiming for a sustainable management of the divergence between supply and demand, allowing knowledge from outside the traditional shipping industry into decision making processes will prove beneficial. Examples of incorporating external knowledge are closer cooperation between knowledge institutes and shipping companies for the sake of better macro-economic forecasting, increased influence of independent board of director members in decision making or allowing sector analyses by consultancies to play a role in decision making. Without a party actively organizing cooperation, a fruitful cooperation will not likely start. The prime candidates for promoting and facilitating such closer cooperation are cluster organisations and shipowner associations.

To keep assuring **safety and security** in a digitalised world is another challenge the maritime sector faces around the world. With an ever increasing amount of maritime equipment and systems that are internet- and computer-based, for instance the Internet of Things (IoT) incorporated in ports and automation processes in shipping, the matter of safety and security has been changed irreversibly. Having security personnel on the ground and in the water is not enough, cybersecurity is needed to be able to guarantee safety and security. However, the lack of awareness of cybersecurity has been widely documented [17]. Even for the International Maritime Organization (IMO), cybersecurity was not an issue until recently [18].

The structural threats to safety and security are crime and terrorism. The maritime sector has been sensitive to crime, smuggling and piracy in particular, since the very first ports arose. Digitalisation brings opportunities to improve inspection techniques. However, that a centrally organized, hackable inspection brings potential hazards is evident. Looking at terrorism, attacks on ports or taking over ships could potentially have a disastrous outcome. Due to the crucial and strategic role ports have in the infrastructure of a country, security from malicious powers, both foreign and domestic, is critical for ports. Shutting down port(s) could paralyze an entire infrastructure system [19]. The same level of caution is necessary in countering automatic ships being taken over by malicious powers, especially when handling CBRN[3] material.

Although a full description of possible countermeasures to combat crime and terrorism in a digital age within the maritime sector is widespread enough to fill a book, three lines of countermeasures can be identified in previously written literature [17][19][20][21][22]. Firstly, increasing awareness for cyber-security. Combatting digital threats is still rather new or non-existent in most parties of the maritime sector. Secondly, assessment of where in the system the main vulnerabilities can be identified. Thirdly, an increase of funding in order to increase cybersecurity. Moreover, not only private companies are recommended to increase maritime cybersecurity funding, but governmental agencies just as much (f.e. IMO, Homeland Security).

Most papers call for an industry-wide approach in combatting threats in a digital age, or at least a more integrated approach than currently is being applied [17][18][21][22]. More specifically, integrated on both a national as well as an international level (f.e. through safety standards set up by the IMO). An integrated approach including the government makes sense since the government is the main institute responsible for public safety. Thus, the government is a stakeholder in combating crime and terrorism. An integrated approach is what is currently missing. Besides minor exceptions, cooperation within the maritime industry between either industry, government or knowledge-institutes (f.e. sharing best practices or collaborating projects) cannot be identified. Incentives to cooperate in cybersecurity or programmes aimed to increase maritime cybersecurity are not widespread. Through a cooperative approach between the parties of the triple helix, the level of knowledge can be increased (f.e. centre of excellence, sharing best practices, increased cybersecurity research). Plus, the financial power of the industry increases with a sector wide integrated approach (f.e. subsidies).

Ecological sustainability is perhaps the most obvious structural challenge for all industries of the maritime sector. When looking at one of those industries, ecological sustainability plays an important role in shipping. It is documented that ships, in terms of carbon-dioxide (CO_2) emissions, are most efficient at carrying large loads over long distances, but because of the sheer mass of the shipping industry, the total impact on the environment is still substantial. A significant decrease of CO_2 emissions in road transport can be identified over the last decades, at the same time, CO_2 emissions of naval transport have hardly changed over the last decades. It shows the complex and structural nature of the ecological sustainability problem. Furthermore, note that the ecological challenge for the maritime sector is twofold, the maritime sector affects the ecological sustainability of both air and water.

Potential consequences of not paying attention to the greening process are twofold. Firstly, efforts regarding the greening process will affect the position of the sector socially. When the maritime sector becomes one of the sectors which impacts the environment the most, the maritime sector will become the centre of social outcry. Secondly, a lack of effort to become more eco-friendly potentially has an economic dimension. Ecological sustainability arguments play a role in the decision which mode of transport to choose. The fact that shipping is more eco-friendly than road-, air- or rail transport is an important factor in choosing naval transport, especially short sea shipping. In the case the carbon footprint of road-, air- and/or rail transport becomes lower than the naval transport's carbon footprint, these arguments in favour of naval transport could disappear. The differences in speed of the greening process of, especially, road transport compared to the greening process in naval transport show that greening the maritime sector is necessary in order for a modal shift to (remain to) make sense.

[3] N.B. Chemical, Biological, Radiological and Nuclear

Solutions for greening the maritime sector are widespread, of which the prime example is a shift to more eco-friendly propulsion of ships. Theoretically, regulation forcing this shift would be a simple solution. Moreover, a trend towards stricter regulations can be identified. For instance, as of 2015, the IMO set up low sulphur zones around Europe and North America [23], plus the EU allocates many subsidies towards eco-friendly initiatives and research. Be that as it may, coercion of these regulations for ecological sustainability within the maritime sector remains difficult due to the pre-eminent international nature of the industry. The big question in solving the ecological sustainability problem, therefore, is which party can and will coerce certain greening measures? Countries are afraid to lose all maritime business when setting up strict greening regulations, while international organizations (f.e. IMO) lack the power to enforce regulations[4]. A movement coming from inside a maritime industry aiming for the greening of their industry is not likely. In general, the maritime industries are too fragmented for the forming of that kind of coalitions/alliances. Again, the shipping industry is a prime example of this. In order for a significant greening process to happen, an entity coercing this greening process is key.

Although the problem of coercion is a central question in solving ecological challenges, this does not mean nothing can be done in the meantime. Firstly, fundamental research is vital is making progress in the greening process. For instance, by investigating potential eco-friendly ways of propelling a ship. Secondly, actively stimulating green initiatives lowers the barrier for industrial players to implement such initiatives. Stimulating can be done on a moral level, an organizational level (f.e. facilitating collaborative projects) and, foremost, on a financial level (f.e. Horizon 2020 or similar member state programmes). Currently, both measures are not being used to its full potential.

The fourth structural challenge is **coping with disruptive developments**. Technological evolutions like autonomous shipping, 3D-printing, artificial intelligence and big data, plus social dynamics like the focus on the aforementioned safety & security and ecological sustainability[5], are of such a disruptive nature that decision making processes need an upgrade. Renewal of decision making processes in the maritime boardrooms is necessary for the maritime sector to remain able to adapt to these recent technological and social developments. The current model of decision making lacks the flexibility to adapt to the exponential nature of these developments. Moreover, these new models of decision making are not only nice to have, but are a necessity to coop with these developments.

Upgrading decision making processes on the side of the industry is already mentioned when discussing further professionalization of the maritime industry in combating volatility of demand. The proposed upgrade for the maritime industry by not solely relying on entrepreneurship, but also allowing external knowledge in the decision making, is very relevant for this challenge as well. Furthermore, upgrading decision making processes on the side on the government is as necessary. The current system in which governments dictate national or international policy for the entire maritime sector is a prime cause of the lack of adaptability to disruptive developments. In order for governmental decision making to gain the necessary flexibility, a closer cooperation between government, knowledge-institutes and industry is needed on both a strategic as well as an operational level. In order to achieve closer cooperation, governments best focus on structural cooperation based on trust and shared values, rather than a suspicious attitude based on rules and contracts.

Towards a triple helix

In the introduction the general idea of a triple helix model was outlined. Now looking more in depth, what are the proposed roles and responsibilities of the three parties within this model?

The ideal role of the government within the maritime sector is that of facilitator of cooperation. This can entail cooperation between industrial players as well as between knowledge-institutes and industry or between government and industry. As an impartial party, the government is the ideal candidate to initiate cooperation, especially between industrial players. Note that this does not exclude the maritime industry from facilitating cooperation, which is a shared role.

The facilitating role of the government does not exclude outside help. Due to their knowledge of the sector, or specific niches of the sector, having partners in the field can have significant value. In order words, governmental programmes including external partners like shipowner associations, cluster organisations and consultancies can actively incentivize cooperation between the parties of the triple helix. Naturally, this entails an organizational part. Examples are providing data and/or IT, optimize regulation in order to stimulate cooperation, mapping the qualities and potential merits of the organizations within the maritime sector and actively bringing these organizations together. Incentivizing brings forth a financial component as well. A bottom-up approach, in which governments recognize and stimulate potential initiatives of added value brought up by the industry, is recommended.

Knowledge institutes provide the maritime sector with a theoretical backbone. In interaction with the industry, it can contribute to increasing the level of knowledge within the maritime sector. By doing this, knowledge-institutes can help in further professionalizing the maritime industry. The maritime sector is pre-eminently a sector which can benefit from a

---

[4] N.B. In the case of the low sulpher zones, the IMO did not establish a fine or sanction, the IMO leaves it up to the State Party.

[5] N.B. These social developments are partly led by technological developments, but social developments independent of technological developments can be identified as well (emotion, social-cultural trends, etc.)

theoretical backbone due to the rather conservative nature of the maritime sector. Generally, the maritime industry consists of firms that carry a long tradition. Knowledge-institutes can help rationalize the debate within the maritime sector through providing a better understanding of the merit of more progressive views.

In aiming at increasing knowledge within the maritime sector, interaction with the industry is important. A much heard complaint from the maritime industry is that knowledge-institutes spend too much time on their own hobbies, without any added value to both fundamental and applied research in the long run. In order for a fruitful cooperation, a willingness of the knowledge-institutes to align their work with the industry is necessary. Academic bubbles will hinder the applicability of the results, a translation to practical needs and a pragmatic vision are vital for knowledge-institutes.

As mentioned before, the innovative capacity of the maritime sector lies primarily within the industry. Therefore, the main role of maritime industry within a triple helix is that of producer/innovator. Furthermore, as mentioned before, the industry can play a role in initiating cooperation as well. To ensure an industry that actively contributes to triple helix cooperation, a change of attitude of the maritime industry on four principles is necessary. Firstly, in order to enable a bottom-up approach, the role of the industry entails an assessment of their own needs. The basis of a cooperation with both knowledge-institutes and government is an industry that gives direction by providing problems that need solving. In other words, the basis is an industry that provides questions that need solving. Secondly, an attitude that focusses solely on short-term results, an attitude which is much more common in the industry than in government or knowledge-institutes, could hinder a fruitful cooperation. In order to get added value out of a cooperation, the maritime industry should broaden their usual timespan for projects. Thirdly, triple helix cooperation does not benefit from long implementation phases or long time to market timespans. The industry should act more hands on, for instance through experiments or pilots. This entails showing a willingness to accept failures more quickly as well. Fourthly, the maritime industry should broaden their scope when cooperating, thereby showing more concern for the common good. Not solely approaching a cooperation as a client or supplier, but as a partner.

The current line of maritime policy that can be identified throughout Europe has its origin at the end of the 1990's[6], which provided the maritime sector a great impulse. It entailed the entry of the tonnage tax, a liberal crew policy and the introduction of cluster organisations. Perhaps it is time for a revival of the cluster organisations as the link between all parts of the maritime network. This entails a renewed valuation of cluster organisations by the three parties of the triple helix, plus an assessment by the cluster organisations of their performance and vision as well.

4 Conclusion

This paper aims at showing the necessity of closer cooperation in combating structural challenges in the maritime sector. Four prominent challenges show the shortcomings of the current lack of incentive to cooperate, plus, show the merit of the triple helix model towards solving those challenges. The essence of the proposed triple helix model is a government implementing incentives for cooperation. These incentives will result in closer cooperation between government, knowledge-institutes and industry, which in turn will lead to the actual merits of an increasing level of available knowledge within the maritime sector, increasing financial power for the industry to take innovative steps and increasing flexibility to adapt to disruptive developments. Note that the proposed triple helix approach does not immediately solve the challenges in the maritime sector. The described challenges within a conservative and fragmented maritime sector are too complicated to be solved by merely a changed view on public-private cooperation. However, closer cooperation is necessary to set the right conditions for combating the challenges.

'Incentivizing cooperation' can be conceived as rather vague, therefore, to be more concrete, three recommendations for governmental policy are drawn up. Firstly, we recommend pro-active governmental programmes allocating funds to hire external partners, thereby actively searching for opportunities for triple helix cooperation that can be of added value. Cluster organisations are the prime candidates for the external partner spot. As an impartial player, cluster organisations are responsible for fine-tuning the dialogue between government, knowledge-institute and industry, plus streamlining the process of acquiring subsidies, especially EU subsidies for research and innovation. Secondly, in order to avoid wasting resources on programmes that do not respond to actual problems, we recommend a bottom-up approach when incentivizing cooperation. Governmental resources should be based on needs and problems provided by the industry. Note that this recommendation does not only appeal to governments. The industry providing their needs and problems (f.e. through shipowner associations) is vital. Thirdly, in order to avoid solely a one-time subsidy, we recommend a process-based approach for promoting and facilitating cooperation by the government. Allocating resources for a longer period of time secures continuity in the proposed closer cooperation.

[6] N.B. The introduction of this line of policy was in The Netherlands on January 1st 1996 and in the EU in 1997. Many countries followed afterwards. This line of policy was introduced by Peeters et al. [15]

References

[1] H. Etzkowitz, 1993. Technology transfer: The second academic revolution. Technology Access Report **6**, 7-9

[2] H. Etzkowitz, L. Leydesdorff, 1995. The Triple Helix: University - Industry - Government Relations: A Laboratory for Knowledge-Based Economic Development. EASST Review **14**, 14 - 19

[3] M. Ranga, & H. Etzkowitz, Triple Helix systems. *Entrepreneurship and Knowledge Exchange*, *7*, 107. (2015).

[4] P.D. Langen, Clustering and performance: the case of maritime clustering in The Netherlands. *Maritime Policy & Management*, ***29*****(3)**, 209-221. (2002).

[5] C. Peeters, A. Lefever, A. Soete, P. Vandendriessche, & H. Webers, De Nederlandse Maritieme Cluster: Literatuuronderzoek en plan van aanpak economische impact studies. *Nederland Maritiem Land 1*. (1997).

[6] M. Shinohara, Maritime cluster of Japan: implications for the cluster formation policies. *Marit. Pol. Mgmt.*, ***37*****(4)**, 377-399. (2010).

[7] N. Wijnolst, *Dynamic European maritime clusters* (**Vol. 30**). IOS Press,(2006).

[8] C. Egloff, U. Sanders, D. Khanna, S. Mohottala, K. Georgaki, G. Stratigis, K. Xu, The New Normal in Global Trade and Container Shipping. (2016).

[9] AT Kearney. (2012). Balancing the Imbalances in Container Shipping.

[10] UNCTAD, (2017), *Review of Maritime Transport 2017*

[11] A. S. Grzelakowski, *The Impact of Freight Markets and International Regulatory Mechanism on Global Maritime Transport Sector.* Marine Navigation and Safety of Sea Transportation: STCW, Maritime Education and Training (MET), Human Resources and Crew Manning, Maritime Policy, Logistics and Economic Matters, 273. (2013).

[12] K.G. Raun, (12-01-2018). *South Korean fund to support construction of 200 new ships.* [Website]. Visited on: 07-03-2018. Found on: https://shippingwatch.com/secure/Services/article10196511.ece

[13] B. Barnard, 24-11-2016. State Aid undermining container shipping recovery. [Website]. Visited on: 07-03-2018. Found on: https://www.joc.com/maritime-news/container-lines/state-aid-undermining-container-shipping-recovery_20161124.html

[14] O. Duru, Motivations behind irrationality in the shipping asset management: Review of fundamental theories and practical challenges. Maritime Business Review, **1(2)**, 163-184. (2016).

[15] C. Peeters, A. Lefever, A. Soete, P. Vandendriessche, & H. Webers, De Nederlandse Maritieme Cluster: Literatuuronderzoek en plan van aanpak economische impact studies. Nederland Maritiem Land 1. (1997).

[16] J. P. Gander, A Simple Model of University-Industry Research Linkages and the Sharing Principle Under Uncertainty (No. **2017_01**). University of Utah, Department of Economics. (2017).

[17] C. R. Hayes, Maritime cybersecurity: the future of national security (Doctoral dissertation, Monterey, California: Naval Postgraduate School). (2016).

[18] Y. C. Lee, S.K. Park, W.K. Lee, & J. Kang, Improving cyber security awareness in maritime transport: A way forward. Journal of the Korean Society of Marine Engineering, **41(8)**, 738-745, (2017).

[19] N. Meyer-Larsen, & R. Müller, *Enhancing the Cybersecurity of Port Community Systems*. In International Conference on Dynamics in Logistics (pp. 318-323). Springer, Cham, (2018).

[20] G. Berner, M. Lutzhoft, & J. Scanlan, Maritime Cyber Security-Hacking on the High Seas. In The 2017 Marine Electrical and Control Systems Safety Conference (pp. 1-7), (2017).

[21] O. Fitton, D. Prince, B. Germond, & M. Lacy, The future of maritime cyber security, (2015).

[22] J. Kramek, The critical infrastructure gap: US port facilities and cyber vulnerabilities. Center for 21st Century Security and Intelligence (2013).

[23] IMO. (23-12-2014). Sulphur limits in emission control areas from 1 January 2015. [Press Release]. Found on: http://www.imo.org/en/MediaCentre/HotTopics/GHG/Documents/sulphur%20limits%20FAQ.pdf

[24] R. Knulst, *3D printing of marine spares. A case study on the acceptance in the maritime industry* (Master's thesis, Open Universiteit Nederland) (2016).

[25] A.J. Mokashi, J. Wang, & A. K. Vermar, A study of reliability-centred maintenance in maritime operations. *Marine Policy*, ***26*(5)**, 325-335. (2002).

[26] M. Christiansen, K. Fagerholt, B. Nygreen, & D. Ronen, Maritime transportation. *Handbooks in operations research and management science*, ***14***, 189-284, (2007).

Main Drivers for Maritime Cluster and Shipping in the Baltic Sea

Riitta Pöntynen[1*], *Anne* Erkkilä-Välimäki[2], and *Hanna* Nieminen[3]

[1]University of Turku, Centre for Maritime Studies, Finland
[2]University of Turku, Centre for Maritime Studies, Finland
[3]Finnish Environment Institute SYKE, Finland

Abstract. One of the main aims of Maritime Spatial Planning (MSP) is to observe and determine factors affecting the development of various future sea uses. As a part of the transboundary MSP research done in the Maritime Spatial Planning for Sustainable Blue Economies (Plan4Blue) project, drivers for the creation of alternative scenarios for Blue Economy in the Gulf of Finland and the Archipelago Sea have been defined. The Blue Economic sectors analysed were energy, the maritime cluster, maritime and coastal tourism, the blue bioeconomy and subsea resources. The drivers affecting the development of blue economy sectors by the year 2050 were identified by experts through Delphi-surveys and workshops. First, the drivers were explored using the PESTEL-classification: political, economic, social, technological, environmental and legal drivers. The main drivers were then analysed and presented in futures tables. The results of this analysis are discussed from the viewpoint of the maritime cluster and shipping. The most important shipping drivers identified are transnational in nature: global economic trends and environmental policies, which connect national and regional MSP with the global scale. Driver analysis supports the identification of the sector's needs in the context of MSP. Challenges encountered in the process are highlighted in this text.

1 Introduction

Marine Spatial Planning (MSP) processes inherently include the envisioning of future developments in marine and coastal areas [1]. Futures research methods facilitate regional knowledge creation and management; in particular, they fulfil the need to strengthen foresight activities and increase visionary capabilities at the regional level [2]. In futures studies, scenarios are widely used as a tool for collective learning, for reframing perceptions and to preserve uncertainty if it is pervasive. A central part of the scenario process is to understand and define the driving forces behind future changes [3].

* Corresponding author: riitta.pontynen@utu.fi

https://doi.org/10.1051/shsconf/20185801024

In this study, the consideration of the driving forces, i.e. drivers, is a part of the examination of the alternative future scenarios concerning the regional blue economy in the Gulf of Finland and the Archipelago Sea in the Baltic Sea (Fig. 1). Here, the drivers are defined as external forces, which the individual actors cannot influence with their own means, but which affect the region's operational environment [4]. They are seen as contextual scenario elements that externally frame the conditions beyond the control of the relevant actors [5, cf. 2.].

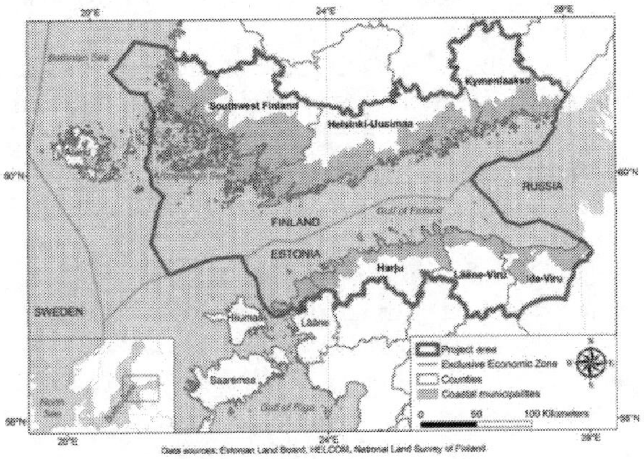

Fig. 1. Plan4Blue project area. [6]

This study explores drivers that may have major effects on the future development of the operational environment of regional blue economics and the use of coastal and sea space. Experts, as representatives of different MSP stakeholders, have extensive knowledge on the driving forces within their field of expertise. Thus, they play a crucial role in defining the drivers of regional change [3, 7]. Their views are explored in this study.

The shallowness, narrow straits and numerous islands, along with the possibility of ice cover in winter, make the Baltic Sea one of the most challenging areas for navigation [8, 9]. In addition, it is one of the most heavily trafficked sea areas in the world, and the Gulf of Finland is among the most congested sea areas of the Baltic Sea. It is anticipated that maritime transport in the Gulf of Finland will still increase, for example, due to the increases in Russian transport activity [8, 9, 10]. The area has large ports with lively passenger and cargo traffic, including increasing oil transport traffic. All this increases the risks associated with maritime transports in the Gulf of Finland and the Archipelago Sea [10]. The increases in maritime transport activities compete for space with the increasing leisure boating and other activities, making MSP processes highly important for the maritime transport and port businesses. Exploring drivers that influence the development of blue business sectors and their development prospects up to 2050 is crucial for scenario building. Understanding the drivers that affect the maritime cluster in the study area is crucial for the success of MSP processes.

The study is a part of the Plan4Blue project, which focuses on the building of cross-border MSP capacity [11]. Among other tasks, alternative future scenarios are produced based on the quantitative and qualitative data produced in the project. [12]

2 Methods

2.1. Defining Blue Economy Sectors

In the first phase of the economic analysis and scenario process, the analysis included all existing blue economy sectors in the study area, to enable the selection of the most important ones. The sectors are based on EU definitions [e.g. 13, 14, 15], and the extent of the analysis was agreed upon with the project partners, considering, in particular, the views of MSP planners. The analysed blue economy sectors are: energy; the maritime cluster; tourism, culture, services for leisure activities; and the blue bioeconomy and subsea resources. These sectors were included in the quantitative economic analysis as well as the rounds of the Delphi-panel and the workshop in Helsinki, both of which took place in 2017.

This paper focuses on the drivers of the maritime cluster. The following subsectors of the maritime cluster were chosen for further investigation in the first phase of the economic analysis and scenario process:

- Fairways: Building, deepening and maintenance
- Ports: Building, expanding port and harbour areas, other water constructions
- Sea and coastal cargo transportation
- Sea and coastal passenger transportation
- Building of ships, clean tech and equipment for the vessels (manufacturing)
- Offshore construction (fixed or floating platforms)
- Warehousing and storage related to cargo transports
- Pilotage and towage of ships
- Demolition of ships
- Warehousing and storage of leisure boats

In the second phase, which took place in 2018, the analysis focused on the sub-sectors identified in the 2017 results.

2.2. Delphi panel and workshops

The Delphi method was selected because it is a structured communication technique based on the anonymity of the participants, allowing them to freely express and change their opinions during the process [e.g., 3]. On the other hand, the workshops offered direct communication and discussion between experts. The aim of the scenario workshops has been to complement and deepen the results of the Delphi questionnaires.

The Delphi studies in 2017 and 2018 were independent two-round questionnaires. The overall aim of both Delphi studies was to define the blue business sectors and drivers that are relevant to the future regional development of the Gulf of Finland and the Archipelago Sea. The 2017 Delphi and Helsinki workshop also explored, quite extensively, the perceptions of experts regarding the future locations and activities of the blue businesses. The drivers were assessed viewing the blue economy as a whole. Based on the Delphi and other economic surveys carried out in the Plan4Blue-project in 2017, the 2018 Delphi and Tallinn workshop focused more on the most important blue business sectors from the perspective of Maritime Spatial Planning, as well as on their sub-sectors.

With the help of an expertise matrix, a cross-boundary and multi-sectoral panel of experts was formed. It consisted of all the relevant blue economy business sectors, both from Finland and from Estonia. The business representatives were mainly from business support organizations and unions, as choosing individual companies might have emphasized their opinions excessively. In addition, there were only few academics in the panel, as the focus was on stakeholders that have concrete interests or activities in the coastal and sea areas, either in terms of business operations or sea-use planning.

In 2017, the panel consisted of 55 members that represented blue economy businesses, public authorities and NGOs from Finland and Estonia. For the Delphi rounds in 2018, the panel composition was checked. New members were recruited to balance the representation towards the business sectors that were the focus of the 2018 rounds. In this phase, the panel had 57 members.

The Delphi studies were performed through on-line questionnaires. In 2017, the 1st questionnaire was composed using the HARAVA tool [16] and sent on the 26th of April. The second questionnaire was sent on the 14th of June, just before the Helsinki workshop. The questionnaire included a mixture of Delphi-type and more traditional survey questions. The two rounds of Delphi in 2017 were answered by 43 panelists. In 2018, the first Delphi questionnaire was sent on February 16th and the second on April 18th. Altogether, 16 replies were received. The questionnaire included statements on the development of the blue economy sectors that were related to the alternative scenarios. The 2017 questionnaires had identical Finnish, Estonian and English versions. The workshops and the 2018 Delphi were in English.

The first workshop was arranged in Helsinki on the 15th and 16th of June 2017, and the second one in Tallinn on the 23rd and 24th of January 2018. The Helsinki workshop was attended by 40 participants, six of whom were Delphi panelists and 24 of whom were other invited experts. Of the experts, 9 participants were from Estonia and 21 from Finland. In the Tallinn workshop, there were altogether 41 participants. Of the experts, 17 were from Estonia and 6 from Finland. Four Delphi panelists attended the Tallinn workshop. In the Tallinn workshop, the working groups were designed to work from the point of view of the different blue economy sectors, whereas in Helsinki, the groups worked from the point of view of the blue economy as a whole.

2.3. The classification and mapping of the drivers

To extensively explore the external forces that may have impacted various blue economy sectors in the study area, the PESTEL approach was adopted. The concept was introduced by Aguilar [17] as the PEST analysis for the Business Environment Scanning Task. PESTEL is a tool to "help form the futures table" and it analyses the impacts of **p**olitical, **e**conomic, **s**ocial, **t**echnological, **e**cologic / **e**nvironmental and **l**egal changes. It is used in scenario processes to chart external factors and drivers, and it originates from economics. The timespan for a PESTEL analysis is usually considered to be 3-10 years, though when combined with a scenario process, it is possible to create structured descriptions of alternative futures, and those drivers that are central for an activity or a sector are explored in more detail [4].

One of the themes of the Delphi-questionnaire included the assessment of the drivers according to the PESTEL classification: an estimation of the major political, economic, social, technological, environmental and legal drivers affecting the development of the blue economy by the year 2050. Prior to the 2017 Delphi study, a set of general drivers that may have regional-level effects in the study area were selected based on the literature and the suggestions of project partners. According to Vuorinen [4], the strength and probability of the drivers, their positive or negative influence and the timespan of a change should be assessed. In the first questionnaire, the positive or negative influence were evaluated considering those drivers that are global megatrends already taking place, or based on political decision-making, thus already being intended to have a clear effect on the businesses in the study area. Regarding other drivers, the experts were asked to evaluate their significance on the human activities in the study area. In the questionnaire, respondents had the possibility to add or comment on the important PESTEL drivers in open questions.

In the Helsinki workshop, drivers were discussed as one of the three working group themes. The world café method [18] was used. Each of the three working groups contributed into three thematic working group sessions. Groups circulated the themes and complemented the previous discussions. The PESTEL drivers from the Delphi study, with some added drivers from the comments and literature, were presented to the drivers group. The participants selected the most important drivers and explained their reasons for choosing them. Based on the Delphi and workshop responses, key drivers for different blue economy sectors, e.g. for the maritime cluster, were identified by the research team. The importance of these drivers and their consequences was the focus of the Delphi-round of 2018 and the Tallinn workshop.

Participants of the Tallinn workshop and the Delphi panelists were asked to identify the three most important drivers for the specified blue economy sectors, e.g. the maritime sector. The Delphi panelists were asked to assess the possible effects of the drivers they selected. In the Tallinn workshop, the participants discussed the consequences and future developments of the most important drivers chosen in the working group.

3 Results and discussion

3.1 Identification of drivers for the blue economy

In the first Delphi study and the Helsinki workshop, environmental policies were viewed as the most important political driver, and their impact was assessed to be significantly positive. Other sectoral policies were considered important drivers as well, and the percentage of respondents was almost the same. Traffic policy was considered a positive factor by the respondents, as well as the industrial policies. The connection between increased political tensions in the Baltic Sea and strengthening EU integration was pointed out at the workshop. The issue of global security was also raised in the discussions, in particular related to tourism.

Both in the Delphi panel and in the workshop, global economy conditions and trends were considered to be the most important economic driver. In Delphi, it was closely followed by the regional economy conditions and trends. Globalization was considered to affect, in particular, the maritime industry, through the increasing demand for maritime transport. Global economy as a driver emphasizes the significance of ports and affects the volume of maritime transport, as well as the volume of shipbuilding orders. Maritime transport also affects the regional level in a positive way. In the case of the maritime cluster, the attitudes of customers and shippers influence the sustainable development of the sector. The aim of achieving a low carbon society was discussed from the perspectives of both economic and policy aspects.

Among the social drivers, ageing, urbanization and attitudes towards blue businesses were considered particularly important. These drivers are linked to trends and consumption habits, and thus to the future transport needs of the area. The selected technological drivers have many connections to the maritime cluster: clean tech innovations for blue businesses, increasing automatization and the overall development of ICT and digitalization. As one of the environmental drivers, the eutrophication of the sea was unanimously considered the most important driver in the Delphi-questionnaire in terms of its effect on the development of the region's blue economy by 2050. In the Helsinki workshop, climate change as a larger phenomenon was proposed to be included as a driver, thus combining many of the environmental drivers presented in the questionnaire. In Delphi, tight restrictions for environmental permits were considered the most important legal driver. In the workshop, environmental regulations and the permit process as a whole, including changes and restrictions, was mentioned the highest number of times.

While the causal relations were not discussed in depth, certain links between these drivers were identified. The level and extent of co-operation in the Baltic Sea Region was considered to be linked with the safety situation and stability of the Baltic Sea. For example, the relationship with Russia affects maritime transport via the ports of Finland and Estonia, and the future of transit cargo passing through Estonia and Finland. Environmental policies have an impact on the fuels used in shipping. Transport policy affects ship owners and ports.

3.2 Drivers of the maritime cluster

In the second phase, in 2018, the Delphi panelists believed there would be environmental benefits through economic development: strong economic growth would also bring economic benefits into that region. The regional economic situation would promote co-operation, and the better the situation, the more the amount of shipping would increase. It was noted that the trends of global economy always act as a foundation for the state of the maritime cluster. The economic situation controls supply and demand. The panelists assessed the environmental regulations and the chosen fuels, which affect the price levels of fuels and ports. The use of low-carbon fuels can support carbon neutrality. Energy efficiency and clean air were also considered important drivers.

The level of co-operation in the BSR – the safety situation and stability of the Baltic Sea area, combined with the regional economic situation as well as environmental regulations and legal practices, were considered extremely relevant to both the amount of cruise passengers in the BSR as well as cargo transport activities. Autonomous shipping was expected to revolutionize shipping altogether, even if the need for a crew would remain e.g. in the archipelago areas. Various climate conditions, for example ice cover and storms, were considered to affect what types of freight vessels can be operated, for example, in the BSR.

In the Tallinn working group, globalization, increasing economic growth and the growth of cargo volume were considered to have an influence on the land infrastructure of the maritime cluster. Larger port areas on land, better infrastructure and port–hinterland connections would be required. Increased ship size would require deeper fairways. It was also noted that global shipping means global solutions, as well. In terms of sustainable development, the profile of vessels may change. However, global economic growth was expected to have a larger effect in Asia and 3rd world countries than in the BSR. Transit cargo would also provide an opportunity for the local operators and service providers. Possibilities for port expansion were discussed in relation to e.g. conservation areas near port areas. Offshore waiting areas would be needed to access the congested ports.

The attitudes of customers and shippers were expected to influence what types of cargo would be transported in the area, but also the ship owners' investments in clean tech and fuels. In the future, customers might be more willing to pay more for high-quality and clean tech solutions. Experts felt that attitudes could possibly be changed through information exchange and motivation building, and these actions should preferably be steered at the EU-level. The results of the choices in the Delphi and the workshop in 2018 are presented in **Błąd! Nie można odnaleźć źródła odwołania.**Table 1 below.

Table 1. The most important drivers for the maritime sector in Delphi (2018/1) and the Tallinn working group (24th January, 2018)

Driver	**Mentions in Delphi n=14, 41 answers**	**Mentions in the Tallinn workshop n=4, 10 answers**
Environmental regulations, legal practices	8	
Conditions and trends of global economy, globalization	6	4
Clean tech / emissions from the maritime cluster (energy efficiency)	6	
Fuels used in shipping (environmental policy)	5	
Regional economic situation	4	
Level of co-operation in the BSR – the safety situation and stability of the Baltic Sea area	4	1
ICT, digitalization	3	2
Attitudes of customers and shippers	2	3
Transport routes	1	
Climate conditions	1	
Other: Development of autonomous ships (without crew)	1	

In the context of globalisation and its consequences, some possible new roles and services for ports were envisioned that would have impact on port planning. For bigger ports, the multiuse of ports, mobility and new services were expected. These would include:

- Energy: serving offshore wind power plants (incl. maintenance)
- Tourism: more coastal cruises
- Cleantech would be situated in ports or in their vicinity
- Circular economy; for example, ports may produce energy from collected material and reuse waste in ports.
- Offshore platforms, also for multiuse by several blue economy sectors, as well as artificial islands, were considered future development options.

Potential synergies between cargo, passengers and energy were identified. New, future cargo types were discussed intensively. For example, in the future waste might become a new type of cargo. If the Arctic route through Finland was developed, there would be cargo through Finland from the north via railways or pipelines. Fossil fuels from the Arctic might be a new cargo type as well. This would create a new North–South transport corridor in the project area in addition to Russian and Eastern transit cargo.

Digitalisation and ICT were discussed as one of main drivers for the maritime sector. Automatization would impact vessels and ports, as well as their safety. Surveillance would be needed for the ports' offshore areas. Although most vessels would operate on automated fairways, there would be a risk for collisions between the autonomous and traditional uses of the fairways.

3.2. Challenges of the process of identifying drivers

In the first phase, the discussion of the long-term future developments at the regional level seemed to be challenging, even for experts. This was, to some extent, unexpected, because while the timeframe of 30–35 years is long in many societal and business issues, it is more familiar in land-use and regional planning, which is the sector that a majority of the experts represented. A longer timeframe is also preferable in MSP processes [e.g. 1].

While the idea of creating several alternative future scenarios and different opinions was emphasized, the Delphi responses and workshop discussions were about the current situation and issues or the familiar near-future changes that were already discussed, for instance, in the media, such as the development of autonomous shipping. Only a few truly visionary suggestions and ideas were presented, but the discussions around them did not expand. However, the discussions around the drivers deepened in Tallinn, which shows that the learning process for creating visions of the future developments takes time and iterations. The fact that the experts were more acquainted with certain issues on the table, each other and the moderators in the Tallinn workshop probably also affected the discussions in a positive way. This would be a similar result as in previous projects, where the transboundary planning is anticipated to benefit from informal situations, such as iterative face-to-face meetings that help create mutual trust and knowledge of the other parties [19, 20]. This is challenging, as the experts are busy people and the consideration of the planning issues requires time and familiarization, as well as the consideration of other stakeholders' views. In any case, stakeholder participation is necessary for the success of the MSP processes [e.g., 21].

It was beneficial to provide lists of the potential drivers in the workshops as a starting point for group discussions regarding their significance and effects on regional blue economics. The complex and multilevel chains of causal relationships formed by the drivers [cf. 17, 22. e.g. 23] were not discussed in depth. A few individual causal relationships were presented briefly in the Delphi studies and the workshops; and in the Tallinn workshop, the participants were encouraged to discuss the consequences of the drivers.

By definition, the contextual environment may only impact issues indirectly , but making the distinction between external factors and factors that, for instance, an individual organization can influence is often difficult [24]. These different levels of hierarchies were also noted by some of the experts in the workshops, but discussion on their connections and interrelations were not accomplished within the strict time-frames of the workshops.

4 Main conclusions

In the regional context, the international politics in the Baltic Sea region and the fluctuations of the global economy affect the general operational environment. In particular, the global economy and globalization directly affect maritime transport and the ports' business, probably even more than regional drivers in the study area. Additionally, environmental regulations for maritime transport are global, although in the Baltic Sea region the requirements are stricter. For the maritime cluster, this also offers possibilities for clean tech business.

The future changes of maritime transport are anticipated to have effects on sea use, and especially on the issues of safety and security in the study area, where other uses of the sea compete for space with maritime transport. Ports may acquire new roles and services if, for example, the study area develops more as a maritime transport corridor for international and Arctic trade. The positions of ports may also become increasingly important in serving different blue economy businesses, which may also have new spatial requirements. Planning decisions and politics will define whether the requirements of maritime transport and ports are favoured or ignored in the future.

The challenge for the experts working with questionnaires and in workshops is to vision the long-term future changes and thoroughly discuss complex and multilevel issues such as drivers. Our study gave hints that in MSP processes, the iterative rounds of discussions between experts are necessary in order to achieve successful long-term management decisions. The complex interrelations between the regional drivers should be studied more as the marine spatial planning processes are developed, as they have a central role in creating an understanding of the various possibilities of future development. The effects of the most important drivers for alternative future scenarios should be assessed through the examination of long-term development of trends and their interactions with maritime businesses. In addition, methods allowing experts from different stakeholder groups to participate in marine spatial planning as a part of the long-term planning, monitoring and management processes should be developed further.

The article is based to Delphi-panel and workshops realized as part of Plan4Blue project.

References

1. P. M. Gilliland and D. Laffoley. Key elements and steps in the process of developing ecosystem-based marine spatial planning. Mar. Policy **32**, 787–796. (2008)
2. T. Uotila, H. Melkäs, V. Harmaakorpi. Incorporating futures research into regional knowledge creation and management. Futures **37**, 849–866. (2005)
3. J. Maness. *Techniques, Methods & Applications in Futures Studies*. EBSCO Publishing: eBook Collection (EBSCOhost). (2012)
4. T. Vuorinen. *Strategiatyökirja. 20 työkalua*. Talentum (2013).
5. R. Milestad, Å. Svenfelt. K.H Dreborg. Developing integrated explorative and normative scenarios: The case of future land use in a climate-neutral Sweden. Futures **60**, 59-71. (2014)
6. T. Nylén. Basemap of Plan4Blue project area. (2017).
7. H. Ritchie, G. Ellis. 'A system that works for the sea'? Exploring Stakeholder Engagement in Marine Spatial Planning. J. Environ. Plann. Man **53**:6, 701-723. (2010)
8. Baltic Lines. Baltic LINes: Shipping in the Baltic Sea – Past, present and future developments relevant for Maritime Spatial Planning. Project Report I. 35 p. (2016):
9. HELCOM BSEP 152. HELCOM Assessment on maritime activities in the Baltic Sea 2018. Baltic Sea Environment Proceedings No.152. Helsinki Commission, Helsinki. 253pp. (2018). Available at http://www.helcom.fi/Lists/Publications/BSEP152.pdf
10. WWF. Future Trends in the Baltic Sea. WWF Baltic Ecoregion Programme 2010). Available at http://wwf.panda.org/what_we_do/where_we_work/baltic/publications/?194764/Future-trends-in-the-Baltic-Sea
11. Plan4Blue project, available at http://www.syke.fi/projects/plan4blue
12. R.Pöntynen, A. Erkkilä-Välimäki, T. Nylén, R. Varjopuro, H. Nieminen, A. Lähteenmäki-Uutela, T. Pohjola, M. Roose, H. Uusitalo, J. Törrönen. *Setting of Blue Growth Scenarios. Qualitative analysis based on expert opinions*. (to be published).
13. European Commission / EUNETMAR. Study on Blue Growth, Maritime Policy and the EU Strategy for the Baltic Sea Region. Final report. (2013). Available at https://webgate.ec.europa.eu/maritimeforum/en/node/3550
14. Blue Growth. Opportunities for marine and maritime sustainable growth. COM (2012) 494 final.
15. Sustainable Blue Growth Agenda for the Baltic Sea Region (2014). Available at service.mvnet.de/php/download.php?datei_id=132314
16. HARAVA tool, available at https://dimenteq.fi/palvelut/harava/
17. F.J. Aquilar. *Scanning the Business Environment*. The Macmillan Company, New York. (1967)

18. J. Brown. The World Café: Living Knowledge through Conversations that Matter. The Systems Thinker. **12**:5. (2001).
19. Ruskule et al. Pan-Baltic stakeholders' dialogue on MSP: Synthesis report from PartiSEApate single-sector workshops held in 2013. (2014). Available at http://www.partiseapate.eu/wp-content/uploads/2014/08/Syntesis-report-from-pan-Baltic-Dialogues-2013.pdf
20. Baltic Scope 2017. Recommendations on maritime spatial planning across borders. Available at http://www.balticscope.eu/content/uploads/2015/07/BalticScope_OverallRecomendations_EN_WWW.pdf
21. R.Pomeroy, F. Douvere. The engagement of stakeholders in the marine spatial planning process. Mar. Policy **32**, 816–822. (2008)
22. G. Tovstiga, M. Aylward. *Strategic Direction. Study Guide*. Henley Business School. (2008).
23. R. J. A. Collins. Graphical Method for Exploring the Business Environment. Available at us-ers.ox.ac.uk/~kell0956/docs/PESTLEWeb.pdf (8.6.2017) (2010).
24. H.A. Wolters, J. Gille, J.M. de Vet, R.J. Molemaker. Scenarios for selected maritime economic functions. Futures **1**:11. (2013)

Multi-use of the sea: from research to practice

Joanna Przedrzymirska[1], Jacek Zaucha[1], Daniel Depellgrin[2], Rhona Fairgrieve[3], Andronikos Kafas[3], Helena Maria Gregório Pina Calado[4], Marta Horta de Sousa Vergílio[5], Mario Cana Varona[6], Marija Lazić[1], Angela Schultz-Zehden[7], Ivana Lukic[7], Eva Papaioannou[8], Martina Bocci[9], Rianne Läkamp[10], Ioannis.Giannelos[10], Aneta Kovacheva[10], and Bella Buck[11]

[1]The Maritime Institute in Gdansk, Długi Targ 41/42, 80-830 Gdańsk, Poland
[2] CNR - National Research Council of Italy, ISMAR - Institute of Marine Sciences Arsenale - Tesa 104, Castello 2737/F, I-30122 Venice, Italy
[3]Marine Scotland Science, 375 Victoria Road, Aberdeen, AB11 9DB, Scotland, UK
[4]MARE – Marine and Environmental Sciences Centre; FCT - University of the Azores, 9501-801 Ponta Delgada, Azores, Portugal.
[5]CIBIO – Research Centre in Biodiversity and Genetic Resources/InBIO – Associate Laboratory; University of the Azores, 9501-801 Ponta Delgada, Azores, Portugal.
[6]University of the Azores, 9501-801 Ponta Delgada, Azores, Portugal.
[7]SUBMARINER Network for Blue Growth EEIG, Kärntener Str. 20, DE–10827 Berlin, Germany
[8]University of Dundee, Nethergate, Dundee DD1 4HN, United Kingdom
[9] Thetis SpA, Castello 2737/f, 30122 Venice, Italy
[10]Ecorys Nederland, Watermanweg 44, 3067 GG Rotterdam, The Netherlands
[11]Am Handelshafen 12, 27570 Bremerhaven, Germany

Abstract. The increasing demand for ocean resources exerts an increasing pressure on the use of ocean space across all European Sea Basins. This underlines issues of compatibility (or conflicts) between different maritime uses as well as between economic activities and environmental protection. The idea of multi-use (MU), as a guiding concept for efficient allocation of compatible activities in the same marine space, can increase spatial efficiency and at the same time provide socio-economic and environmental benefits. However, its transition from a concept to real-world development is facing several barriers. Based on analysis of five European sea basins done under the Horizon 2020 MUSES project (Multi-Use in European Seas), this paper aims to clarify the concept of MU by discussing: 1) the definition in the literature and practice so far, and; 2) how existing regulatory and planning regimes are supporting and challenging the development of several MUs (considered as the most promising). The analytical methodology developed for the MUSES project relied on data collected via desk research and semi structured interviews with key stakeholders (e.g. industry, regulators), over the period of seven months. The semi-quantitative analysis of data conducted, identified the commonalities and differences among countries in respect to each of the analyzed MUs. The paper points out priorities for the MU development in different sea basins and recommends initial steps to overcome existing barriers, whilst maximizing local benefits. This paper is a starting point towards a broader scientific debate on: (i) what could be the role of management policies (like for instance maritime spatial planning - MSP) in supporting and fostering MU concept development, (ii) what are technical and technological challenges for technically advanced MUs, (iii) how added values of MUs concept (e.g. benefits for local economies, positive impacts on environment) could be enhanced.

Acknowledgement

Analyses presented in this paper were financed by the European Union's Horizon 2020 research and innovation programme under Grant Agreement no 727451. The authors would like to thank the MUSES project consortium and all project partners for their valuable contributions to the project. The MUSES project is funded by the EU Framework Programme for Research and Innovation Horizon 2020. Grant number: 727451. More information can be found at https://muses-project.eu/.

1. Introduction

Multi-use (MU) at the sea is a relatively new research topic that has emerged from three distinctive sources. The first one is research and innovation, that prompts out development of new technologies offering novel ways of exploitation of sea resources and improvement of its conservation measures. Several research projects [1] have been conducted e.g. to investigate the possibility of establishing multi-purpose off-shore platforms, serving needs of off-shore energy production, mariculture or sea tourism and even regular navigation. Some combinations have been researched even more extensively such as off-shore wind energy and aquaculture and substantial documentation exists in this field (e.g. [2], [3]). The second source is business itself, e.g. tourism in the Mediterranean region considering new opportunities such as pescatourism [4] or underwater culture heritage. This provides possibilities for combinations such as tourism and fishery or tourism and protection of underwater artifacts. The third source is scarcity of space. The emergence of maritime spatial planning [5-13] made it evident that marine space is not abundant and it should be treated as a scarce resource [14]. Therefore, MSP urges for sparing use of the sea space [15] and MU is among the considered solutions. Some scholars see MU as an analogue to the economies of scales that drive terrestrial spatial development[16].

2. The essence of Multi-Use

The discussions and developments of MU of marine resources in the political and academic arena have generated a variety of terms to describe the context. Each nomenclature is trying to capture and convey important information about the particularities of their investigated scenario. As a result many differing terms have emerged during the last 15 years for the same concept idea: co- and translocation, multi- and multi-functional use, co-use, secondary and additional use and coexistence to name a few.

The information conveyed by these terms can cover every dimension from legal and business relationships of users to even temporal and physical aspects of the multi-use relationship.

According to a definition elaborated within the MUSES project, 'multi use' (MU) is considered as a "joint use of resources in close geographic proximity". The term is an umbrella term covering a multitude of use combinations in the marine realm and representing a radical change from the concept of exclusive resource rights to the inclusive sharing of resources by one or more users. This can involve either a single user or multiple users. The use means distinct and intentional activity through which a direct (e.g. profit) or indirect (e.g. nature conservation) benefit is drawn by one or more users. The user means individual, entity or group that intentionally benefits from a given resource, and the resource is understood as a good or service that represents a value to one or more users (e.g. biotic, such as fish stocks; or abiotic, such as ocean space) and can be exploited through either direct (e.g. fishing) or indirect (e.g. nature conservation) uses [1].

It is often difficult to differentiate between genuine MU and the mere coexistence of several uses. For example, ships and fish use the same seawaters. However, this should not be considered as a MU, even though the condition of lack of exclusivity is fulfilled. It's doubtful whether this joint use of resources is intentional (rather than coincidental) and is beneficial to both parties. Also diving in wreck sites should not be considered as a MU. However, diving in the intentionally prepared and maintained underwater sanctuaries is considered as a MU, because it is based on conscious decisions and provides benefits to both users.

Recognizing the multitude of possible multi-use scenarios in European seas, two essential types of MUs are defined:

a) Multi-use of geographical, human, biological resources
b) Multi-use of technical resources (marine infrastructure & platforms)

The first type means that multi-use of marine resources refers mainly to the geographical connection of resource uses to create benefits for society and single actors. An example of such a multi-use is the combination of offshore wind and tourism through boat tours viewing the offshore wind farm [17].

The second type means even closer (functionally and geographically) integration of uses to create even more added value than a side-by-side scenario. This closer integration looks for synergies in integrating the operations and implementation of offshore activities and can start by e.g. the simple sharing of the use of offshore supply vessels to reduce individual operations costs. The synergistic integration of activities culminates in multi-use platforms. MU offshore platforms are engineering solutions, designed to incorporate modules of other compatible activities (e.g. TROPOS Project). A fully integrated multi-component and multi-purpose offshore platform serves as a main infrastructure shared by two or more ocean uses (e.g. H2Ocean project designed a platform coupling renewable energy harvesting + hydrogen generation + aquaculture + environmental monitoring)[18].

In terms of sequence in which the development occurs, two scenarios of MU creation are considered as presented in Figure 1.

Joint development of uses		Staggered development of uses
MU where two (or more) combined uses (from the blue growth sector i.e. aquaculture or offshore wind) are applying for licenses at the same time	VS	One existing (traditional) use is already in place and the new (emerging) one is coming in MU where one sector is already in place (e.g. underwater heritage protection) and is being combined with the new use (e.g. tourism)

Fig. 1. Two possible scenarios for the sequence of multi-use developments. Source: own elaboration SUBMARINER and AWI

3. Multi Use as research subject

Several types of MU combinations have been researched in a wide variety of possible MU combinations, all of them at different stages of their maturity and feasibility. The list of combinations (Tab.1) was compiled after identifying combinations that have been analysed by past projects. A total of 26 case studies analysed in past projects (e.g. MARIBE, MERMAID, H2Ocean and TROPOS) have resulted in 11 uses considered as MU.

Table 1. MU combinations identified in the international projects

Project	Use	Co-Uses
EU funded projects		
COEXIST Project ID 245178	Fisheries and aquaculture	Other coastal activities (stakeholder)
H2Ocean Project ID 288145	Wind and Wave energy	Aquaculture, Hydrogen (stored and shipped to shore as green energy carrier)
MARIBE (Marine Investment for the Blue Economy - Baltic, North Sea, Atlantic, Caribbean, Mediterranean) Project ID 652629 (collected results from all other finished EU multi-use projects)	Caribbean: Aquaculture	Tourism, Wave energy, Desalination
	Mediterranean: Aquaculture	Tourism
MERMAID (Baltic, North Sea, Atlantic Mediterranean, Lead: DTU) Project ID 288710	Atlantic: Offshore wind and wave energy	Maritime transport,
	Mediterranean: Wave energy	Leisure , Aquaculture , Maritime transportation
	North Sea: Wind energy	Aquaculture (seaweed and shellfish), Tourism
	Baltic: Wind farm	Passive Fisheries, Aquaculture (fish and seaweed)
ORECCA (Offshore Renewable Energy Conversion platforms – Coordination Action) Project ID 241421	Offshore Renewables	Aquaculture (biomass and fish), Monitoring of the sea environment (marine mammals, fish and bird life)
TROPOS (Mediterranean, Tropic, Sub-tropic, Lead: PLOCAN) Project ID 288192	Maritime transport (offshore port and base	Fisheries (service station, storage), Aquaculture (fish), Energy (solar and ocean wave),

	of logistic service for energy sector)	Leisure activities (floating hotel, underwater observation facility, scientific tourism, diving base, yachting services)
MARINA Platform Project ID 241402	Wind Energy	Wave Energy
National funded projects		
Project	**Use**	**Co-Use**
AquaLast (Germany – Lead: AWI; University of Applied Sciences Bremerhaven, Fraunhofer, Weswerwind, TKB) **(AWI)**	Offshore Wind Energy	Aquaculture (loading on offshore support structures, such as wind turbine foundations, caused by mussel longlines)
Biological and technical feasibility study of marine aquaculture in the Thorthonbank area, Belgium: Co-use of space with offshore wind farms (Belgium - University of Ghent, SINTEF Ocean)	Offshore Wind Energy	Aquaculture (farming of blue mussel)
Coastal Futures (Germany – Lead: University of Kiel; AWI, GKSS) **(AWI)**	Offshore Wind Energy	Aquaculture (integrated coastal zone management for the integration of aquaculture into wind farm areas)
Flandres Queen Mussel (FIOV) (Belgium - Stichting voor Duurzame Visserijontwikkeling -SDVO, ILVO)	Offshore Wind Energy	Aquaculture (development of floating buoys with mussel ropes for spat collection)
Gulf of Mexico OOA (USA – University of Texas)	Offshore Oil Platforms	Aquaculture (multi-use of offshore fish cultivation in combination with offshore Oil & Gas)
Integrate the offshore wind technology with aquaculture –development of fish farm equipment for offshore conditions (Norway - Statoil, SINTEF Ocean and Lerøy Seafood Group)	Offshore Wind Energy	Aquaculture (fish farming of salmon)
KOREA Co-Location (South Korea – Lead: Korea Electric Power Cooperation Research Institute (KEPCO); Korean Institute of Ocean Science and Technology - KIOST)	Offshore Wind Energy	Fisheries (passive fisheries), Aquaculture (seaweed production for biomethane and bioproducts in wind farms)
Mosselkweek in Belgische windmolenparken – Mussel production within Belgium Wind Farms (Belgium – Lead: University of Ghent; ILVO, AWI, SINTEF, et al.)	Aquaculture	Wind energy, Maritime energy
MytiFit (Germany – Lead: AWI; Engel Netze, LAVES) **(AWI)**	Offshore Wind Energy	Aquaculture (mussel fitness, infestation of parasites, and selection of hard substrates for multi-use)
NutriMat (Germany – Lead: IMARE; Greim Fish Consulting, AWI, University of Applied Science Bremerhaven, WeserWind, Louis Schoppenhauer GmbH & Co. KG)	Offshore Wind Energy	Aquaculture (use of fouling organisms of offshore platforms for fish feed in land-based aquaculture)
Nysted Sea Wind Farm Mussels (Belgium – DTU)	Offshore Wind Energy	Aquaculture (investigation on the possibility to multi-use for longline mussel farming)
Ocean Forest (Norway – Leroy Seafood Group, Bellona Foundation)	Aquaculture (multi-trophic) Energy	Aquaculture (bio-mass production for energy generation)
Offshore-Aquaculture (Germany – Lead: AWI; Terramare)	Offshore Wind Energy	Aquaculture (investigations of the settlement and growth of bivalves and macroalgae in the German Bight to test its feasibility for offshore multi-use)
Offshore Site Selection (Germany – Lead: AWI; Thünen, University of Rostock, Kutterfisch, WindMW, Deutscher Fischereiverband, Skretting)	Offshore Wind Energy	Aquaculture (offshore site selection for IMTA in co-use of offshore wind farms)
Open Ocean Use (OOMU) (Germany – Lead: IMARE; EWE, University of Hannover, Thünen Institute, Bard Engineering, Kutterfisch, Frosta, AWI)	Offshore Wind Energy	Aquaculture (investigation on integrating an offshore fish cage into tripile foundation)
Roter Sand Project (Germany – Lead: AWI)	Offshore Wind Energy	Aquaculture (development of system design for the use of offshore environments for the cultivation of species for aquaculture and bioextraction)

SOMOS – Safe production Of Marine plants and use of Ocean Space (The Netherlands – Lead: Wageningen University; TNO)	Offshore Renewable Energy	Aquaculture (Seaweed farming)
Stichting Noordzeeboerderij (The Netherlands – Hortimare, Schuttelaar and Partners)	Offshore Wind Energy	Aquaculture (development of a seaweed technology and mass algal production)
WINSEAFUEL (France - French National Research Agency)	Offshore Wind Energy	Aquaculture (seaweed mass production for biomethane and bioproducts in wind farms)

Source: own elaboration by SUBMARINER and AWI

Figure 2. illustrates that combinations may differ in terms of their potential/feasibility and time of appearance. Some of them are very probable in the near future, some may be possible in several years' time, and others are not likely to occur at all. However, this matrix is indicative of the complexity of the MU research.

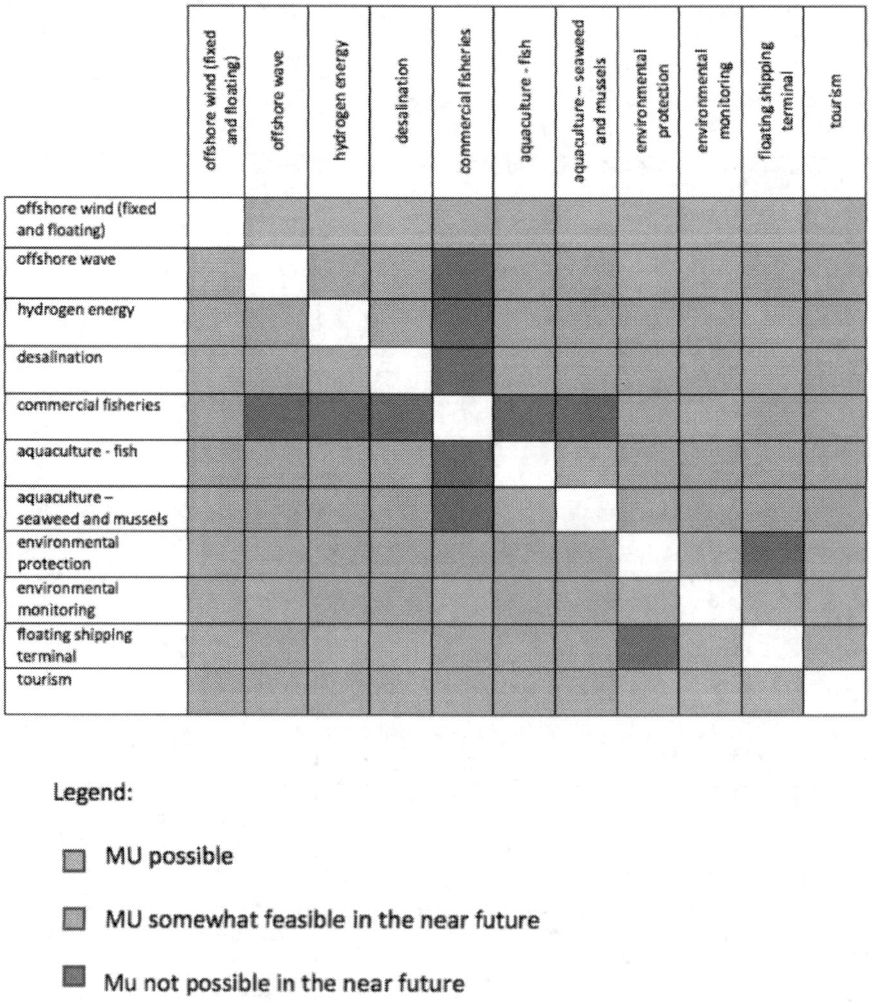

Fig. 2. Feasibility of multi-uses combining two uses. Source: own elaboration by AWI and SUBMARINER

4. The researched sea basins

It is evident that MUs might differ in the EU sea basins due to their specific features facilitating development of some uses and hindering others. Five distinctive sea basins are defined in the EU sea waters if the outermost regions are not included: the North-Eastern Atlantic (EA), the North Sea (NS), the Baltic Sea (BSR), the Mediterranean Sea (Med) and the Black Sea (BS) (Fig 3). Each of these sea basins is characterized by different physical conditions resulting in different uses of sea resources. However, despite obvious differences, several common trends important for MU development are observed: 1) sectors dominating in the given sea basin seem to strongly influence development of MU, 2) environmental assets tend to have a more important role in allocation of the sea space to particular uses, 3) local and regional economic development is a driving force for local MU initiatives.

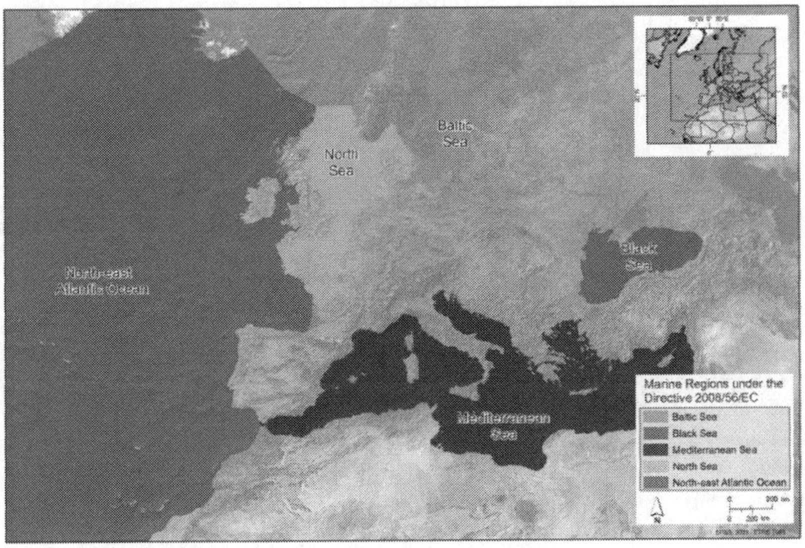

Fig. 3. Sea basins analysed under MUSES project (drawing on [1])

A brief comparison between sea basins is provided in Table 2. Evidence is clear that some physical conditions (wind potential) support multi-use based on wind farms in the North Sea and Eastern Atlantic, whereas high temperature signifies an importance of tourism and possibility to combine tourism with other activities in the Black Sea and Mediterranean Sea. Low salinity hinders mussels aquaculture in the Black Sea and the Baltic.

Table 2. Overview of prevailing physical conditions in the analysed sea basins

Sea Basin	EU Countries involved	Area [km^2]	Physical characteristics						Notes
			Wind	Waves	Tides	Temperature [°C]	Salinity [psu]	Depth [m]	
EA	Portugal (PT), Spain (ES), France (FR), Ireland (IR), United Kingdom (UK)	?	Excellent	Powerful	Strong	Surface: between 7 and 15 Deep waters: between 5.5 and 7.5	35 or higher	5000 (ocean)	Part of the world ocean. EU countries have jurisdiction over large maritime spaces.
NS	Denmark (DK), Germany (DE) Belgium (BE), France (FR), United Kingdom (UK), Netherlands (NL),	570,000	Excellent	Strong (in comparison to the open ocean smaller speed and the larger amplitude)	Strong	Average: 17 in the summer and 6 in the winter	25 - 34.5	average 90	Partially enclosed by land but directly connected with Atlantic Ocean, intensively used for various economic sectors.
BSR	Germany (DE), Denmark (DK), Sweden (SE), Finland	415,266	Good	Moderate	Weak	Surface: between -0.5 to +20	18 (west) – 0	average 54	Completely enclosed by land, connected with the North Sea through Danish

	(FI), Estonia (EE), Lithuania (LT), Latvia (LV), Poland (PL)					depending on the season	(north-eastern)		straights, intensively used for shipping and fishery.
MED	France (FR), Spain (ES), Malta (MT), Cyprus (CY), Croatia (HR), Slovenia (SI), Greece (GR), Italy (IT),	2,505,000	Moderate and good	Varies	Weak	Surface: 21-28 in the summer, 10-17 in the winter	36.5 - 39	?	Completely enclosed by land, connected with the Atlantic Ocean through the Strait of Gibraltar, intensively used for shipping, tourism and fishery with growing importance of aquaculture.
BS	Romania (RO), Bulgaria (BY)	436,000	Moderate	Moderate	Weak	Summer: up to 30 (surface) and 8.5 (deep waters)	17-18	Average 1253	Completely enclosed by land, connected with the Mediterranean Sea through the Dardanelles and the Bosporus, intensively used for shipping, tourism and fishery.

Source: own elaboration by MIG

Availability of space is also a relevant factor influencing the development of MU. In small sea areas where space is scarce, MU might be seen as an opportunity to use space in a more efficient way. In the oceans and other deep sea areas, MU might be driven mainly by the economic benefits of such an approach (e.g. offshore MU platforms) rather than spatial efficiency.

5. The most relevant multi-use combinations for each sea basin

Analyses were conducted at three geographical scales:

- Scale 1 – Intra-country scale: within single country;
- Scale 2 – Basin/sub-basin scale: sum of findings from all countries within a basin or sub-basin;
- Scale 3 – Trans-boundary scale: two or more countries.

The analyses were conducted with the use of various research methods including desk-based review of relevant regulations (international to local levels), project reports and case studies, scientific reports, workshops and interviews with stakeholders associated with marine planning in general and MU in particular. Relevant data was collected at country-level and results aggregated and analysed at sea basin level.

The stakeholders' preferences for individual MUs were revealed in the course of in depth interviews. The stakeholders' opinions were confronted with the previous desk research findings and related to the sectoral experience with MU development and to some extent also with the policy will in promotion of MU. As the result, the combinations were prioritised taking into account that at least one sector has been already existent (and preferably demonstrated some MU experience) and some policy will was in place (e.g. for tourism, fishery and aquaculture in the Mediterranean sea basin). The top three combinations per sea basin are presented in Table 3.

Altogether, in all five sea basins, 14 MU combinations have been identified as existing or having potential. Six of them have been selected as the most relevant in the sea basins at least in one of the sea basins (Table 3).

Table3. The most relevant MUs selected in the sea basins analysis

Note: blue number indicates the number of countries within the sea basin in which the given MU exists, orange number indicates the number of countries in which the given MU has potential as one use is already in place.

	MU name	EA	NS	BSR	MED	BS
MU1	Offshore Wind and Aquaculture	1/2	3/1	1/3	1/1	-
MU2	Offshore Wind and Tourism	1/1	1	3/2	-	-
MU3	Offshore Wind and Fisheries	1	4	1	-	-
MU4	Aquaculture and Tourism	3/1	-	1	3/3	2
MU5	Fisheries and Tourism and Environmental Protection	3	-	1	5/3	2
MU6	Underwater Cultural Heritage and Tourism and Environmental Protection	3	-	4/2	1/4	2

Source: own elaboration by Maritime Institute in Gdańsk

Out of eight in-depth analysed MUs, the most frequent (in terms of appearance as the existing in EU countries, see table 3) have been the three related to tourism (MU2, MU4 & MU6). The combination of Fisheries and Tourism and Environmental Protection has been tested or established in ten countries and within three sea basins in which tourism is a driving force for blue growth. Also, the combination of Underwater Cultural Heritage and Tourism and Environmental Protection has often occurred (ten countries) in four out of five sea basins. The third most frequent MU is Aquaculture and Tourism – existing (according to the MUSES categorisation)| in six member states of Southern Europe located in the Mediterranean and East Atlantic sea basins. Thus, in the policy supporting MU, tourism as a MU driver should be properly considered.
The fourth MU in terms of practical deployment is Offshore Wind and Aquaculture, that has been tested or exists in six countries (though in some cases energy is mainly a supplement to existing aquaculture with no ambition to produce energy for sale).

However, in terms of future development, the picture looks quite different. The biggest expectations are formulated by stakeholders towards Offshore Wind and Aquaculture as well as Underwater Cultural Heritage and Tourism and Environmental Protection. Both MU combinations exist or have development potential (with one use already in place) in 13 and 16 countries respectively. Both of these MU combinations have been prioritised in four sea basins, however the first type of MU is not so prominent for the Black Sea and the second one for the North Sea basin. This can be explained at least partially by physical characteristics of these sea basins and their policy specificities in terms of blue growth. Wind energy is not a priority in the Black Sea whereas in the North Sea underwater cultural heritage is not regarded as a development driver (i.e. the following sectors take a lead in blue economy: commercial fisheries, oil and gas production, shipping and maritime transport, tourism and offshore renewable energy development).

6. Conclusions for further research

Selection of the most important MUs for each sea basin seems only a top of an iceberg. There is a need for further research in order to make the MU concept operational. To summarize the findings from this paper, the following topics need further detailed research:

1. Researching MU in the context of resilience of marine ecosystems, since so far the economic and planning perspective prevails in researching MU.
2. Establish the economic value of different combinations which might be challenging due to important externalities related to them. Such research can change opinions of stakeholders and give space for new priorities in relation to MU deployment.
3. Better understanding of stakeholders' opinions with regard to MU. For instance in the Mediterranean sea basin, combinations related to tourism were prioritised as the most relevant due to their prevalence and importance for almost all countries in the given sea basin. Whereas combinations related to offshore wind scored high due to high probability of France to increase investment in multitrophic aquaculture combined with floating wind turbines that might offer EU breakthrough for this MU. Those peculiarities deserve more in depth analysis.
4. Analysing possible deployment paths of the most promising MUs in the sea basins, in particular the combinations related to offshore energy and tourism as driving sectors. The assistance should be tailored to the maturity level of the supported combinations and the size of barriers hindering their development. Also sea basin specificities must be taken into consideration. The support must be adjusted to the macro regional needs. Casting support for MU development requires prior understanding of the reasons behind prioritisation of some uses by macro regional experts.

All of the aspects mentioned above call for further research. In order to support the MU approach in a conscious way - i.e. to move from research to practice – a different approach seems necessary. Previous research has been focusing on the technical aspects of MU deployment. To complement this research, the social aspects need to be further investigated, using behavioral economics, business anthropology and other fields of social science. Applied research covering stakeholders' motivations and attitudes, describing drivers and barriers and identifying feasible policy solutions is essential for successful MU deployment in the future.

References

1. J. Zaucha, M. Bocci, D. Depellegrin, I. Lukic, B. Buck, M. Schupp, M. Caña Varona, B. Buchanan, A. Kovacheva, P.K. Karachle, et al. (2017) Analytical Framework (AF) – Analysing Multi-Use (MU) in the European Sea Basins. Edinburgh: MUSES project
2. T. Michler-Cieluch, G. Krause, B. H. Buck (2009) Reflections on integrating operation and maintenance activities of offshore wind farms and mariculture.Ocean & Coastal Management **52(1): 57-68**
3. B. H. Buck, R. Langan (eds.) (2017)Aquaculture Perspective of Multi-Use Sites in the Open Ocean. The Untapped Potential for Marine Resources in the Anthropocene. Springer International Publishing, 404 p
4. W. Piasecki, Z. Głąbiński, P. Francour, P. Koper, G. Saba, A. Molina García, V. Únal, P.K. Karachle, A. Lepetit, R. Tservenis, Z. Kızılkaya, K.I. Stergiou (2016). Pescatourism—A European review and perspective. *Acta Ichthyol. Piscat.* **46 (4): 325–350**.
5. A. Schultz-Zehden, K. Gee, K. Scibior (2008). Handbook on Integrated Maritime Spatial Planning. Berlin: S.PRO, 98 p.
6. F. Douvere, C.N. Ehler (2009). New perspectives on sea use management: Initial findings from European experience with marine spatial planning. *Journal of Environmental Management*, **90(1): 77–88**
7. H. Calado, K. Ng, D. Johnson, L. Sousa, M. Phillips, F. Alves (2010). Marine spatial planning: Lessons learned from the Portuguese debate. *Marine Policy*, **34: 1341–49**
8. Jay S., Flannery W., Vince J., Liu W.-H., Xue J.G., Matczak M., Zaucha J., Janssen H., van Tatenhove J., Toonen H., Morf A., Olsen E., Suárez de Vivero J.L., Rodríguez Mateos J.C., Calado H., Duff J., Dean H. (2013). Coastal and marine spatial planning. W: Chircop A., Coffen-Smout S., McConnell M. (red.). Ocean Yearbook. Leiden: Brill (Ocean Yearbook; 27): 171–212
9. J. Zaucha (2014a). The Key to governing the fragile Baltic Sea. Maritime Spatial Planning in the Baltic Sea Region and Way Forward. Riga: VASAB, 110 p.
10. J. Zaucha (2014b). Sea basin maritime spatial planning: A case study of the Baltic Sea region and Poland. *Marine Policy*, **50: 34–45**
11. A. Barbanti, P. Campostrini, F. Musco, A. Sarretta, E. Gissi (red.) (2015). Developing a Maritime Spatial Plan for the Adriatic-Ionian Region. Venice: CNR-ISMAR, 255 p.
12. W. Flannery, G. Ellis, M. Nursey-Bray, J.P. van Tatenhove, C. Kelly, S. Coffen-Smout, R. Fairgrieve, M. Knol, S. Jentoft, (2016). Exploring the winners and losers of marine environmental governance/Marine spatial planning: Cui bono?/"More than fishy business": epistemology, integration and conflict in marine spatial planning/Marine spatial planning: power and scaping/Surely not all planning is evil?/Marine spatial planning: a Canadian perspective/Maritime spatial planning – "ad utilitatem omnium"/Marine spatial planning: "it is better to be on the train than being hit by it"/Reflections from the perspective of recreational anglers... *Planning Theory and Practice*, **17: 121–151**
13. A. Schultz-Zehden, K. Gee, (2016). Towards a multi-level governance framework for MSP in the Baltic. *Bulletin of the Maritime Institute in Gdańsk*, **31(1): 34–44**
14. J. Zaucha, (2009). Planowanie przestrzenne obszarów morskich. Polskie uwarunkowania i plan pilotażowy. Gdańsk: Instytut Morski w Gdańsku, 149 p.
15. K. Gee, A. Kannen, B. Heinrichs (2011). BaltSeaPlan Vision 2030: Towards the sustainable planning of Baltic sea space. Hamburg: BaltSeaPlan, 46 s.
16. J. Zaucha (2018) Gospodarowanie przestrzenią morską. Warszawa: Wydawnictwo naukowe Sedno
17. V. Westerberg, J. Bredahl Jacobsen, R. Lifran, 2013. The case for offshore wind farms, artificial reefs and sustainable tourism in the French mediterranean. Tourism Management **34: 172-183**. doi.org/10.1016/j.tourman.2012.04.008.
18. M. Stuiver , K. Som, P. Koundouri , S. van den Burg, A. Gerritsen, T. Harkamp, N. Dalsgaard, F. Zagonari, R. Guanche, J-J Schouten, S. Hommes, A. Giannouli, T. Söderqvist, L. Rosen, R. Garção, J. Norrman, C. Röckmann, M. de Bel, B. Zanuttigh, O. Petersen, F. Møhlenberg (2016). The Governance of multi-use platforms at sea for energy production and aquaculture: challenges for policy makers in European seas. Sustainability **8(4):333**. doi: 10.3390/su804033
19. S. Davoudi, J. Zaucha, Brooks E. (2016). Evolutionary resilience and complex lagoon systems *Integrated Environmental Assessment AND Management*, **12(4): 711–8**

SHS Web of Conferences **58**, 01026 (2018) https://doi.org/10.1051/shsconf/20185801026
GLOBMAR 2018

ECA compliance and enforcement – legal regime for ships

Dorota Pyć[1,*]

[1] Maritime Law Department, Faculty of Law and Administration, University of Gdańsk

Abstract. This paper addresses the importance of the prevention air pollution from ships in terms of international legal compliance and enforcement. In addition to effectively enforcing international norms, standards and procedures, flag States have an obligation to prevent polluting ships from sailing. The International Convention for the Prevention of Pollution from Ships (MARPOL) has been described by the International Maritime Organization as "the most ambitious international treaty covering maritime pollution ever adopted". Annex VI of the MARPOL allows for the establishment of Emission Control Areas (ECAs). ECAs have been implemented to reduce the amount of air pollution from ships. The new ECA regulations entered into force on the 1st of January 2015. First and foremost, this paper examines some aspects of the implementation of the new ECA regulations on shipping industry and control air emissions in the Baltic Sea ECA. Moreover, it outlines that Port State Control and Flag State Control are commitment to monitor and enforce compliance through inspections and penalties.

1 Introduction

Emissions generated in the combustion of marine fuels entering the air pollute the environment and have a significant negative impact on marine ecosystems. These emissions are a type of pollution from ships. Most often, in the legal doctrine pollution from ships (vessel-source pollution) is divided into: accidental, operational pollution and operational pollution by emissions (air pollution) [1]. Air emissions from ships have an impact on the processes of eutrophication and acidification of the seas and consequently threaten human health and the environment as an integrated whole. In particular, this applies to semi-enclosed seas as the Baltic Sea, which is also the special area and particularly sensitive sea area [2].

Reduction of air emissions from ships and the contribution of these emissions to global air pollution and environmental quality is one of the main objectives of the International Maritime Organization (IMO). IMO decided to create a policy to prevent air pollution in 1991 [3]. At that time, work was commenced on the development of a new annex to the International Convention for the Prevention of Pollution from Ships (MARPOL). The Marine Environment Protection Committee IMO (MEPC IMO) at the conference convened in London on September 1997 (Air Pollution Conference) presented the text of Annex VI Provisions on the Prevention of Air Pollution from Ships to the MARPOL, which was

* dpyc@prawo.ug.edu.pl

adopted at the time together with the NOx Technical Code [4], which is mandatory for the application of Annex VI to MARPOL [5].

In 2008, IMO adopted amendments to Annex VI to the MARPOL (the 'revised MARPOL Annex VI') containing regulations for the prevention of air pollution from ships. The revised MARPOL Annex VI entered into force on 1 July 2010 and Article 1(6)(b) of Directive 2012/33/EU introduced one of the main changes of the revision, namely the mandatory use of marine fuels with a maximum 0,10% sulphur content in SOx-ECAs as of 1 January 2015, into the European Union law (Article 6(2) of Directive (EU) 2016/802). Similarly to the revised MARPOL Annex VI [6], Article 1(6)(c) of Directive 2012/33/EU also established the use of fuels with a maximum sulphur content of 0,50% in all other waters falling under the jurisdiction of a Member State ('European waters') outside the SOx-ECAs as from 1 January 2020 (Article 6(1) of Directive (EU) 2016/802) [7].

It should be emphasized that, the European Union is significantly interested in pollution prevention for many years, taking action to improve instruments for monitoring the air pollution from ships [8]. The main objective of the EU is to reduce the negative effects of the emission of air from ships and to ensure the proper functioning of the internal market in the field of shipping, EU ports and abatement technologies. In this respect, specific purposes include: (1) ensuring alignment of the EU law with the latest international standards and procedures for marine fuels, including advanced technical standards and technologies; (2) the definition of additional or alternative methods to limit the negative impact of emissions from shipping on human health and the environment; (3) strengthening the implementation of the monitoring and implementation system of EU law.

The purpose of this paper is assessment of the significance of selected legal instruments aimed at reducing harmful health effects and environmental damage caused by sulphur dioxide (SO2) emissions resulting from combustion of certain types of liquid fuels, as well as other air emissions from ships in the Baltic Sea area in the context of legal compliance. In addition, this article reviews recent developments at the Polish domestic law to promote compliance by ships with the ECA regime and enforcement at national level.

Although compliance and enforcement terms are sometimes used interchangeably, there are generally no major doubts as to what they mean in international law. Compliance is used in an international context while enforcement is used in a national one. In public international law, compliance is understood as the taking of effective action by state to meet an international obligation. Implementation of the international obligations is prerequisite for compliance. Legal compliance means that the state has made compliance of the domestic law with the international law (rules, standards and procedures). Proper implementation of legal standards should result in law enforcement in practice. Compliance refers to the situations in which a state is with regard to its obligations under MARPOL Annex VI. Enforcement refers to set of actions, which state (e.g. maritime administration) should take to ensure implementation of ECA standards. Enforcement of the MARPOL Annex VI can be exercise through ship inspections (flag state inspection or port state inspection) to ensure that ships fulfill technical norms and monitoring ship compliance and by punishing non-compliance ships [5].

There are some instruments to promote compliance *inter alia* inspections, reporting and monitoring compliance. Inspections and fuel sampling are actions to be taken at national level to enforce compliance with the ECA regime. In order to assess the significance of these selected legal instruments for the control of air emission standards, a normative analysis was used. In addition to dogmatic research, empirical research has become particularly important in order to justify legal decisions by European Union institutions.

Reporting on compliance with air emission standards from ships and inspections of compliance with sulfur content standards in marine fuels are international instruments introduced into domestic law by the State-Parties to Annex VI of MARPOL and required by the European Union law. Both of these instruments are used to monitor compliance with international standards. Reporting as well as inspections allows assessing the degree of compliance of obligated entities to comply with certain rules, standards and procedures.

2 Selected legal instruments for the control of air pollution from ships

The definition of the Emission Control Area (ECA) is contained in Regulation 2 of MARPOL Annex VI. The ECA means an area where the adoption of special mandatory measures for emissions from ships is required to prevent, reduce and control air pollution from NOx or SOx and particulate matter or all three types of emissions and their attendant adverse impacts on human health and the environment. Emission Control Areas shall include those listed in, or designated under, regulations 13 and 14 of Annex VI.

Effective preparation is essential to safeguard a global level-playing-field for operators and to reduce air pollution from ships across the world. For ships operating outside designated the Emission Control Areas, IMO has set a limit for sulphur in fuel oil used on board ships of 0.50% m/m (mass by mass) from 1 January 2020. This will significantly reduce the amount of sulphur oxide emitted from ships and should have major health and environmental benefits for the world, particularly for populations living close to ports and coasts. Ships must be issued with an International Air Pollution Prevention Certificate (IAPP) by their Flag State. This certificate includes a section stating that the ship uses fuel oil with a sulphur content that does not exceed the applicable limit value as documented by bunker delivery notes or uses an approved equivalent arrangement. Port and coastal States can use Port State Control to verify that the ship is compliant.

2.2 Reporting on compliance with air emission standards from ships

Reporting on compliance with air emission standards from ships and preparing for the enforcement of the global 0.50% sulphur cap are to be an active support increasing the sustainability of maritime transport in the European Union.

The Council Directive 1999/32/EC relating to a reduction in the sulphur content of certain liquid fuels was amended several times. Last time it has been changed by Directive 2012/33/EU of 21 November 2012 as regards the sulphur content of marine fuels [7]. In order to ensure consistency of the Directive, it was codified in Directive (EU) 2016/802 relating to a reduction in the sulphur content of certain liquid fuels [7].

Article 14(1) of Directive (EU) 2016/802 requires EU Member States to submit each year by 30 June a report to the Commission outlining compliance with the sulphur standards of the liquid fuels falling within the scope of the Directive on the basis of checks carried out in the preceding year. Article 14(1) furthermore requires the Commission to evaluate the need for further strengthening of the relevant provisions of the Directive and to make any legislative proposals to that effect. On the basis of these national reports and on the notifications regarding the non-availability of marine fuels, the Commission shall each year report on the implementation of the Directive.

The EU Member States are obliged to take action against fuel suppliers that have been found to deliver marine fuel that does not comply with the specification on the bunker delivery note and to maintain a publicly available register of suppliers of marine fuels to increase transparency. Member States are also required to control marine fuel suppliers by taking samples of their products upon delivery to a ship in case of recurrent reports of alleged quality issues ('letters of protest') (Article 4 of Implementing Decision (EU) 2015/253) [9].

2.3 Mandatory frequency of inspections and fuel sampling

Mandatory frequency of inspections and fuel sampling ensure an equal enforcement burden among Member States and a level-playing-field for the European Union ports and operators. Physical sampling of marine fuels being used by ships for the purpose of verifying their sulphur content should be carried out either by analyzing the sealed 'bunker sample' or the 'representative sample' accompanying the bunker delivery note and available on-board, or by obtaining and analyzing a 'fuel spot sample' drawn from the ship's fuel service system [9].

To ensure an equal enforcement burden among Member States and a level-playing-field for the EU ports and operators, Directive 2012/33/EU empowered the Commission to define a binding frequency of inspections and sampling of marine fuels (Article 13(4) of Directive (EU) 2016/802), which was laid down in the Implementing Act [10].

The EU Member States have to complement the documentation check by sampling and analyzing the sulphur content of the fuels of 20 to 40% of the inspected ships depending on whether the Member State is (partly) bordering a SOx-ECA or not [9].

2.4. Polish Legal Perspective

The provisions of the Polish Act on prevention pollution from ships from 1995 [11] ensure compliance with Annex VI of the MARPOL and implementation of the requirements of the Directive 2016/802 [7].

According to the Polish Act on prevention pollution from ships (Article 13c) [11], the use of marine fuel which does not meet the requirements or additional qualitative criteria are prohibited on ships of the Polish nationality and ships of foreign nationality in Polish maritime areas. Emission abatement methods may be used on ships of the Polish nationality or ships of foreign nationality in Polish maritime areas as an alternative to using marine fuels that meet the requirements and additional qualitative criteria. Emission abatement method means the use of any fitting, material, device to be fitted in a ship or other procedure, alternative fuel, or method, ensuring the reduction of sulphur dioxide emission that is verifiable, quantifiable and enforceable. The minimum criteria to be met by emission abatement methods are laid down in Annex II to the Council Directive 1999/32/EC (as amended) [7]. On ships which use emission abatement methods, sulphur dioxide emission abatement should be achieved on an ongoing basis to at least the level equivalent to the value of sulphur dioxide emission that would be achieved using marine fuel that meet the requirements and additional qualitative criteria. Equivalent emission values for sulphur dioxide are laid down in Annex I to the Directive 1999/32/EC [7]. For ships at berth in ports, the use of onshore power supply systems, if available in a given port, can be allowed as an alternative solution to emission abatement methods.

The Act specifies the competences of the director of the maritime office in the field of keeping records of fuel suppliers. The director is required to keep the register of suppliers of marine fuel, operating in Polish maritime areas (Article 13e). Before the first delivery, the supplier is obliged to notify the director of the maritime office competent for the place of delivery about the fuel supply activity in Polish maritime areas. The notification should include the name, seat, address for correspondence and telephone number of the supplier, a copy of the fuel trading license, and should specify the types of supplied fuels. The supplier must inform the director of the maritime office about the change of data within 30 days from the date of the change.

By 31 March each year, the supplier is obliged to submit the report from activity in the previous year, including at least the information about the quantity and type of fuel sold in individual ports, taking into account the sulphur content, to the director of the maritime office competent for the place of delivery. If the supplier ends fuel supply activity in Polish maritime areas, the supplier must notify the director of the maritime office competent for the place of delivery in writing about it, within 30 days from the completion of the activity. The director of the maritime office provides the list of suppliers available to the public in the Public Information Bulletin at the website of maritime office. The list includes names of suppliers, their seats, and addresses for correspondence, telephone numbers and specification of the types of fuels supplied.

According to the Article 13f of Polish Act on prevention pollution from ships the supplier is obliged to deliver fuel that meets the requirements for sulphur content of marine fuel, including the method of its labelling as well as additional qualitative criteria for marine fuels supplied to ships, to protect the environment against pollution from ships, laid down in the provisions specified in Polish implementing regulations.

Detailed obligations of the supplier and the master of the ship regarding fuel delivery to the ship are specified in the MARPOL Convention. If the master of the ship of the Polish nationality outside Polish maritime areas cannot purchase marine fuel with sulphur content allowed in a given maritime area, the master shall notify about this the director of the maritime office competent for the home port of the ship and the competent authorities of the port of destination. If the master of the ship in Polish maritime areas cannot to purchase marine fuel with sulphur content allowed in those maritime areas, the master shall notify about this the director of the maritime office competent for the area where the ship is, the administration of the country of the ship's nationality and the competent authorities of the port of destination.

The director of the Polish maritime office, as a governmental maritime authority, is competent to perform inspections of sulphur content of marine fuel on ships (Article 13g). The inspection regarding the sulfur content of marine fuel is carried out in accordance with art. 3-6 of Decision 2015/253 (Implementing Act) [10] and includes: checking each time fuel delivery documents containing information of the fuel supplier with sulfur content in marine fuel, including checking whether sealed marine fuel samples accompanied by these documents are signed by the representative of the vessel accepting the fuel, and checking the machine log and oil record book regarding entries documenting operations change of fuel; for at least 40% of inspected vessels downloading marine fuel samples: (a) from the on-board fuel circulation system, (b) from the vessel's fuel tank, if technically possible, (c) using sealed marine fuel samples accompanying fuel delivery documents, transferred to the ship by the fuel supplier in accordance with regulation 18 of Annex VI to the MARPOL Convention or (d) during its delivery to the ship - and testing the sulfur content in it.

In the case of taking marine fuel samples on board in accordance with the provisions of Decision 2015/253 (Implementing Act) [10], the ship's representative shall keep one sample for a period of not less than 12 months from the date of its collection. As a result of an inspection, that a ship does not comply with the requirements for the sulfur content of marine fuel, the director of the maritime office shall request the master of that vessel: (1) if possible, take measures to adapt marine fuel to the implementing applicable regulations; (2) the presentation of: (a) a list of measures taken to comply with the requirements concerning the sulfur content of marine fuel in a given sea area, (b) proof that an attempt has been made to purchase marine fuel with a sulfur content permissible in a given sea area in accordance with the itinerary and, if the fuel was not available at the place where it was planned to be purchased, an attempt was made to locate alternative sources of such fuel, and that, despite such attempts, it was not possible to purchase such fuel.

Proof of attempted acquisition of marine fuel with the content of acceptable sulfur in a given sea area is considered a written statement from the fuel supplier about the lack of availability of such fuel in a given port. The director of the maritime office shall take into account the ship's voyage plan, including the planned date of leaving the port.

Information about the results of the ship inspection carried out in the field sulfur content in marine fuel include: (1) the date of the inspection carried out; (2) vessel identification data: identification number given by the International Maritime Organization, name, affiliation, reconnaissance signal, gross tonnage (GT), type of propulsion, emission reduction methods, if different from those based on the use of low-sulfur fuels; (3) the names of the fuel tanks owned by the ship and their determination capacity; (4) specification of the type of combustion device and its power specified in kW; (5) data on marine fuel delivered on a ship obtained from fuel delivery documents or on on the basis of the journal entries: date of fuel delivery, fuel type, quantity in m^3, sulfur content in %, name of the fuel supplier and name of the fuel delivery port; (6) data from the International Air Pollution Prevention Certificate (IAPP): name of the authority that issued the certificate, date of issue, date the name of the body that carried out the periodic inspection confirming the validity of the certificate, and the date and place of the last periodic inspection; (7) data on the marine fuel sample taken: fuel type, number in ml, sulfur content in%, sampling source, sampling method, sample identification number, name of the port in which the sample was taken, and date of sample analysis and laboratory name analysis of the sample; (8) non-compliance with the requirements regarding the content of sulfur in the fuel has been found shipping, including type of deficiencies, and actions taken to achieve compliance; (9) other data obtained during the inspection.

The director of the maritime office draws up a report from inspections each year. The report contains results of analysis of sulphur content and is submitted to the President of the Office of Competition and Consumer Protection by 30 April of the following year. The report should include in particular the information about: total number of analysed samples by type of fuel; fuel quantity used for the analysis each time and the determined average sulphur content; number of inspections on ships; average sulphur content of marine fuels used in the area of competence of the director of the maritime office, calculated based on performed analyses and information provided by fuel suppliers. The minister responsible for maritime economy should ensure that the information referred to in Article 7 of Decision 2015/253 (Implementing Act) [10] is forwarded to the European Commission by 30 June each year.

The shipowner whose ship polluted the marine environment when in Polish maritime areas in breach of the Act in connection with ship operation is liable to a fine being the equivalent of 1,000,000 reference units called Special Drawing Rights (SDR) defined by the International Monetary Fund. This penalty can be also imposed on the shipowners of Polish nationality ships which polluted the marine environment in breach of the Polish Act on prevention pollution from ships outside Polish maritime areas, in connection with ship operation (Article 36).

If a shipowner fails to exercise the obligations stipulated by the law or violates prohibitions established by the law by using fuel with impermissible sulphur content on a ship shall be liable to a fine of 600,000 SDRs. If the shipowner applies emission reduction methods on board, without the approval required for them, shipowner shall be subject to a penalty of up to SDR 600,000. If the shipowner does not pass the results of tests of emission abatement methods in accordance with the prescribed procedure, shipowner shall be subject to a penalty of up to 100,000 SDRs. (Article 36aa).

The supplier who fails to fulfil the obligation defined in Article 13e(2)–(5) and Article 13f when supplying fuel shall be liable to a fine of 50,000 SDRs (Article 36c).

In accordance with Article 37, if a master or any other crew member *inter alia*: (1) fails to exercise the obligations stipulated by the law and fails to keep the records of each instance of switching to a different fuel in line with the requirements of the MARPOL Convention; (2) does not have nautical delivery documents on board in the ship's fuel tanks or documents do not contain all information required in accordance with the MARPOL Convention; (3) does not carry on-board marine fuel samples in the ship's fuel tanks or marine fuel samples are not sealed or unsigned (non-compliance with MARPOL); (4) does not have the required documents (Article 13g); (5) does not have a valid permit for testing emission reduction methods or does not meet the legal requirements; (6) uses emission reduction methods on a ship in a manner that does not ensure reduction of emissions required by law - shall be liable to a fine whose amount does not exceed 20 times the average salary in the national economy for the preceding year, published by the President of the Central Statistical Office.

3 Conclusions

The International Maritime Organization plays a leading role in promoting legal instruments of a preventive nature. Legal measures are essential to prevent pollution from ships. In 1997, the Annex VI "Regulations for the prevention of air pollution from ships" was added to MARPOL Convention to minimize emissions form ships [12]. Since then, the system of legal instruments aimed at reducing harmful emissions from ships has been considerably strengthened. The IMO and the European Union strive to achieve international legal compliance, i.e. coherence of the norms of national law with international law. In 2012 the European Commission adopted more detailed provisions on the content and format of the annual reports of Member States. The Directive 2012/33/EU has introduced many changes , e.g. the mandatory use of low sulphur marine fuels in the Baltic Sea designated as Sulphur Oxides Emissions Control Areas, took effect only on 1 January 2015.

The EU has an important role to play in compliance reinforcement through the mandatory frequency of inspections and fuel sampling, as well as monitoring compliance and the development of reporting mechanisms of the EU Members States' conduct. Reporting on compliance with air emission standards from ships aims at exchanging information between the Commission and the EU Member States. Reporting is closely related to monitoring compliance. Frequency is determined by the annual number of individual ships calling in port of the EU Member State. According to Implementing Act [10], all Member States may carry out inspections of the ship's documentation on at least 10% of the individual ships calling in their ports. Member States have to complement the documentation check by sampling and analysing the sulphur content of the fuels of 20 to 40% of the inspected ships depending on whether the Member State is bordering a SOx-ECA or not. The frequency of sampling in the SOx-ECA areas is higher than that in other European waters.

Polish law is in compliance with international law contained in Annex VI of the MARPOL Convention as well as the European Union law on the prevention of air pollution from ships. In the last few years, the Polish Act on Prevention Pollution from Ships has been amended several times. The Act was last amended in 2017. International legal compliance is undervalued, however at national level particular attention must be paid to proper implementation and law enforcement.

References

1. E.J. Molenaar, Coastal State Jurisdiction over Vessel-Source Pollution, Kluwer Law International, The Hague/Boston/London, p. 19 (1998)
2. A. Lähteenmäki-Uutela, S. Repka, T. Haukioja, T. Pohjola, How to recognize and measure the economic impacts of environmental regulation: The Sulphur Emission Control Area case, Journal of Cleaner Production **154**, 553-565 (2017)

3. IMO Res. A.719(17)
4. MEPC.177(58)
5. M. Fitzmaurice, The International Convention for the Prevention of Pollution from Ships (MARPOL), [in:] IMLI Manual on International Maritime Law, vol. III Marine Environmental Law and Maritime Security Law, gen ed. D.J.Attard, ed. M. Fitzmaurice, N. A Martínez Gutiérrez, R. Hamza, Oxford University Press, p.55, pp. 61-63 (2016)
6. http://www.imo.org/en/OurWork/Environment/PollutionPrevention/AirPollution/Pages/Index-of-MEPC-Resolutions-and-Guidelines-related-to-MARPOL-Annex-VI.aspx
7. Council Directive 1999/32/EC relating to a reduction in the sulphur content of certain liquid fuels has been substantially amended several times, lastly by Directive 2012/33/EU of 21 November 2012 as regards the sulphur content of marine fuels, OJ L 327 of 27.11.2012, p. 1, which entered into force on 17 December 2012 and had to be transposed by 18 June 2014. To preserve clarity and rationality following the different substantial amendments of Council Directive 1999/32/EC, it was codified as Directive (EU) 2016/802 of 11 May 2016 relating to a reduction in the sulphur content of certain liquid fuels, OJ L 132 of 21.5.2016, p. 58.
8. J. Harrison, Atmospheric Pollution of the Marine Environment [in:] IMLI Manual on International Maritime Law, vol. III Marine Environmental Law and Maritime Security Law, gen ed. D.J.Attard, ed. M. Fitzmaurice, N. A Martínez Gutiérrez, R. Hamza, Oxford University Press, pp. 190-191 (2016)
9. Report from the Commission to the European Parliament and the Council on implementation and compliance with the sulphur standards for marine fuels set out in Directive (EU) 2016/802 relating to a reduction in the sulphur content of certain liquid fuels, European Commission Brussels, 16.04.2018 COM(2018)188 final.
10. Implementing Act: Commission Implementing Decision (EU) 2015/253 of 16 February 2015 laying down the rules concerning the sampling and reporting under Council Directive 1999/32/EC as regards the sulphur content of marine fuels, OJ L 41 of 17.2.2015.
11. Polish Act on Prevention Pollution from Ships, 16.03.1995, consolidated text: Journal of Laws 2017, item 2000
12. D. Pyć, Uwagi *de lege ferenda* o statusie Morza Bałtyckiego jako obszaru kontroli emisji tlenków azotu ze statków morskich, Prawo Morskie **t. XXVIII**, 235-247 (2012) www.journals.pan.pl/delibra/journal/107532?language=en

External effects in the maritime sector leading to new economic perspectives.

Magdalena Rozmiarek, and *Ad* Faasse.

Abstract. Our present economic system is facing some real challenges. It is a so-called linear system like "take-make-use-lose". Sustainable economic development requires valuing (environmental) resources correctly and factoring them into production and consumption decisions. Having a focus on the economic production process, external effects do not seem to fit in the linear system. So-called external effects are not included in economic models, business decisions etc. External effects like the Sustainable Development goals and legal actions in court are of growing importance for economic developments ranging from port development to design of ships and logistics. The distinction Economic Capital, Social Capital and Natural Capital is helpful in creating consciousness about the concept of circular economy. Nature provides tools, systems and knowledge from 3.8 billion years of evolution that will enable us to enter and develop new circular economic developments, to develop real "green ports"; to combine divergent functions in a smart way.

Keywords externalities, legal actions, biomimicry, sustainable economic growth, circular economy, doughnut economics, green port development.

Introduction

Koos Biesmeijer [1] distinguished three types of Capital: Economic Capital (labor, capital, raw materials); Social Capital (labor and e.g. regulations & law) and Natural Capital (all living organisms, ecosystems, as well as wind, sun, gold etc.) This division means to show the interrelatedness between different, though not separate elements of our present economic concepts and thinking.

Each distinguished capital forms a system with its own dimension, doctrines etc. The real challenge and power is or will be in drawing hyphens between the different systems and create a more circular economic system by doing so.

Economic Capital

The science "Economy" originates from the Greek words "oikos" and "nomos" which in the basics means the home economics. In other words, how to manage all the things that relate to a living; running a family c.a. From origin a social science, economics evolved as a metric concept without a value system and with the fiction of infinite growth. Adam Smith's "invisible hand of the market" seemed only focused on the economic production process and did not consider "external effects" like politics, law, nature, or in his case the (unpaid) care by his mother.

Our economic system is a linear system best defined as a "through-put-system" [2] or "Earth is inexhaustible – so take all you want. There will be no shortage of Earth resources if markets are permitted to do their job."[3 4]

The monetary (or mathematical) value of our economy is – on macro level – expressed in the concept of GDP, as the total value of everything produced by all people and companies in the world. This excludes explicitly the value nature provides. GPD is concentrated around (certain aspects of) human behavior. GDP is made of a set of elements as there are: personal consumption, expenditures plus business investment, plus government spending

[1] Leiden University. Oration teaching assignment Functional Biodiversity "Natural Capital". (March 9th 2018)
[2] Utrecht University. Farewellspeech Herman Wijffels. (October 3rd 2016)
[3] K. Raworth "Doughnut Economics, Seven ways to think like a 21st Century Economist" p 70. (2017). ISBN 978-1-847-94138-1
[4] Though: Frankfurter Allgemeine Zeitung: "BMW tries to secure the supply of cobalt and lithium for the upcoming 10 years" (14.02.2018)

plus export minus imports. GDP has been criticized for being complex, not being optimal and for not including "costs" of externalities in the economic system.

In his famous speech for the University of Kansas Robert Kennedy said that ..."GDP counts napalm and nuclear warheads but does not allow for the health of our children…etc.[5] D.J. Thampapillai in his text on Environmental Economics: "Clearly, the natural environment is a critical component of the economic system, and without the natural environment, the economic system would not be able to function. Hence, we need to treat the natural environment in the same way as we treat labor and capital; that is, as an asset and a resource" [[6]]

"Externalities occur where the use of a resource by one party imposes costs or benefits on others, but these impacts are not factored into economic decisions. As a result, economic agents – individuals, firms or governments – do ***not face the full costs/benefits of their actions on society***. Externalities can be either positive or negative, depending on whether actions produce unpriced beneficial or detrimental effects – positive externalities will tend to result in under-provision of the good or service, whereas negative externalities will lead to over-provision. For example, in the absence of regulation, sewage companies discharging effluent into waterways (cf. the Baltic Sea. AF) will not face the full social cost of their activities – in terms of recreational and other benefits foregone and/or the cost to society to remediate the damage – leading to degradation of the environment beyond the economically efficient level. Conversely, the pollination of plants by bees kept for their honey is a positive externality, which cannot necessarily be captured by beekeepers, leading to under-provision of this service compared to the economically efficient level." [[7]]

In the example of the sewage company discharging (polluted) effluent into waterways, Defra (Department for Environment, Food and Rural Affairs/UK) speaks about social costs. In our opinion these costs are of economic value as well, the private property affected by polluted waterways will be less valuable, health can be affected by polluted water, which causes medical costs etc. This example shows that social costs are interlinked with economic costs like lost value in real estate, health costs etc. Those costs are often divided over several parties, excluded by the sewage company, nevertheless they are simply there and not to be ignored.

Sustainable economic growth requires valuing (environmental) resources correctly and factoring them into production and consumption decisions. Underestimating or not adequately valuing these (costs or) benefits leads to overuse.[[8]]

If we make this definition of externalities applicable to business cases, business plans or other economic decisions on a large scale, we consider that (nearly) all business cases we know about do not integrate 'externalities' like clean air, clean water, contributing to biodiversity etc. As stated before by Kate Raworth it is about "take-make-use-lose.[[9]]

Social Capital

Social Capital is a system, a network of regulations, laws, work and social relations all around. We share our view on economics with the Rotterdam based Goldschmeding foundation[[10]] which means economics is not about money, it is about relations, relations between human beings, and the interrelatedness with nature in the broadest way.

Part of this (socio-political) system is formed by the Sustainable Development Goals (SDG) from the United Nations and the Paris Agreements in 2015. The SDG's together form a framework of – at first sight – separate topics. These topics are interlinked and interact with each other, they are to be distinguished and not to be separated. The SDG's express the desire of the world community for clean air, clean water, avoidance of pollution; physical and mental health, a decent work environment, good relations with the surrounding areas, biodiversity, clean seas etc.

It is the task of governments/companies to show their responsibility by integrating those targets in policies and business plans/decisions. The SDG's – though likely being considered as externality - are of growing importance for all economic sectors. Especially for the maritime economy the SDG's 8, decent work & economic growth; SDG 9, Industry, innovation & infrastructure; SDG 11, sustainable cities & communities; SDG 13, climate action; SDG 14, life below water; SDG 15, life on land; are to be considered.

[[5]] Robert F. Kennedy Speeches, University of Kansas (John F. Kennedy Presidential Library and Museum) (March 18th, 1968)

[[6]] S. Beder "Economy and Environment: Competitors or Partners". University Wollongong http://uow.edu.au/~sharonb

[7] Defra, Everett, Ishwaran, Ansaloni & Rubin "Economic growth & Environment PB 13390" (March 2010) p.31

[[8]] Defra, Everett, Ishwaran, Ansaloni & Rubin Economic growth & environment. PB 13390 (March 2010) p. 30

[[9]] K. Raworth "Doughnut Economics, Seven ways to think like a 21st Century Economist". Random House Business Books (2017). ISBN 978-1-847-94138-1

[[10]] Goldschmeding Foundation Rotterdam. www.goldschmedingfoundation.org.

They might be looked at as blocking economic growth, or prohibiting economic activities, in our opinion they are a framework and possibilities for a new and more circular economy, adding more quality to life and the environment. This requires a change in mindset, the acceptance of the necessity to change our behavior, the willingness to develop transition pathways, and creativity. The power of a different public mindset is shown explicitly in the Netherlands in the public noise about the 50 % pay-rise of the CEO of the ING-bank. The Supervisory Board decided to allow the CEO this pay-rise because they concluded that in the international context he was considered to be under-payed. The public and political noise, as well as accountholders leaving the bank caused the rollback of this decision for reasons of reputation damage control and to restore confidence in the bank. The conclusion is justified that reputation and reliability is difficult to express in monetary value but as a social asset/externality of importance for sound economic development. [11] The public noise inspired politicians to work on legal (administrative) instruments to avoid future pay rises like this in the (system)bank sector.

How SDG's are related to specific areas can be seen in investigating problems in and around the Baltic sea. The Baltic Sea faces a lot of challenges and problems. An integrated and coordinated cross-sectoral and cross-governmental approach as well as a shared vision is needed to create solutions for the three primary threats in the Baltic Sea as there are; eutrophication, contamination and overfishing. Two scenarios were developed, "Clear Waters" and "Shipwrecked". Choosing for and working on the Clear Water scenario could amount to 550.000 jobs (in tourism, agriculture and fishing industry) and €32 billion in value added by the year 2030. This demonstrates that the health of the Baltic Sea is not only an environmental concern but an important economic and social one as well.[12] "The Baltic Sea region has the potential to become a global leader in technology-driven environmental solutions. Reframing the challenges as "economic" will create positive benefits of a "Clear Water" scenario."

A special spotlight of attention deserves functioning of the civil and international law system. Mission statements and annual reports of – most times – larger and international organizations refer to or state that legal requirements will be met. In general, this refers to legal obligations prohibited by criminal law and/or imposed by administrative law. Both legal systems are dominated by government and therefore under political influence. [13] Law-enforcement is the monopoly of the government and depends on the priorities set by the governing parties. It requires a government organization with its own attributed powers, priorities, budgets, expertise (or the lack of it) etc.

An interesting and for the moment unique criminal case is the following: The Dutch shipowner "Seatrade" is summoned for the criminal court in Rotterdam[14] for demolishing 4 of their ships on the beaches in the far East. Demolition on those beaches causes large environmental damage, is bad for workers health and safety, and waste is dumped without proper treatment. The public prosecutor accuses the ship owner of violation of the European waste directive, by dumping hazardous waste (components like bunker oil, asbestos, PCB's) outside Europe. " As court proceedings are going on there is no verdict yet. The prosecution expects more similar criminal cases to follow.

The system of (international) private/civil law does not require such governmental meddling. It depends on the plaintiff's discretion. Civil law suits can be called to help in situations where the balance between economic activity and social capital and natural capital is at stake. Or, in other words, where economic activity affects the interests of the other party. Even when it is about nature itself, civil legal actions are possible. Most times this is not included in business decisions.

To protect their own (or nature's) interest, civilians, action groups and even local governments use civil law to force the other party to comply with the law and/or to forbid causing damages or to order (punitive) damages related to private and even public interests. In some parts of the world even nature has legal standing in court.[15]

A general legal principle is that behavior, though not always specified in detail and being detrimental to "the other party" in the circumstances of the event, may constitute liability for damages. And, the execution of rights of one party could be prohibited or limited by the rights of the other party.

Several big cities in the USA filed cases in courts against "big oil" for the damages climate change constitutes to these cities. The damage claimed consists of all the measures that must be taken, to prevent cities and citizens for the changing climate conditions, heavy storms etc. Impacts of weather extremes can lead to large-scale disasters

[11] Compare: The Facebook and Cambridge Analytica scandal caused a major reduction (10 billion US $) in Zuckenburgs stock value.
[12] Boston Consulting Group "The economic case for revitalizing the Baltic Sea". March 2014.
[13] F.e. the Trump administration opens ways for monopolists by cutting in the prohibitions and granting permits to establish pipelines in protected areas, by allowing drilling in vulnerable places like the arctic.
[14] Criminal Court Rotterdam, (15.03.2018) State versus Seatrade.
[15] A brook with Legal Rights: The rights of Nature in Court http:/scholarship.law.georgetown.edu/facpub/1906 (2016); New Zealand's Whanganui River is a person under domestic law and Ganges River (India) was granted human rights. The Conversation. (June 19th 2017).

if organizations across sectors and related communities incur physical damage, losses and/or disruption of their routine functioning and are unable to cope with the event effectively [16]

A Peruvian farmer [17] appealed his case in the Regional Court of Appeal in Hamm, Germany, suing for 0,5 % of his (climate) damages caused by the German based energy company RWE. The amount of damages sought is based on the fact that RWE is responsible for around 0,5 % of global historical industrial emissions. It is "……not about "guilt" it is about taking responsibility for what the company has done and what they continue to do…"

The US District Court of California [18] ruled:

"..Plaintiffs state law nuisance claims are premised on the theory that – despite long-knowing that their products posed severe risks on the global climate – defendants (Royal Shell; Chevron; Exxon; BP ConocoPhillips) produced fossil fuels while simultaneously engaging in large scale advertising and public relations campaigns to discredit scientific research on global warming, to downplay the risks of global warming, and to portray fossil fuels as environmentally responsible and essential to human well-being. The plaintiffs were allowed to advance the litigation. This decision of the court does not mean (yet) that the claims will be granted, but that the plaintiffs have a justified case in court.

These judicial findings are relevant for a worldwide as well as local or regional legal approach of polluters. Ignoring these upcoming "signs of noise" in economic decisions or business plans will be economically disastrous. On the other hand, they are strong incentives for changing economic modelling and behavior.

Natural Capital

Natural Capital is essential for economic development as it allows companies to make products or deliver services. "Nature" is not to be exchanged for money and/or subsidiaries. Natural capital is the stock of all resources, we could use for improving our existence. Natural Capital is to be cherished, and preferably be strengthened." [19]

In economic terms: we should use the interest on the capital, instead of consume/exhaust our capital. National law and international treaties shape the framework for the protection of the "vulnerable nature" as indicated in the prior text.

By introducing the concept of Natural Capital Biesmeijer draws attention to the wealth of nature, and the way nature serves our economy. This is not done by extracting things from nature but by learning from nature and adding something to the eco-system, like what happens in nature, e.g. ".. a whale contributes to life in the ocean by diving deep, chasing krill, supplying poop as food for other organisms and working on the oxygen system of the world by reducing carbon dioxide. But…. Whales are killed for so called scientific research, with no specific results, other than a piece of whale flesh in a certain cuisine. [20]

Einstein once said: "Look deep into nature, and then you will understand everything better". In line with this statement the upcoming science Biomimicry is an approach to innovation that seeks solutions to human challenges by emulating nature's time-tested patterns and strategies[21]. "Innovation inspired by nature", Biomimicry [22] is a rather new shoot to the trunk, and important for economic development.

In her book Dr. Benyus shows a deeper look into nature and displays the best natural solutions for our today problems. From the waterproof fibers of a spider (five times stronger than steel) to photo-syntheses in plants, to organizing patterns in nature. For example, the way a sharkskin is developed is of importance for new ship design. The specific texture of the skin does not allow for fouling or parasites, it allows a shark to develop speed with less energy. So why not develop ship's skin in this way? [23]

Wageningen University conducted a research about the value of "green" in urban developments in Amsterdam.[24] One of the conclusions was that, depending on the location and the surrounding area the value of ***green*** might even have a higher monetary value than the existing real estate. "If housing is realized at the cost of "green" the question arises whether or not social prosperity will increase" The recommendation for the City of

---

[16] Martina Linnenluecke e.a. "Extreme weather events and the critical importance of anticipatory adaptation and organizational resilience in responding to impacts." Business Strategy and the Environments (2011). Doi 10.1002/bse.708

[17] Reuters (Nov, 10. 2017)

[18] US District Court of California. 27.02.2018 judge William Alsup. No. C17-06012WHA

[19] Koos Biesmeijer. Oration "Natural Capital" Leiden University (March 9th 2018)

[20] Produced by Chris Agnos. https://youtu.be/M18HxXve3CM. How whales change the climate

[21] Joint Global Conference Biomimicry and Bio Inspired Innovation November (11th – 13th 2016) Utrecht University

[22]Janine M. Benyus Biomimicry, Innovation inspired by Nature. (2002) HarperCollings Books ISBN 978-0-06-053322-9

[23] Compare: Bloomberg Technology "Boeing copies Flying Gees to save fuel" (08.08.2017)

[24] Wageningen University & Research"Growth versus Green" in dutch "groei versus groen". (March 2018) Report 344. ISBN 978-94-6343-758-5 DOI https://doi.org/10.18174/443008.

Amsterdam was to act sensibly by taking into account "green" in decision processes about where and how to realize housing". 'Green' creates values, in monetary terms as well as in leisure. Living environment, sustainability, SDGs etc. are important aspects for developments.

The situations at stake were about urban developments. This conference is about the maritime sector, part of which is about port development. Having regard to the above mentioned "movements" it is at the least interesting to consider these findings in port developments, and the relation between ports and the hinterland.

Circular economy

The concept of circular economy is much discussed. It seems that some 114 definitions exist, ranging from vague, to theoretical dreams etc.[25] Most times the concept of circular economy is used as a buzz-word and in politically correct statements, assuming that when we e.g. collect our garbage and recycle the fractions we do a good job in circularity. The best way to learn what circularity means is to investigate nature, as circularity is to be found there.[26]

Circularity starts with design, how to design in such a way that no damage is done to the environment, there is optimal use of the design, and after use, no harm to the environment. Circular economy is considered a concept for the future, with new business models, new financial models and may be a solution for the upcoming lack of raw materials.

For being able to really think and work "circular" a new mindset is needed, an innovative approach to doing business and to design our economy. The scientific world needs to (re-) organize itself in a transdisciplinary way to be able to contribute to cohesively and integrated transition. Therefore, we need people with a deep knowledge in a discipline and the capacity to integrate this knowledge in outcomes.[27] Learning from nature is one of the solutions to be recognized and explored though it can – at the moment - be regarded as "terra incognita"

Dayna Baumeister [28] formulated life's principles. All of them fit in the concept op circular economy. Some of them are now worth drawing attention to as they contain interesting concepts: "evolve to survive" > ensure enduring performance; "adapt to changing conditions" > appropriately respond to dynamic contexts; "be locally attuned and responsive" > fit into and integrate with the surrounding environment; "integrate development with growth" > invest optimally in strategies that promote both development and growth. These concepts are applicable to develop "circular economy" and they fit into the point of view prof. Wijffels previously suggested, namely reorganizing the scientific world into a transdisciplinary way and integrated transition.

Closing remarks

The underlying message is a call for change, awareness of the (deplorable) state of our planet and a call for cooperation and integration. The World Economic Forum 2018 put a series of important topics on the agenda like shutting all coal fired power stations by 2021 (Macron); How to do business with doughnuts (Raworth); Global cooperation, not walls (Merkel); How companies must adapt ; How investors can help save the planet, energy transition etc.

Broadening the context of the economic science will offer interesting opportunities for (circular) economic development. The examples mentioned in this text are inspiring and open ways to new economic opportunities. Being aware and integrate planning and working with Economic, Social and Natural Capital will lead to an economy that will be more distributive and generative.

Externalities like the SDG's and the upcoming series of legal actions around climate change could be considered as incentives for transforming the present economic concepts into a more circular approach. The city council of Paris recently adopted a motion to study the feasibility of suing oil companies for climate damages.[29] So, it seems to be time to include "externalities" into economic models and decision making. As for the Baltic Sea, the "Clear Water" scenario is interesting, feasible, worth developing.

There is a lot of interesting and inspiring work to be done with new economic perspectives.

No time to waste.

---

[25] Kirchherr, Reike, Hekkert "Conceptualizing the circular economy: An analysis of 114 definitions" https://doi.org/10.1016/j.resconrect.2017.09.005.

[26] Albert Einstein quotes: "Look deep into nature, and then you will understand everything better."

[27] Utrecht University School of Economics. Farewell speech Herman Wijffels, prof. Sustainability and Social Change (October 3rd 2016)

[28] Biomimicry Resource Handbook. A seed bank of best practices. Biomimicry 3.8 Missoula MT USA.

[29] Conseil de fevrier 2018 (23 janvier2018) "Voeu pour un Paris decarbone. "a l'instar de New York, etudie la faisabilite d'assigner en justice les petroliers. (Like the example given by the city of New York, to study the feasibility of suing oil companies for climate damages.)

References

1. S. Durston, T. Baggerman, Changing understanding of reality, Amsterdam University Press (2017).
2. K. Raworth Doughnut Economics, seven ways to think like a 21 st Century Economist Random House Business Books London SW1V 2 SA, (2017).
3. J. Benyus, Biomimicry Innovation inspired by Nature, HarperCollins Publishers NY, (1997).
4. D. Baumeister, Biomimicry Resource Handbook , Biomimicry 3.8 Montana USA , (2014).
5. Hajer, Nilsson, Raworth c.s, Beyond Cockpit-ism: Four Insights to enhance the Transformative Potential of the Sustainable Development Goals. MDPI Sustainability (2015), **7(2)** 1651-1660 doi:10.3390/su7021651.
6. Linnenluecke et al., Extreme weather Events and the Critical Importance of Anticipatory Adaptation and Organizational Resilience in Responding to Impacts. published on line Wiley Online Library DOI 10.1002/bse.708 (20100
7. S. Beder, Economy and Environment: Competitors or Partners, Prof. Sharon Beder's Publications http://www.uow.edu.au/~sharonb.
8. "Growth versus Green" Wageningen University & Research, ISBN 978-94-6343-758-5 DOI https://doi.org/10.18174/443008, (March 2018).

Dealing with Maritime Economy at the Local Level. Trends along the Coastal Poland

Barbara Szejgiec-Kolenda [1,*], and *Jacek* Zaucha [2]

[1]Institute of Geography and Spatial Organization, Polish Academy of Sciences, Twarda 51/55, 00-818 Warsaw, Poland
[2]University of Gdańsk, Faculty of Economics, Armii Krajowej 119/121, 81-824 Sopot, Poland; Institute for Development, Czyżewskiego 6/1, 81-706 Sopot, Poland

Abstract. Although the interest in the concept of 'blue economy' has grown rapidly in recent years, the most studies assess maritime activities' size, scope, basic trends or position in national economy mostly at the basin/national, sometimes even regional level and there has been little research taking into account the local dimension of maritime economy. This is partly due to a lack of appropriate statistical data concerning maritime economy. The aim of the analysis is to define and describe the local maritime economy in Poland as well as to establish its importance for various territories. This study considers the challenges that maritime local studies face in terms of data availability and provides a research path that is to some extent complementary to analyses at the national and regional level. It explores a two-step approach to measure and evaluate a maritime local economy in Poland in 2016: (1) a more general countrywide attempt to identify the problem; (2) addressing the local dimension of blue economy in the spatially limited area (coastal regions). Moreover, the approach allows to identify territorial differences (functional region types), the extent to which these activities differ among local economies and the pathways for maritime economy structures transitions along the coastal Poland.

1 Introduction

An increase in interest in maritime economy along with growing potential for deriving benefits from the seas and the oceans wealth contributed to an increased intensification of efforts related with planning and strategic governance of maritime areas [cf. 1-4]. As a result, as of 2015, the Maritime Policy of the Republic of Poland up to 2020 (with perspectives up to 2030) [5] was adopted, whose main directions of development were determined earlier by the Green Paper on Marine Policy for the EU [6], as well as by the EU Strategy for the Baltic Sea Region [7].

In practical terms, there appeared demand for facilitating cohesion between scientific knowledge and governance practice to strengthen the decision-making processes and governance of economy and natural environment of seas and oceans, as well as of coastal

* Corresponding author: b.szejgiec@twarda.pan.pl

zone. The Blue Economy, by providing considerable developmental opportunities for maritime regions and contributing to more rapid development of national economies, caused an increase of information requirements that are indispensable for public debate and development of national and international policies for sustainable sea governance. Over the quite short period of time the information gaps appeared on many dimensions: in terms of spatial approach "on land" (local, regional and national dimension of maritime economy), from a substantive perspective (identification of maritime sectors), from a subjective perspective (marine stakeholders), in the context of the definitional debate (measuring the marine economy) and in terms of tool-based approach (availability of coherent, comparable and adequate data, choice of measurement methods) [8].

The study comes within a contemporary discussions of maritime economy. The aim of the paper is to indicate specificity of defining the maritime economy on a local basis, and elaboration of measurement methods and evaluation of local maritime economy in Poland. Moreover, the study takes into consideration identification of spatial disparities, as well as searching for regularities in economic space of coastal regions in regard to blue economy in the context of usability of obtained findings in the process of governance.

2 Theoretical and conceptual framework

2.1 Theoretical concepts of marine economy

Studies aimed either at evaluation of economic activity associated with the oceans, or at delimitation of maritime functional regions or at development of other typologies of marine space are in the initial phase. Although one may try to describe the term "blue growth" explicitly as a broad economic use of seas' and oceans' resources, such as fisheries, shipping, tourism and coastal protection [9], the very terminology associated with seas and oceans is applied on a variety of ways in different parts of the world. There occur, for example, such terms as "ocean industry, "marine economy", "marine industry", "marine economy", "marine activity" and "maritime sector" [10].

Analogically, the scope of the marine economy is defined on multiple ways, depending on the particular countries and regions, which may pose problems while quantifying the economy's marine sectors. As shown by the example of United States, UK, France, Australia, China, Japan and South Korea, Park and Kildow [10] point out that disparities in understanding of marine economy are noticeable and may depend on the importance of the particular sectors in a given country. The common sectors of marine economy comprise such categories as: fisheries, marine mining, offshore oil and gas, ship and boat-building, marine manufacturing, marine construction, marine transportation, ports, marine tourism, ocean renewable energy, marine bio industry; and public sectors such as: education, national defence, R&D, and public administration. Drawing on the aforementioned scope of marine economy sector, Parker and Kildow [10] break down maritime activities into activities that: (a) explore and develop ocean resources, (b) use ocean space, (c) protect the ocean environment, (d) use ocean products as a main input and (e) provide goods and services to ocean activities.

As far as the research studies in EU countries are concerned [11], the proposal is to take into account the following marine sectors: living resources, non-living resources, ship and boat building, construction, transportation, tourism and recreation, public administration, education and R&D and others, with making use of industries classified with the NACE Rev.2 four-digit numerical code. Suris et al. [11] break down the marine activities into three groups: completely marine, mainly marine and partially marine, pointing out at the same time that certain dose of flexibility ought to be preserved in defining the scope of maritime economy for the particular countries.

Similarly Colgan [12] distinguishes two concepts of economic activities falling under the framework of marine economy sector: (1) the ocean economy perceived as "economic activity which derives all or part of its input from the ocean" (p. 5), which is a function of both industry and geography" (p. 5), i.e. a significant part of activities takes place within the coastal regions, but some part maybe located outside those regions (e.g. seafood retailers), and (2) the coastal economy that is an economic activity in the coastal region and "incorporates a broader set of economic activity" (p. 5).

2.2 Dimensions of local marine economy in practice

Different approaches to concepts and definitions of maritime economy cause that the basic difficulty lies in delimitation of maritime economy. The EU public statistics omits to single out the sector of maritime economy. Inasmuch as, in Poland at the national level, the adequate estimates are carried out by the Polish Central Statistical Office [13]; one can hardly find relevant studies at the regional and local level, which comes along with limited access to statistical materials. Also there is a risk that defining maritime economy as a set of selected types of business activities may lead to both underestimation (registering in accordance with the prevailing kind of business activity, utilization of the enterprise method) as well as overestimation of the actual values (difficulty with distinguishing from types of business activities the particular ones that are strictly related to maritime economy). The information gap in the system of public statistics (both in Poland and in other EU countries; cf. 1) results from the inflexible or rigid system of data collecting, based on traditional breakdown into types of business activity, which makes it impossible to aggregate quantitative data with regard to sectors developing at variance with traditional breakdown (e.g. maritime economy or the so-called silver economy). It is necessary, in that case, to adopt lots of assumptions that as a result may lead to inaccurate conclusions.

The current study draws on EU experiences [cf. 14] which indicate selected types of business activity classification (NACE Rev.2 – PKD 2007) as referring to marine economy (Tab. 1). To some degree, also the concepts of marine economy are made use of, as proposed by Suris et al [11] and Colgan [12], basing the scope of marine economy on spatial scope of analysis (country vs. coastal areas). The Table 1 shows the types of activities, marked by grey background, when the analysis is restricted to maritime communes.

Table 1. Listing and grouping the types of business activities for the purposes of identification of maritime economy sector

Types of activities by NACE Rev.2 – PKD 2007 classification (sectors and codes)	Sector of maritime economy
Marine (03.11) and freshwater (03.12) fishing Marine (03.21) and freshwater (03.22) aquaculture	1 – fishing
Extraction of crude petroleum (06.10) and of natural gas (06.20) Support activities for petroleum and natural gas extraction (09.10) and for other mining and quarrying (09.90) Operation of gravel and sand pits; mining of clays and kaolin (08.12) and of salt (08.93)	2 – mining and energy
Processing and preserving of fish, crustaceans and molluscs (10.20) Wholesale of other food, including fish, crustaceans and molluscs (46.38)	3 – food industry
Building of ships, boats and floating structures, including pleasure and sporting boats (30.1, 30.11, 30.12) and Repair and maintenance of ships and boats (30.15), machinery (33.12) and dismantling of wrecks (38.31)	4 – building of ships and boats
Works related to construction of water projects (42.91, 42.99)	5 – construction of water projects

Sea and coastal passenger and freight water transport (50.10, 50.20) and inland passenger and freight water transport (50.30, 50.40) Service activities incidental to sea transportation (52.22, 52.29) Warehousing and storage and cargo handling in sea ports (52.10, 52.24)	6 – water transport
Research and experimental development on biotechnology and natural sciences and engineering (72.11, 72.19)	7 – research & development
Hotels, holiday and other short-stay accommodation, amping grounds and campsites (55.10, 55.20, 55.30) Restaurants and other eating places (56.10) Rental and leasing of recreational and sports goods and water transport equipment (77.21, 77.34) Tour agents activities (79.11)	8 – tourism

Source: own elaboration based on [8].

2.3 A proposal of evaluating local marine economy

The term 'maritime economy' is used to denote economic activity (not economic value), being measured through the number of business establishments (entities of national economy included in the business register – REGON) [cf. 1]. This approach is determined largely by availability of data at the local level. Statistical data concerning the number of entities were obtained from the REGON database. For a tourism-related group, the number of facilities and accommodation places available in tourist accommodation facilities was used as an index value (this stems from the fact that lots of entities providing accommodation are not registered in the REGON database). The use of REGON database entails certain limitations concerning the accuracy of conclusions. Spatial attribution of information concerns the place of registration rather than the actual place where business activity is carried out. The database covers also firms that ceased to operate, i.e. those that have not deregistered. These and other weaknesses of REGON database are well described in the literature of the subject [e.g. 16, 17]. However, it seems that in spite of the aforementioned drawbacks the database allows for comparing the intensity/volume of maritime economy between communes (gmina, LAU 2), since the similar scale of error maybe assumed with regard to all of the investigated communes, except for large cities – where undoubtedly greater number of larger firms is registered, including those that operate outside their area. That is why the REGON database does not always ensure the accuracy of estimations concerning the importance of maritime economy in large urban centres.

To evaluate maritime economy at the local level, particularly to identify regions in which maritime activities play a particularly important role, as well as to conduct comparisons inside the coastal regions, the index method was employed which is regarded to be a traditional research method in socio-economic geography [18]. The basic index that is helpful in differentiating between communes' economic space falling within the framework of maritime sector is the number of economic entities per 10 thousand working age population.

Additionally, in order to enhance the knowledge on the situation existing and processes going on in the coastal regions' economic space, as well as on the basic structures of maritime economy and their evolution, the classification methods were made use of. To carry out classification of communes, multidimensional comparative analysis concerning importance and structure of maritime economy in maritime communes was utilized, with a special focus on development of tourism – classification of facilities performed by means of Ward's method based on the set of features. Two analyses were carried out: for the year 2016 (static analysis) and for the years 2011-2016 (dynamic analysis).

For the purposes of classification of communes, variables are utilized that as far as possible comprehensively characterize maritime economy, while ensuring completeness and comparability of data. The adopted concept of study includes three main areas of maritime economy analysis, i.e. (a) based on static approach – importance of sector (entities in maritime economy per 10 thousand working age population, share of entities in maritime economy in relation to the total number of entities), sector's structure (share of entities in the particular sectors of maritime economy in relation to the number of entities in maritime economy), and tourism potential (number of accommodation places in total per 10 thousand population, share of year-round accommodation/hotel accommodation in relation to the total number of accommodation) as the important components of blue economy in Poland; as well as (b) based on dynamic approach (2011-2016 – dynamics of maritime economy sector (dynamics of entities in marine economy, changes in share of entities in marine economy in relation to entities in total), durability of maritime economy structure (change in share of entities in particular sectors of maritime economy in the number of entities in maritime economy), as well as changes in tourism development in communes (dynamics of the number of accommodation places in total, changes in share of year-round accommodation/hotel accommodation in relation to the number of accommodation places in total) representing potential structural changes of marine economy.

Two spatial scopes of study were made use of, to which the separate scopes of marine economy sectors were attributed (both narrow and broad approach to marine economy; cf. Tab. 1 – section 2.2). The former one covers the territory of whole country broken down into communes (LAU 2), and the latter level of analysis was limited to coastal communes. Delimitation of coastal regions constitutes a compromise between different interactions and relations between sea and land as well as objectives of the study itself. In European studies carried out at the regional level, those statistical regions (NUTS 3) which have at least half of their population within 50 km of the coast (according to Eurostat definition) are recognized as coastal regions. For the purposes of the current study it was assumed that the communes lying directly by the sea, as well as those not lying by the sea, but located within 10 km-wide belt from the sea are regarded as coastal communes, after the "Study of land-use conditions for Polish coastal regions" [19]. According to this approach, in total, coastal regions comprise 68 LAU 2 units (communes) (spatial scope against the background of the whole country – cf. Fig. 1).

3 Results

3.1 Polish local marine economy

The findings obtained from the countrywide analyses show evident clustering of maritime economy especially in maritime communes (Fig. 1). In this context, it comes as a surprise that there are so many maritime economic entities operating in inland regions of Poland. This is due to taking into account, under the framework of that sector, economic establishments involved in the construction of water-engineering facilities, as well as those that carry out maritime-related research. The research studies go to the benefit of large cities. Nonetheless however, by using this approach, it was possible to establish that in all of the maritime communes the number of economic entities exceeds 10, while in the areas characterized by outstanding tourist values and in large seaside cities that number exceeds 50 per 10 thousand working age population.

Inclusion of the measurement of importance of maritime economic sector at the local level, covering the whole country, makes it thus possible to prove the thesis that a marked spatial diversification in location of maritime-related sector of economic entities stems from a varying degree of relation, under the maritime economy activity, to the seaside location (sectors benefiting from the location rents, as well as those independent of/dependent in a small degree upon location in the seaside areas, e.g. part of scientific research, food processing, etc.). On the other hand, it reveals certain weaknesses in operationalization of the maritime economy concept based on a statistic system of economic activity classification and on necessity to adopt numerous assumptions.

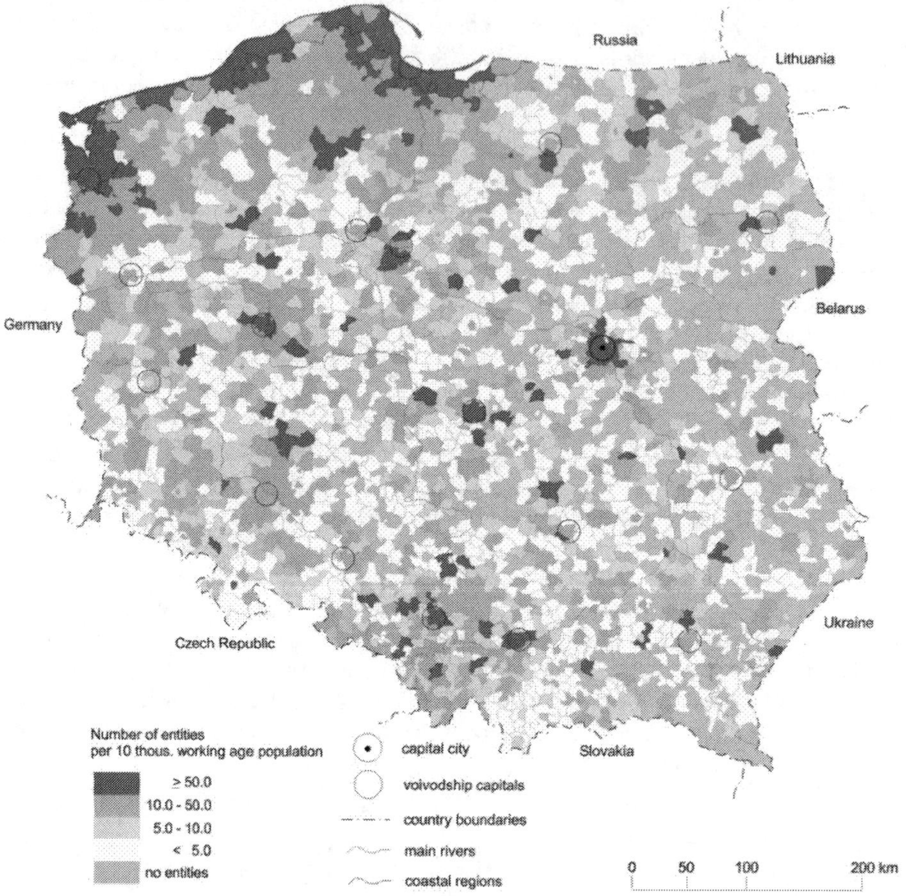

Fig. 1. Number of REGON entities potentially related to marine economy per 10 thousand working age population, as of the end of 2016.
Source: own elaboration based on [8].

3.2 Coastal Poland local marine economy

In order to find solutions to part of the problems posed by definitional approach to the maritime economy scope at the local level, perhaps it would be necessary to limit the spatial analysis exclusively to seaside areas. That is why analogical studies on the importance of maritime economy sector locally were carried out with regard to seaside communes. Sector of maritime economy (in a broader object-based approach including inter alia tourism, warehousing, cargo handling) shows considerable spatial differentiation in regard to seaside communes (Fig. 2). As regards quantitative-based approach, the majority of entities is concentrated in large coastal cities (Gdańsk-Gdynia-Sopot and Szczecin) and in communes adjacent to them.

The exception is the tourist sector located along the coast in smaller municipalities and towns, and occurring in rural and urban-rural communes and small coastal towns: fishing, fishing industry and the production of ships and boats appearing in many different communes. In total, approximately 22 thous. entities of maritime economy were located in coastal areas of Poland in 2016.

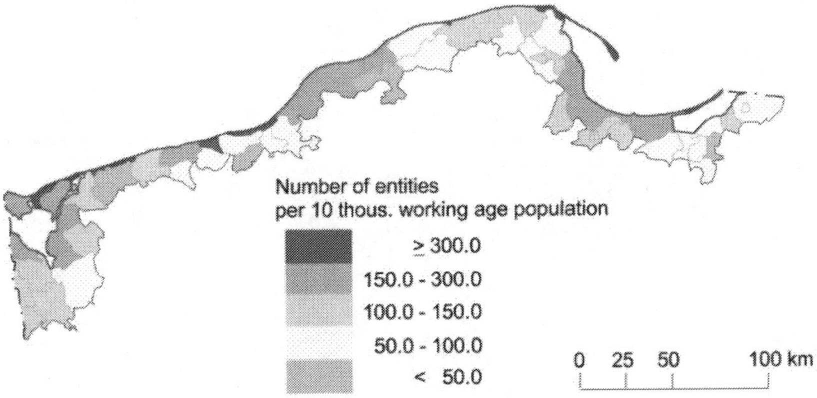

Fig. 2. Number of REGON entities related to marine economy per 10 thousand working age population, as of the end of 2016
Source: own elaboration based on [8].

Spatial analysis of the number of economic entities per 10 thousand working age population indicates that maritime economy is of a relatively great importance to small towns and rural communes (Hel, Jastarnia, Władysławowo, Sztutowo, Krynica Morska, Łeba, Mielno, Ustronie Morskie, Rewal, Dziwnów, Kamień Pomorski). The number of maritime economic entities exceeds 300 per 10 thousand working age population. On the other hand, there occur communes – none of which is lying directly by the sea – with a relatively small number of entities. Against this backdrop, large cities, due to their larger and more diversified economic base, are in the group of areas characterized by an average intensity/volume of maritime economy.

3.3 Types of maritime communes based on importance and structure of marine economy

3.3.1 Static typology

Based on the use of static method of classification, 5 groups of communes have been identified. Group 1 consists primarily of large multifunctional centres together with their surrounding zones that specialise in "higher-order" sectors of maritime economy: ship and boat building, water engineering constructions, water transport, including even, to a relatively high degree, scientific research. Also the tourism infrastructure, in opposition to units from groups 3-5, is made up chiefly of hotel accommodation base, which partly results from the fact that it provides also business-oriented services. However, it needs to be stated that the marine sector is just one of many sectors that operate within these centres. Group 2 is made up of communes, similarly as in the case of group 1, with diversified character of marine sector which includes mainly tourism (prevalence of hotel facilities), ship and boat building as well as fishery. The marine economy plays a noticeable role in economy of these communes, but its share is lower than in communes representing 4 and 5 types. In turn, group 3 consists of communes in which, as compared to other groups, the marine economy does not play a particular part in economy; these communes are predominantly small multifunctional centres. Group 4 is made up of communes that have moderate significance for marine economy and average tourist-oriented services. An important role is played especially by these units that are also centres associated with fishery and fishing industry, as well as with ship and boat building. Group 5, the least numerous one, consists of units with strict tourism specialization; these are the communes with the largest concentration of marine economic entities – in which the marine sector to greatest extent is linked to tourism (mostly monofunctional in character) – food and beverages establishments, tourist traffic services and primarily holiday centres (often seasonal). In addition to tourism, fishery is a sector of great importance. Groups 2 and 3 are made up of units that distinguish themselves by differentiated location in relation to coastal line – with small exceptions these are the communes which are not immediately situated on the coast of the Baltic Sea. Group 4 and 5 includes communes that are located alongside the Baltic coast. Figure 3 presents location of particular groups.

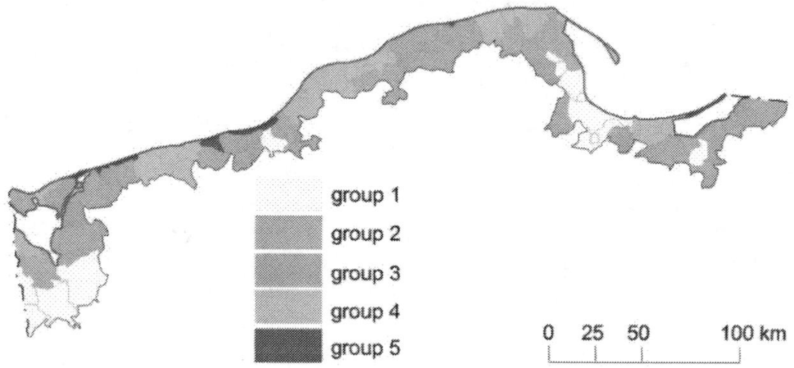

Fig. 3. Typology of maritime communes in regard to maritime economy criteria (static analysis)
Source: own elaboration based on [8].

3.3.2 Dynamic typology

Thanks to the use of classification method based upon dynamic approach, similarly as in the above-mentioned case, 5 groups of communes have been identified; however, it must be stated that their numerical distribution is even more unequal than the one elaborated on the basis of static approach. The most numerous group (1) constitute communes characterised by an average growth of maritime economy, as well as by stable structure of that industry sector. A characteristic feature of units belonging to group 2 is a relatively high dynamics of marine sector (second position in the ranking after group 5). The changes are additionally influenced by structural transformations, a decrease in significance of fishery as well as of construction of water engineering and tourist facilities, and simultaneously by a growing importance of ships and boats building. An extremely high dynamics of accommodation places is accompanied by an increased seasonality of the tourist accommodation infrastructure that follows from the rapid growth in number of seasonal places. Group 3 is made up of units which, under conditions of stabilised importance of maritime industry sector as well as of its structure, are characterised by the highest increase of entities in absolute numbers. In the area of tourism, it should be noted that there occurs a drop in share of year-round accommodation places (similarly as in group 2). These communes are characterised by high levels of coastal tourism development (high increases in absolute numbers, but with small relative increments). Also, they distinguish themselves by a high percentage of continuing businesses. Group 4 comprises units with a stable maritime industry sector, however, with a tendency towards decreased importance of that industry sector (including tourism sector). Nevertheless a large percentage of companies continues to enjoy a stable situation in the area of undertaken activity. Considering the structure of the industry sector, it is the only group in which a slight increase of fishing importance is noted. In turn, group 5 comprises communes where the maritime sector, as measured in terms of economic entities number, is developing at the most dynamic pace. Also, as it is observed, there occur such processes as the qualitative improvement of accommodation places (with a simultaneous decrease of their overall number), the highest growth of year-round accommodation places in hotel facilities, which is partly due to their extremely low initial level, and the largest increase of entities in the tourism sector. At the same time, however, tourism is losing its importance in favour of those activities that are associated with ship and boat building as well as water transport.

Figure 4 presents map showing division of maritime communes based on dynamic approach. Spatial distribution of communes, performed by means of Ward method, reveals some regularities, namely a relatively rapid growth of maritime economy in communes where previously this sector of economy never had much significance, also in the communes that are not located at the immediate location on the Baltic shore. St the same time, it reveals that there is an ongoing stabilization in development of that industry in typical tourist communes and in large multifunctional centres together with their surrounding zones that specialize in "high-order" maritime economy sectors.

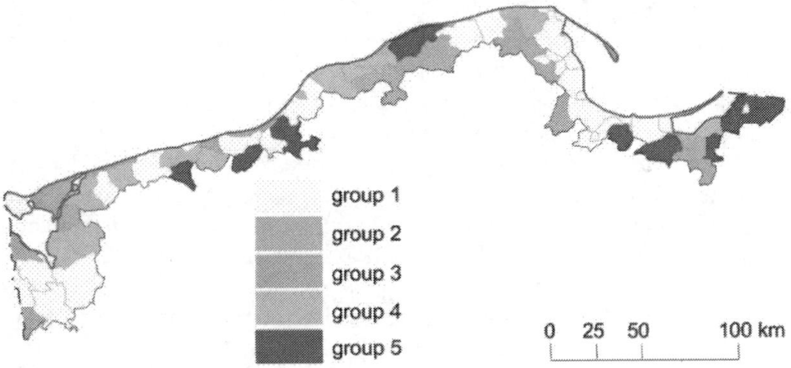

Fig. 4. Typology of maritime communes in relation to marine economy criteria (dynamic analysis) Source: own elaboration based on [8].

4 Conclusions

Identification of the importance of maritime economic sector (in a narrow sense) at the local level in Poland, based on the index of the number of economic entities per 10 thousand working age population, tends to denote: (a) a high level of heterogeneity of Poland's economic space, (b) a concentration of strong links with maritime economy in the areas situated especially along the coastal line (traditionally predisposed toward stronger links with that sector), (c) trends towards concentrating business activity in the sectors connected with maritime economy also in the agglomeration areas (marine activity independent of location in the direct vicinity of the Baltic sea), and last (d) that the large part of Poland's space has no links whatsoever with maritime economy (this concern in the main the peripheral areas located in east Poland, as well as the so-called inner peripheries). The similar studies in foreign literature deal more often with attempts to identify disparities in maritime economy at the regional rather than the local level. Hence the utilized set of features is based primarily on indicators concerning GDP (or GVA), labour productivity within the marine sector or the level of employment in marine economy sectors. Nevertheless their results reveal that regional disparities are significant among regions and also that regional sectoral specializations are developing [12, 20, 21], the ocean economy plays a greater role in absolute terms predominantly in urban areas, however, it has greater significance for less developed local economies (including rural areas) [1], and, in certain cases, gaps between maritime economies of varied coastal regions are gradually decreasing [21].

Analysis of the maritime sector (in a broader approach as compared to analysis of the area of the whole country), limited spatially to coastal regions, reveals strong spatial differentiations between communes in terms of significance of maritime economy, both among economic entities (in addition to communes characterized by a very high index of economic entities per 10 thousand population – small administrative units that have coastal location, there are communes that have a relatively insignificant links with maritime economy, including also those that are situated along the Baltic coast), and in terms of accommodation offer (in this case spatial distribution is to a larger extent related to the distance from the shore – seaside municipalities are among the most tourist-wise developed localities). The large heterogeneity of maritime areas in terms of development of maritime economic sector is also noticeable with reference to the European Atlantic areas (the highest level of heterogeneity is noted in case of Portugal regions), where at the coastal NUTS 3 level, to measure and compare the importance of the maritime sectors, a synthetic index value was utilized (the EcIMar index) [22].

The index method, applied for evaluation of the level and significance of maritime economy on a local scale, can be further developed on a number of aspects: (a) broadening the list of indices with the ones that concentrate on a sectoral structure of maritime economy or (b) paying attention to dynamics of change in maritime economy. Spatial classifications provide more opportunities for locally-oriented studies of the marine economy sector. The presented typologies have innovative character; in the Polish literature of the subject, one can hardly find any cases of communes' classifications where the maritime economy would serve as a principal classification criteria. This indicates that the economic dimension revolving around the junction between land and sea has been greatly undervalued by Polish researchers. The presented findings can be compared to classifications covering either all Polish communes [23], or communes that have a specific economic profile, e.g. rural communes [e.g. 24], or tourist communes [e.g. 25, 26]. The typology proposed in the current study provides, however, other, greatly important aspects to the knowledge of maritime communes. For example, a group of communes identified by Śleszyński and Komornicki [23] with other non-agricultural functions can be broken down into communes that have evident tourist profile (group five) and those where multi-functional maritime economy is developing (group four). While with regard to communes having other functions, the typology pinpoints e.g. those with greater (group two) and those with lesser (group three) significance for maritime economy. In the foreign literature, there have been sporadic studies that use hierarchical agglomerative clustering for determining the regional regularities within blue economy. For example, Fernandez et al. [27] based on the use of Ward's method grouped regions of Atlantic European Arc, indicating, as a result, mostly nation-driven maritime clusters, which attests to the fact, that such studies ought to be carried out within the particular countries, aiming at analysis of intrastate trends. Summing up, the initial assumptions for the typology employed and choices made concerning the necessity to separately classify dynamic and static phenomena have been confirmed to be correct. Dynamic classification brings completely new knowledge, being in addition an innovative way of looking at the dynamics of changes taking place in maritime communes in the context of marine economy.

Identification of territories, characterized both by a high as well as low importance of maritime economy, may be useful in determining the economic potential of the aforesaid sector at the local level and in pursuing the policy of improving competitiveness. Blue Growth and spatial planning in the seaside regions, including location of strategic investments with specific reference to a place-based approach. At the same time, a much greater emphasis should be put on a comprehensive approach to the subject-matter of local maritime economy.

The paper presents the outcomes of the research financed by the Polish National Science Centre under the project "Economy of maritime space" (2015/17/B/HS4/00918).

References

1. C. S. Colgan, The ocean economy of the United States: Measurement, distribution, & trends, Ocean & Coastal Management, **71**, 334-343 (2013) DOI: http://dx.doi.org/10.1016/j.ocecoaman.2012.08.018
2. L. D. Rodwell, S. Fletcher, G. A. Glegg, M. Campbell, S. E. Rees, M. Ashley, E. A. Linley, M. Frost, B. Earll, R. B. Wynne, L. Mee, P. Almada-Villela, D. Lear, P. Stanger, A. Colenutt, F. Davenport, N. J. Barker Bradshawk, R. Covey, Marine and coastal policy in the UK: Challenges and opportunities in a new era, Marine Policy, **45**, 251-258 (2014) DOI: http://dx.doi.org/10.1016/j.marpol.2013.09.014
3. J. Zaucha, Sea basin maritime spatial planning: A case study of the Baltic Sea region and Poland, Marine Policy, **50**, 34-45 (2014) DOI: http://dx.doi.org/10.1016/j.marpol.2014.05.003
4. M. R. Keen, A.-M. Schwarz, L. Wini-Simeon, Towards defining the Blue Economy: Practical lessons from pacific ocean governance, Marine Policy, **88**, 333-341 (2018) DOI: http://dx.doi.org/10.1016/j.marpol.2017.03.002
5. *Uchwała nr 33/2015 Rady Ministrów z dnia 17 marca 2015 r. w sprawie Polityki morskiej Rzeczypospolitej Polskiej do roku 2020 (z perspektywą do 2030 roku)* (2015)
6. *Green paper towards a future maritime policy for the Union. European vision for the oceans and seas* (European Commission, Brussels 2006)
7. *Communication from the Commission to the European Parliament, the Council, the European Economic and Social Committee and the Committee of the Regions concerning the European Union Strategy for the Baltic Sea Region* (Brussels, 2009)
8. J. Zaucha, *Gospodarowanie przestrzenią morską* (Instytut Rozwoju, Sopot, 2018, to be published)
9. M. G. Burgess, M. Clemence, G. R. McDermott, C. Costello, S. D. Gaines, Five rules for pragmatic blue growth, Marine Policy, **87**, 331-339 (2018) DOI: http://dx.doi.org/10.1016/j.marpol.2016.12.005
10. K. S. Park, J. T. Kildow, Rebuilding the Classification System of the Ocean Economy, Journal of Ocean and Coastal Economics, 2014, **1**, 4 (2014) DOI: https://doi.org/10.15351/2373-8456.1001
11. J. C. Surís-Regueiro, M. D. Garza-Gil, M. M. Varela-Lafuente, Marine economy: A proposal for its definition in the European Union, Marine Policy, **42**, 111-124 (2013) DOI: http://dx.doi.org/10.1016/j.marpol.2013.02.010
12. C. S. Colgan, *Measurement of the ocean and coastal economy: theory and methods* (National Ocean Economics Project, USA, 2004)
13. *Rocznik Statystyczny Gospodarki Morskiej* (Główny Urząd Statystyczny, Urząd Statystyczny w Szczecinie, Warszawa, 2016)
14. T. Brodzicki, J. Zaucha, *Analiza "niebieskiego wzrostu", polityki morskiej oraz strategii UE dotyczącej Regionu Morza Bałtyckiego. Analiza dla Polski* (Dokumenty robocze Instytutu Rozwoju, **4**, 2013)
15. D. P. Shaw, S. Kidd, L. McGowan, S. A. Jay, *ESaTDOR: European Seas and Territorial Development, Opportunities and Risks, Final Report* (Luxembourg, ESPON, the European Observation Network for Territorial Development and Cohesion, 2013)

16. P. Śleszyński, Studium szacunku liczby i rozmieszczenia pracujących w Krakowie (Biuletyn KPZK PAN, **243**, 2010)
17. D. Celińska-Janowicz, Rejestry podmiotów jako źródła danych w analizach lokalizacji działalności gospodarczej w mikroskali, Wiadomości Statystyczne, **1**, 27-43 (2016)
18. Z. Chojnicki, *Dylematy kwantyfikacji geografii*, [w:] Z. Chojnicki (red.) *Metody ilościowe i modele w geografii* (PWN, Warszawa, 9-15, 1977)
19. J. Zaucha, M. Matczak, J. Pardus, J. Faściszewski, J. Rekowska, S. Dendewicz, L. Jastrzębski, D. Nowoświecka, M. Nowotarska, J. Smutek, T. Jaksina, R. Woziński, M. Izdebski, P. Kaszczyszyn, H. Boniecka, A. Gajda, M. Adamowicz, T. Łączkowski, E. Kuzebski, P. Margoński, T. Nermer, W. Pelczarski, I. Psuty, M. Rakowski, L. Szymanek, U. Kowalczyk, M. Kalinowski, B. Szwankowska, W. Kuszewski, M. Szymańska, B. Hac, J. Gajewski, M. Kałas, J. Kapiński, R. Wróblewski, P. Brzeska, L. Kruk-Dowgiałło, M. Gorczyca, I. Kordala, T. Kuczyński, M. Michałek, M. Olenycz, A. Osowiecki, P. Pieckiel, J. Przedrzymirska, I. Rakowska, T. Komornicki, *Studium Uwarunkowań Zagospodarowania Przestrzennego Polskich Obszarów Morskich* (Gdańsk, Instytut Morski w Gdańsku, 2015)
20. K. Morrissey, C. O'Donoghue, The Irish marine economy and regional development, Marine Policy, **36**, 358-364 (2012) DOI:10.1016/j.marpol.2011.06.011
21. L. Baiqiong, X. Min, W. Jing, X. Sumei, Regional disparities in China's marine economy, Marine Policy, **82**, 1-7 (2017) DOI: http://dx.doi.org/10.1016/j.marpol.2017.04.015
22. J. Fernández-Macho, P. González, J. Virto, An index to assess maritime importance in the European Atlantic economy, Marine Policy, **64**, 72-81 (2016) DOI: http://dx.doi.org/10.1016/j.marpol.2015.11.011
23. P. Śleszyński, T. Komornicki, Klasyfikacja funkcjonalna gmin Polski na potrzeby monitoringu planowania przestrzennego, Przegląd Geograficzny, **88**, 4, 469-488 (2016)
24. J. Bański, M. Mazur, Classification of rural areas in Poland as an instrument of territorial policy, Land Use Policy, **54**, 1-17 (2016) DOI: 10.1016/j.landusepol.2016.02.005
25. T. Lijewski, B. Mikułowski, J. Wyrzykowski, *Geografia turystyki Polski* (PWE, Warszawa, 2008)
26. S. Sobotka, Próba wyznaczenia obszarów o najwyższym stopniu wykorzystania turystycznego jako przesłanki do regionalizacji turystycznej Polski, Turyzm, **24/2**, 35-43 (2014)
27. J. Fernández-Macho, A. Murillas, A. Ansuategi, M. Escapa, C. Gallastegui, P. González, R. Prellezo, J. Virto, Measuring the maritime economy: Spain in the European Atlantic Arc, Marine Policy, **60**, 49-61 (2015) DOI:http://dx.doi.org/10.1016/j.marpol.2015.05.010

SHS Web of Conferences **58**, 01029 (2018) https://doi.org/10.1051/shsconf/20185801029
GLOBMAR 2018

The competitiveness of ferry and air transport in the Baltic Sea Region: infrastructure approach

Marta Mańkowska[2], *Dariusz* Tłoczyński[2]

[1] Faculty of Management and Economics of Services, University of Szczecin, 8 Cukrowa St., 71-004 Szczecin, Poland, e-mail: marta.mankowska@wzieu.pl
[2] **Presenting author**: Faculty of Economics, University of Gdańsk, 119/121 Armii Krajowej St., 81-824 Sopot, Poland, e-mail: dariusz.tloczynski@ug.edu.pl

Abstract. The main aim of the article is attempt to get the answer to the question about the impact of services offered by point elements of transport infrastructure on meeting the expectations and preferences of passengers in various modes of transport. Based on the presented objective, the assessment of the competitiveness of air and ferry transport in the service of passenger traffic was carried out through taking into the consideration selected elements affecting the quality of services offered by airports and ferry terminals. One of the key factors of the competitiveness of ports is time availability. The main objective of the research undertaken in the article was therefore to assess the competitiveness of seaports and airports in terms of their time availability. The article uses the case study method of selected airports and ferry terminals located in the Baltic Sea Region (BSR). The comparative analysis covered the time competitiveness of port ground access. The results of the research allow to make a general conclusion that ferry terminals have better ground access for individual car journeys and airports are becoming more competitive in the case of the arrival and departure function of public transport. This indicates that the system of ground access for seaports and airports is adapted to the segment of passengers having the largest share in transport by the analyzed transport branches, i.e. the motorized passengers in ferry transport and non-motorized passengers in air transport.

1 Introduction

The phenomenon of competition between sea transport (ferry shipping) and air transport intensified in the 1990s. The weakening of the competitive position of ferry transport contributed primarily to the introduction in 1999 of the abolition of duty-free and tax-free retailing on board ferries operating on ferry lines connecting the EU member countries [1-3]. On the other hand, the increase in the competitiveness of air transport was related to the development of low-cost airlines supported by the liberalization of the EU air transport sector and the implementation of the open sky policy in the EU [4-8]. As a result of the changes, the function of ferry shipping changed from a typical passenger to a passenger and cargo shipment. The outcome was the conversion of passenger ferries and the introduction of new ro-pax ferries with an extended lane metres at the expense of the passenger space.

These factors changed the structure of the international passenger transport market in the Baltic Sea Region (BSR), which was traditionally dominated by ferry transport. The increase in cost competitiveness and the time advantage of air transport, while the competitiveness of ferry shipping decreased due to abolition of duty-free sale, contributed to the takeover of some of the passengers of the ferries by air transport. The few surveys undertaken so far in the area of competitiveness of air and sea transport focused on the analysis of the demand side, including in particular identification of passengers' preferences as regards the choice of modes of transport. There is, however, no comparative research in the area of inter-branch competitiveness of air and sea transport from the point of view of the quality of services provided by seaports and airports. One of the key elements influencing the quality of airport services is their time of port ground access.

The main aim of the article is attempt to get the answer to the question about the impact of services offered by point elements of transport infrastructure on meeting the expectations and preferences of passengers in various modes of transport. The aim of the research undertaken in the article is to assess the competitiveness of air transport and ferry shipping in servicing international passenger traffic in the BSR through the prism of time competitiveness of port ground access. The results of the research undertaken in the article do not exhaust the issue of competitiveness of the analyzed transport branches from the point of view of the quality of services provided by seaports and airports. Nevertheless, they can be an introduction to further in-depth research in this area, in particular those undertaking a comparative analysis of other elements affecting the quality of port services in passenger traffic.

2 Literature overview

The concept of competitiveness is analyzed on many levels. It is the ability to achieve market objectives in a competitive environment [9]. M. E. Porter [10] defines competitiveness as the ability to create conditions conducive to the development of international competitiveness of enterprises within specific market segments. In the area of passenger transport, competition is examined primarily from the point of view of satisfying the customer's preferences (travel mode choice behavior). The issue of travel mode choice behavior was taken, among others in studies by McFadden, Jackson & Jucker, Ye & Wang, Hunecke, Scheiner, Paulssen, De Vos et. all [11-18]. In the inter-branch transport system, the subject of many studies is the competitiveness between high-speed train and air transport [19-23]. These studies confirm that cost and time are the two main factors that passengers take into account when travelling.

These factors also play a decisive role in the decisions of ferry passengers and air transport. Research in this area is relatively rarely undertaken [24-27]. They show that the main factor in the competitiveness of air transport in relation to ferry shipping is lower cost and the possibility of taking a car. In contrast, less time consuming in comparison with the ferry travel is a major factor in the competitiveness of air transport. There is also no comprehensive research in the literature on the assessment of the competitiveness of ferry and air transport through the prism of the quality of services provided by seaports and airports. Such research is undertaken rather in the inner-branch transport system. For example, Pantouvakis [28], in relation to seaports identified six factors (service, safety and security and cleanliness, communication, guidance and information) that effectively describe the multidimensional construct of port-service quality. Important factors of competitiveness of airports and sea terminals are resource factors that M.E. Porter describes as factor endowment [10]. They are connected, among others, with physical resources such as geographic location and time availability.

In the transport literature, when analyzing the accessibility issues among others the connectivity factor is used [29-31]. Doganis and Denis [32-34] define this as the number of indirect connections that meet the minimum and maximum time requirements for passenger service in the port, taking routing into account. Veldhuis, Burghouwt, Kroes, Danesi, Malighetti and Paleari with their teams [29-30, 35-38] interpret connectivity similarly. In their opinion connectivity is defined as the number of direct and indirect connections, so that the passenger has a guaranteed stay in the transfer port, with the appropriate quality of services provided. The time availability of seaports and airports from the point of view of the port ground access is of significant importance in the broadly understood air- and seaport connectivity [39-44]. The main factors affecting the time availability (connectivity) are: accessilibity coefficient, time coordination of airport operators' service, air travel time, quality of service in the transit port, type of connection, maximum number of travel stages, type of connectivity (through the hub port or through other ports) [45-46]. Tłoczyński [47], analyzing the problems of the airport ground access system, distinguishes three paths of the analyzed problem:

• in terms of time needed for access to the airport,

• in terms of travel costs, car parking (these are the features that can be dimensioned in terms of money),

• competition between the systems that make up the airport / ferry links system (the passenger then uses all possible instruments to value the way of reaching the sea- or airport).

However, Koster, Kroes and Verhoef [48] proposed a model that defines the ideal time needed to reach the port, so that the passenger is neither too early nor too late, at an acceptable cost. As a result of the development of the passenger access system to seaports and airports, the importance of the region increases, then operators are more willing to launch new connections, increase the frequency of existing connections, and this in turn affects further development of the region [49].

Taking into account the above, the main issue raised in the article is the issue of the competitiveness of seaports and airports in handling passenger traffic through the prism of time competitiveness of airport and seaport ground access.

3 Methodology

The subject of research are selected airports and ferry terminals located in the main BSR cities. As an area of the BSR, an area covering territories belonging to countries with direct access to the Baltic Sea, including through the Danish straits, was adopted. In the analyzed area, the most important airports and seaports were selected. The main factors of selection were:

1. The volume of serviced international passenger traffic.
2. Representativeness of most countries in the region (apart from Poland and Germany, capitals of the Baltic countries were adopted).
3. Existence of alternative air and ferry connections.

The analysis excluded ferry terminals serving domestic traffic (e.g. within the Norwegian coast) and very short relations for which there are no alternative air connections, e.g. from the Swedish city of Helsingborg to the Danish city of Helsingør as well as the Danish town of Rødby to the German town of Puttgarden [27]. According to the criterion of representativeness of BSR countries, from the group of airports and ferry terminals, those serving the capitals of particular BSR countries were selected. The exceptions are Poland and Germany, for which airports and ferry terminals located in the coastal regions - Gdańsk, Gdynia and Rostock - were accepted. The indicated origin ferry ports and airports are served by air and ferry connections. The final group of transport hubs (origin ports) that were examined are: Copenhagen, Gdansk and Gdynia, Helsinki, Klaipeda, Oslo, Riga, Rostock,

Stockholm and Tallinn. The Multiple Case Study Design method was used in the analysis of selected airports and ferry terminals [50]. Airports and ferry terminals were subjected to a comparative analysis according to the following criteria:

• location in relation to the city center (km),
• directional structure of served connections (origin ferry ports and airports),
• frequency of connections (per week),
• time availability to the city center (minutes) separately for individual car and public transport (urban and rail).

Most of the analyzed points in which passengers start their journey (6 out of 9) were served by more than one ferry terminal, and two of them by more than one airport. In order to compare the estimated time of transfer on the route from the ferry terminal to the city center and from the airport to the city center in the case of the accessibility of more than 1 transport point for a specific transport branch (ferry terminal or airport), only the shortest (minimum) and longest (maximum) travel time were taken into account. In the analysis of the time availability of selected terminals and airports, the data generated by google.maps were used primarily. This data was taken for car traffic and public transport (including rail) in the same time and period (January 2018). In the absence of data on Google Maps for public transport in Gdansk, Klajpeda and Riga, data from the internet portals dedicated to public transport timetables were used [51-53].

4 Results

4.1. The comparative analysis of ferry terminals and airports (location, traffic volume and connections)

When making a comparative analysis of the analyzed airports and ferry terminals in the scope of passenger service, the sum of all transport points serving the city was taken into account. The analysis of the data shows that in the majority of the cities studied, a higher volume of passenger traffic is served by airports than ferry terminals. The exception is Tallinn and Rostock, where larger volumes of traffic are generated by ferry terminals. The most balanced branch structure of passenger traffic is served in Helsinki and Klaipeda. (Figure 1).

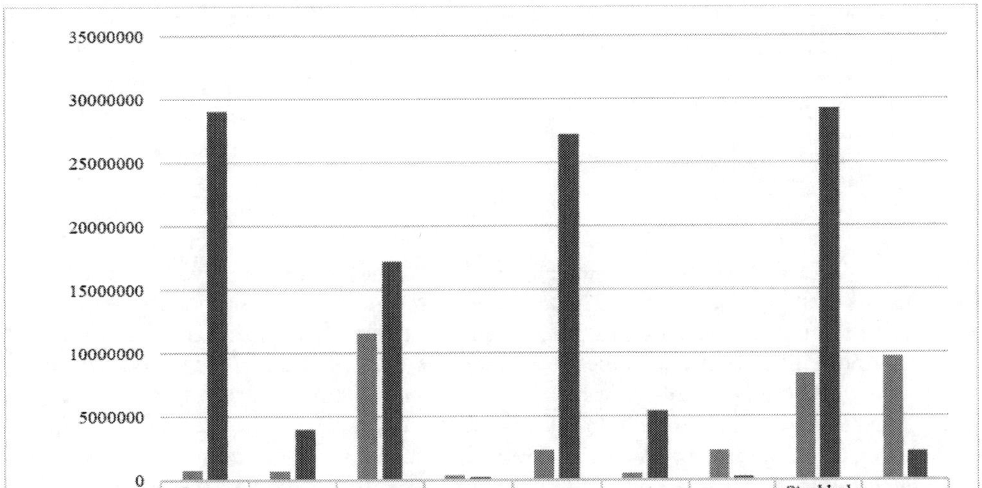

	Copen-hagen	Gdansk	Helsinki	Klajpeda	Oslo	Riga	Rostock	Stockhol m	Tallinn
Ferry ports	749300	703700	11565000	350670	2328245	510157	2300000	8333748	9663258
Airports	29043287	3963783	17184661	232630	27221368	5401243	250200	29235426	2221615

Fig. 1. The volume of passenger traffic served by the examined airports and ferry terminals in 2016 [54-63].

In the surveyed group of origin ports, the largest number of separate ferry terminals operates in Helsinki, 6 ferry terminals are located there. The analysis excluded Hansa Terminal, which handles the Helsinki-Travemünde route with the prevailing cargo traffic. Three ferry terminals also serve Stockholm and Tallinn. In the case of Tallin, ferry services for ferry operators of Viking Line, Eckerö Line and Moby Line SPL's were separated into two separate terminals according to departures and arrivals (successively terminal A and terminal B). The third terminal D serves only Tallink Silja ferries. Two ferry terminals also serve Oslo, Klaipeda and Gdansk. In the case of Gdańsk, Gdynia Ferry Port was accepted as the second terminal. Other analyzed cities such as: Copenhagen, Riga and Rostock are served by one ferry terminal. Specification of the analyzed ferry terminals and the distance from the city center is presented in Table 1.

Table 1. The distance of the analyzed ferry terminals from the city center [64]

destination	**distance from the city center (km)***
Copenhagen	4,5
Gdańsk Westerplatte Ferry Terminal	10
Gdynia Ferry Port	36,5
Helsinki Katajanokka Terminal	3
West terminal 1	3
West terminal 2	4
Makasiini Terminal	2,5
Olympia terminal	3
Klajpeda International Ferry Terminal (IFT)	13
Central Klaipedos Terminal (CKT)	3,5
Oslo DFDS Seaways and Stena Line Terminal	2
Color Line Terminal	3,5
Riga	1,5
Rostock	2,5

Stockholm	
Viking Line terminal (Stadsgården)	14,5
Värtahamnen port	10
Ferry terminal Frihamnen	6
Tallin	
Passenger terminal A i B	1,5
Passeneger terminal D	2,5

* the estimated distance was measured for car traffic and given to an accuracy of 500 meters.

13 out of 19 analyzed ferry terminals were located not more than 5 km from the city center, and as many as 9 of them are located not more than 3 km from the city center. The nearest ferry terminals are located in Oslo, Riga and Tallinn. At a distance of more than 5 km, but no more than 10 km from the city center, there are 3 of the 19 analyzed terminals. Only 3 of the analyzed terminals are located more than 10 km from the city center. These include Gdynia Ferry Port, Stockholm (Viking Line terminal) and Klaipėda (International Ferry Terminal). In the case of the ferry terminal in Gdynia, the distance from the city center is explained by the fact that the city of Gdańsk was accepted as the destination, not Gdynia (the distance between the center of Gdynia and the terminal is 5 km).

For the purposes of comparative analysis, the most important airports serving passenger air traffic, the location of which coincides with ferry terminals, were chosen. With regard to transport, most airports are located in the area of Stochkolm (Arlanda, Skavsta and Bromma) and in Oslo (Gardermoen, Sandefjord-Torp and Moss-Rygge), three air transport points. However, during the analyzed time period (January 2018), the airport serving the Oslo Moss-Rygge region was closed. However, for the remaining agglomerations, one airport was assumed for the needs of research, despite the fact that in close proximity to Danish Copenhagen Airport operates the Swedish port of Malmö-Sturup, serving low-cost traffic (distance 51 km). There are no premises related to cooperation between these ports, and one city from each country was selected for the research. Including Malmo as an origin port would cause inconsistency with the research assumptions. The distances between the particular ports and the city center are presented in Table 2.

Table 2. The distance of the analyzed airports from the city center [64]

destination	distance from the city center (km)*
Copenhagen	7
Helsinki	20
Oslo	
Oslo Gardermoen	49,5
Oslo Torp	119
Stockholm	
Stockholm Arlamda	41,5
Stockholm Skavsta	106
Stockholm Bromma	8
Riga	10,5
Talinn	7
Klajpeda	8
Gdańsk	16,5
Rostock	45

* the estimated distance was measured for car traffic and given to an accuracy of 500 meters.

The examined airports are located at a much greater distance from city centers than ferry terminals. Of the 12 examined airports, the four nearest airports were within 10 km of the city center (airports in Tallinn, Copenhagen, Klajpeda and Stockholm Bromma). The remaining 8 airports are located at a distance of more than 10 km. The airports of Oslo Torp and Stockholm Skavsta are located farthest from the city centers (over 100 km).

Unlike airports, the analyzed ferry terminals support only BSR traffic, which is a natural feature of ferry shipping and results from its regional character [27]. The most popular ferry route is the Helsinki-Tallinn line with the frequency of weekly flights at 94. According to the duration, this route is one of the longest in the BSR (long duration voyages of over 11 hours, [65] and the journey lasts, depending on the ferry operator, 16-17 hours. The second Baltic route with the highest frequency of sailings is the Rostock-Gedser route, which is classified as a short-term route (short duration voyages of up to 2 hours). The travel on this route lasts 1h 45min. Connections with relatively high frequency of sailings per week (over 10) also include ferry services on the Rostock-Trelleborg, Stockholm-Turku, Gdynia Karlskrona and Helsinki-Stockholm routes. However, the lowest frequency sailings include connections with the Baltic with the Russian Federation, i.e. services from Helsinki, Stockholm and Tallinn to St. Petersbug. This is due to their typically tourist nature (ferry cruise, round trip) with a cruise time of over 62 hours from Tallin to St. Petersburg.

Table 3. Structure and frequency of connections in the analyzed ferry terminals [66]

ferry ports	destinantion	frequency (weekly)
Copenhagen	Oslo	7
Gdańsk i Gdynia	**Nynäshamn (Stockholm)**	3
	Karlskrona	18
Helsinki	Sztokholm	7
	Tallin	94
	St. Pettersburg	2
	Sztokholm	15
Klaipeda	Karlshamn	7
	Kiel	7
Oslo	Copenhagen	7
	Frederikshavn	7
	Kiel	7
Riga	Sztokholm	7
Rostock	Trelleborg	33
	Gedser	68
Stockholm	Helsinki	15
	Turku	28
	Tallin	8
	Ryga	7
	St. Petersburg	1
Tallin	Helsinki	94
	Sztokholm	8
	St. Petersburg	1

The analyzed airports play a different role in the transport system. Copenhagen, Stockholm Arlanda, Oslo Gardermoen and Helsinki are typical hub ports, others play a role in regional service (Gdańsk, Torp, Skavsta, Riga and Talinn) and local service (Bromma, Klaipeda and Rostock) (see Table 4). This means that the nature of ports enforces differentiation of the network of connections. While the network of connections from ports located in the BSR to hub ports may be one of the elements of the journey (airline travels through the "hub and spoke" system, connections between regional and local ports have a typical direct character - "point to point").

In the group of the analyzed airports, the largest share of connections within the BSR in the total number of connections (taking into account their frequency) have Stockholm Bromma airports (88.5%), Klaipeda (84%), Tallinn (70.9%) and Stockholm Arlanda (50.6%). The lowest proportion of connections within the framework of the BSR in general are carried out on Oslo Gardemoren and Stockholm Skavsta airports (less than 15%). Rostock airport does not make any air connections to BSR ports, which is largely due to its local nature and proximity to international airports in Berlin (catchment area overlap).

Table 4. Connections of the examined airports under the BSR [55-63]

airports	destination	airports BSR	total connection during week	total connection during week to BSR	BSR/total
Copenhagen	95	13	2373	791	33.3
Helsinki	104	15	1739	636	36.6
Oslo Gardermoen	111	11	2324	324	13.9
Oslo Torp	29	3	128	33	25.8
Stockholm Arlamda	132	24	2346	1188	50.6
Stockholm Skavsta	35	2	115	17	14.8
Stockholm Bromma	15	12	592	524	88.5
Riga	68	9	663	209	31.5
Talinn	25	5	199	141	70.9
Klajpeda Palanga	5	3	25	21	84.0
Gdańsk	59	10	332	95	28.6
Rostock	8	0	27	0	0.0

4.2. The comparative analysis of ferry terminals and airports through the prism of time competitiveness of ground access

The comparative analysis of transfer time from the analyzed ferry terminals and airports to the centers of the cities served by them was carried out in two variants:

- for transfers carried out individually by road transport, assuming the possibility of renting a passenger car at an airport or travelling by ferry with a passenger car,
- for transfers made using public transport (urban and rail).

In the case of the availability of various forms of public transport, the comparative analysis takes into account those with the shortest duration, i.e. rail links to airports in Copenhagen, Oslo Gardemoren, Stockholm Arlanda and Gdańsk. The results of the comparative analysis are presented in Figures 2 and 3.

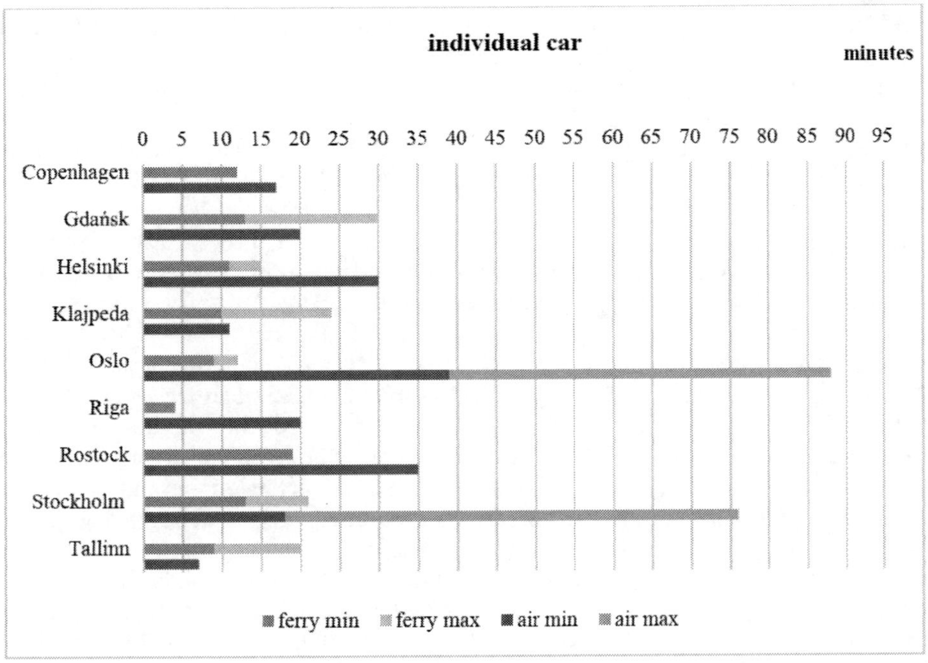

Fig. 2. Comparison of travel time by car from the city center to the ferry terminal and airport [51-53, 64].

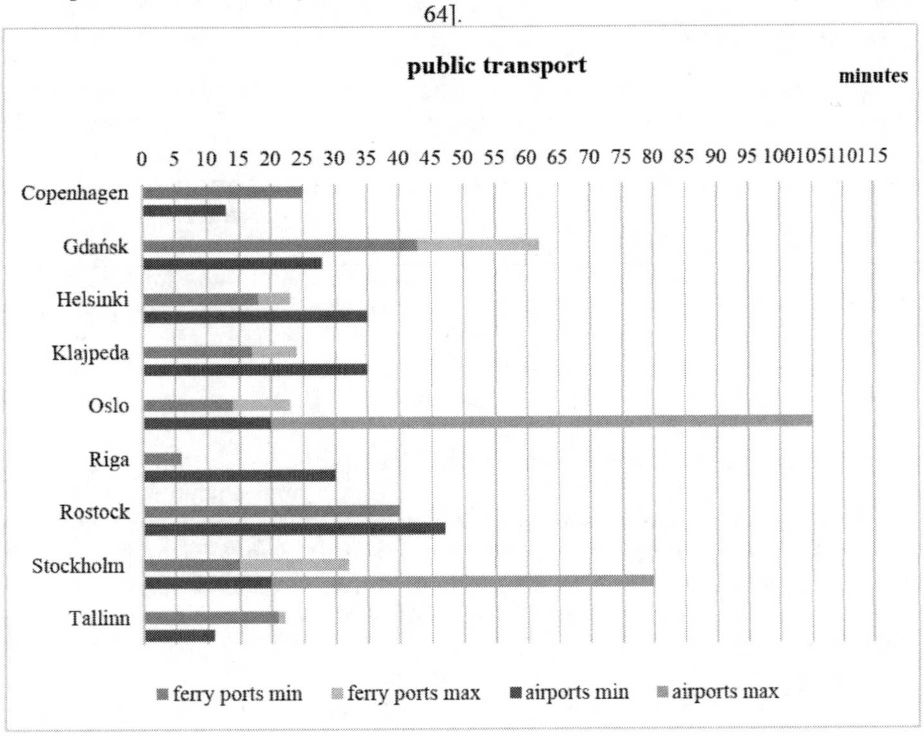

Fig. 3. The comparison of public transport travel time from the city center to the ferry terminal and airport [51-53, 59-60, 63-64, 67-71].

In most of the analyzed origin ferry ports and airports, a shorter estimated time of arrival by car from the transport point to the city center was noted in the case of ferry terminals. The longer maximum time of arrival from the ferry terminal to the center in relation to the transfer from the airport occurs in Tallinn, Klaipeda and Gdańsk. Besides, apart from Tallinn, it concerns transfers only from points further located of ferry terminals. The largest difference in the estimated transfer times concerns Stockholm and Oslo. This is due to the remote location of airports to the city centres.

The comparative analysis carried out for public transport showed that for some origin ports, the time competitiveness of ferry terminals relative to airports in terms of transfer time to the city center (Copenhagen) – decreased, or conversely- the time competitiveness of transfers from ferry terminals at the expense of longer transfers from airports (Klaipeda) increased. In the case of ferry terminals in Rostock and Stockholm, as in the case of car travel, the time of transfer to the city center by public transport is shorter than access from alternative airports. Nevertheless, it was relatively prolonged while reducing the transfer time from alternative airports. In the case of Gdańsk, the difference between the shorter time of arrival from the airport to the city center and the time of transfer from the ferry terminal, as compared to car travel, deepened. The situation hardly changed in the case of Helsinki, Oslo and Riga. It should be emphasized that dedicated railway connections to airports, which are available in some of the analyzed origin airports, significantly shorten the transfer time to the city center. In the case of Oslo Gardemoren and Copenhagen, this is over 62% in relation to bus connections, and Stockholm Arlanda more than 57%. Only in the case of Stockholm Skavsta travel by train takes longer than public transport.

5 Conclusions

The results of the comparative analyzes of the time competitiveness of airport and seaport ground access in the BSR allow to make a general conclusion that ferry terminals have better ground access for individual car journeys. This corresponds to the preferences of ferry passengers which were identified in the other research, who consider it the best option to travel with a car [24, 27]. Ferry terminals in terms of time competitiveness of ground access strengthen their position of ferry shipping in servicing the segment of motorized passengers. The longer time of reaching the destination by public transport results, in the majority of the analyzed cases, due to the lack of dedicated connections by public transport from ferry terminals to the city center (especially railway).

However, airports are becoming more competitive in terms of ground access in the case of the arrival and departure function of public transport. The research additionally indicated that the transfer time of public transport from the analyzed airports to the city center is definitely reduced due to the functioning of airport rail links. As a result, airports are able to meet the main demands made by passengers (time, immediacy, frequency, lack of congestion) thanks to more time-based railway transfers [72]. It should be pointed out that the lower availability of individual transport to airports is mainly due to the large distance between airports and the city center and high fees for the use of parking infrastructure. The airports to a greater extent meet the expectations of individual passengers who make transfers using public transport than ferry terminals.

Summing up, it should be stated that the system of ground access for seaports and airports is adapted to the segment of passengers having the largest share in transport by the analyzed transport branches, i.e. the motorized passengers in ferry transport and non-motorized passengers in air transport.

6 References

1 P. Freathy, & F. O'Connell, Strategic reactions to the abolition of duty free:: Examples from the European airport sector. *Euro. Man. J.*, **18**, 638-645 (2000)

2 A. Gebauer, C.W. Nam, & R. Parsche, Lessons of the 1999 abolition of intra-EU duty free sales for the new EU member states. *CESifo Econ. St.*, **51**, 133-157 (2005)

3 T. Wergeland, Ferry passenger markets. *The Blackwell Comp. to Marit. Econ.*, 161-183 (2012)

4 A. Graham, Liberalization, regional economic development and the geography of demand for air transport in the European Union. *J. Tran. Geog.*, **6**, 87-104 (1998)

5 P. Morrell, Air transport liberalization in Europe: the progress so far. *J. A. Tran. World Wide*, **3**, 42-60 (1998)

6 A. Micco & T. Serebrisky, Competition regimes and air transport costs: The effects of open skies agreements. *J. Int. Econ.*, **70**, 25-51 (2006)

7 X. Fu, T. H. Oum, & A. Zhang, Air transport liberalization and its impacts on airline competition and air passenger traffic. *Tran. J.*, 24-41 (2010)

8 K. Button, The impact of US–EU "Open Skies" agreement on airline market structures and airline networks. *J. A. Tran. Man.*, **15**, 2, 59-71 (2009)

9 K. Moszkowicz, M. Moszkowicz, Konkurencyjność i innowacje – gdzie szukać szansy?. *Tożs. Str. Przed. – modele i doświadczenia*. M. Moszkiewicz (ed.), 110, (House Publ. Technical Wrocław University, 1997)

10 M.E. Porter, *The Competitive Advantage of Nations*, 70-72 (Macmillan Press Ltd, 1986).

11 D. McFadden, *Conditional logit analysis of qualitative choice behavior* (1973)

12 W.B. Jackson & J.V. Jucker, An empirical study of travel time variability and travel choice behavior. *Tran. Sc.*, **16**, 460-475 (1982)

13 Y.L. Ye & Y.S. Wang, Research on travel mode choice behavior in Shanghai-Hangzhou transport corridor. *J. China Rail. Soc.*, **32**, 13-17 (2010)

14 M. Hunecke, A. Blöbaum, E. Matthies & R. Höger, Responsibility and environment: Ecological norm orientation and external factors in the domain of travel mode choice behavior. *Env. Beh.*, **33**, 830-852 (2001)

15 J. Scheiner & C. Holz-Rau, Travel mode choice: affected by objective or subjective determinants?. *Transportation*, **34**, 487-511 (2007)

16 A. Vij, A. Carrel & J.L. Walker, Incorporating the influence of latent modal preferences on travel mode choice behavior. *Tran. Res. Part A: Pol. Prac.*, **54**, 164-178 (2013)

17 M. Paulssen, D. Temme, A. Vij & J.L. Walker, Values, attitudes and travel behavior: a hierarchical latent variable mixed logit model of travel mode choice. *Tran.*, **41**, 873-888 (2014)

18 J. De Vos, P.L. Mokhtarian, T. Schwanen, V. Van Acker & F. Witlox, Travel mode choice and travel satisfaction: bridging the gap between decision utility and experienced utility. *Tran.*, **43**, 771-796 (2016)

19 J.C. Martín, C. Román, J.C. García-Palomares & J. Gutiérrez, Spatial analysis of the competitiveness of the high-speed train and air transport: The role of access to terminals in the Madrid–Barcelona corridor. *Tran. Res. Part A: Pol. Prac.*, **69**, 392-408 (2014)

20 D. Albalate, G. Bel & X. Fageda, Competition and cooperation between high-speed rail and air transportation services in Europe. *J. Tran. Geog.*, **42**, 166-174 (2015)

21 T. D'Alfonso, C. Jiang & V. Bracaglia, Would competition between air transport and high-speed rail benefit environment and social welfare?. *Tran. Res. Part B: Meth.*, **74**, 118-137 (2015)

22 W. Xia & A. Zhang, High-speed rail and air transport competition and cooperation: A vertical differentiation approach. *Tran. Res. Part B: Meth.*, **94**, 456-481 (2016)
23 W. Rothengatter, High-speed rail and air travel complementarity: the case of Germany. *High-Speed Rail and Sustainability,* (Routledge, 123-134, 2017)
24 E. Sambracos, *Air-sea transport: competitiveness or complementarity in the Greek insular market.* (*Inter. Con. A. Tran. A. evol. in the 21st Century,* 3-4, 2001)
25 K. Rigas, Boat or airplane? Passengers' perceptions of transport services to islands. The example of the Greek domestic leisure market. *J. Tran. Geo.*, **17**, 396-401 (2009)
26 K. Rigas, E. Sambracos, & A. Gatzoli, Air and sea transport: Competition strategies under normal and economic crisis environments. *SPOUDAI-J. Econ. Bus.*, **61**, 65-84 (2013)
27 M. Mańkowska, *The concept of development of passenger ferry services in the Baltic sea region in terms of the growing interbranch competition.* (17th Inter. Conf. Tran. Sc. Slovenia: Mar. Tran. Log. Sc., 2015)
28 A. Pantouvakis, Port-service quality dimensions and passenger profiles: an exploratory examination and analysis. *Mar. Econ. & Log.*, **8**, 402-418 (2006)
29 P. Malighetti, S. Paleari & R. Redondi, Connectivity of the European airport network:"self-help hubbing" and business implications. *J. A. Tran. Man.*, **14**, 53-65 (2008)
30 S. Paleari, R. Redondi, P. Malighetti, A comparative study of airport connectivity in China, Europe and US: which network provides the best service for passengers?. *Tran. Res.*, **42**, 198-210 (2010)
31 G. Burghouwt, R. Redondi, Connectivity in Air Transport Networks. An Assessment of Models and Applications. *J. Tran. Econ. Pol.*, **47**, 1, 37-41 (2013)
32 N. Dennis, Airline hub operations in Europe. *J. Tran. Geog.*, **2**, 219-233 (1994)
33 N. Dennis, Scheduling strategies for airline hub operations. *J. Tran. Man.*, **1**, 131-144 (1994)
34 N. Dennis, R. Doganis, Lessons in Hubbing, *A. Bus.*, **3**, 42-47 (1989)
35 J. Veldhuis, The competitive position of airline networks. *J. A. Tran. Man.*, **3**, 181-188 (1997)
36 G. Burghouwt, J. Veldhuis, The competitive position of hub airports in Transatlantic market. *J. A. Tran.*, **11**, 1, 106-130 (2006)
37 J. Veeldhuis, E. Kroes *Dynamics in relative network perfornance of the main European hub airports,* (Proceeding of the Eur. Tran. Conf. Cambridge, 9-11 September 2002)
38 A. Danesi, Measuring airline hub timetable co-ordination and connectivity: definitione of a new index and application to a sample of European hubs. *Euro. Tran.*, **34**, 54-74 (2006)
39 M.L. Tam, W.H. Lam & H.P. Lo, Modeling air passenger travel behavior on airport ground access mode choices. *Transportm.*, **4**, 135-153 (2008)
40 D.A. Tsamboulas, & A. Nikoleris, Passengers' willingness to pay for airport ground access time savings. *Tran. Res. Part A: Pol. Prac.*, **42**, 1274-1282 (2008)
41 M.L. Tam, W.H. Lam & H.P. Lo, Incorporating passenger perceived service quality in airport ground access mode choice model. *Transportm.*, **6**, 3-17 (2010)
42 R.C. Jou, D.A. Hensher & T.L. Hsu, Airport ground access mode choice behavior after the introduction of a new mode: A case study of Taoyuan International Airport in Taiwan. *Tran. Res. Part E: Log. Tran. Rev.*, **47**, 371-381 (2011)
43 T. Budd, T. Ryley, & S. Ison, Airport ground access and private car use: a segmentation analysis. *J. Tran. Geog.*, **36**, 106-115 (2014)

44 I. Gokasar & G. Gunay, Mode choice behavior modeling of ground access to airports: A case study in Istanbul, Turkey. *J. A. Tran. Man.*, **59**, 1-7 (2017)
45 G. Burghouwt, & R. Redondi, Connectivity in air transport networks: an assessment of models and applications. *J. Tran. Econ. Pol.*, **47**, 1, 35-53 (2013)
46 D. Tłoczyński, *Konkurencja na polskim rynku usług transportu lotniczego*, (University of Gdańsk Publisher, 2016).
47 D. Tłoczyński, *The analysis of transport accessibility in airports as the part of the competitiveness in regions. Case study of Polish Airports*, (17th Inter. Conf. Tran. Sc. Slovenia: Mar. Tran. Log. Sc., 2015)
48 P. Koster, E. Kroes, E.T. Verhoef, Travel Time Variability and Airport Accessibility, *Tinb. Inst. Disc. Pap.*, **061/3**, 7 (2010)
49 A. Reynolds-Feighan, P. McLay, Accessibility and attractiveness of European airports: A simple small community perspective. *J. A. Tran. Man.*, **12**, 315 (2006)
50 R.K. Yin, Case Study Methods., *Handbook of complementary methods in education research* J. L. Green, G. Camilli, & P. B. Elmore (Eds.). 111-122 (2006)
51 Stops.lt downloaded from www.stops.lt (10.01.2018)
52 Rīgas satiksme. downloaded from saraksti.rigassatiksme.lv (10.01.2018)
53 Moovit. downloaded from moovitapp.com (10.01.2018)
54 ShipPax Information, (Sweden, 2017)
55 Avinor. downloaded from https://avinor.no (10.01.2018)
56 Copenhagen Airport. downloaded from www.cph.dk (10.03.2018),
57 Gdansk Airport. downloaded from www.airport.gdansk.pl (10.02.2018)
58 Finavia airports. downloaded from www.finavia.fi (10.01.2018)
59 Palanga Airport. downloaded from www.palanga-airport.lt (10.01.2018)
60 Riga Airport. downloaded from www.riga-airport.com (10.01.2018)
61 Rostock Airport. downloaded from www.rostock-airport.de (10.01.2018)
62 Swedavia airports. downloaded from www.swedavia.se (10.01.2018)
63 Tallin Airport. downloaded from www.tallinn-airport.ee (10.01.2018)
64 Google.maps. downloaded from: www.google.com/maps (10.01.2018)
65 I. Kotowska, *Short Sea Shipping in the light of the idea of sustainable development of transport* (Szczecin: Sc. Publ. Hou. Mar. Univ. 2014)
66 Direct Ferries. downloaded from www.directferries.com (10.01.2018)
67 Tripadvisor. downloaded from https://www.tripadvisor.com (10.01.2018)
68 Visitcopenhagen. downloaded from https://www.visitcopenhagen.com (10.01.2018)
69 Visitoslo. downloaded from http://www.visitoslo.com (10.01.2018)
70 Visitstockholm. downloaded from https://www.visitstockholm.com (10.01.2018)
71 ZTM. downloaded from www.ztm.gda.pl (10.01.2018)
72 D. Tłoczyński, M. Wołek, *Komplementarność i substytucja w pasażerskim transporcie kolejowym i lotniczym. Procesy integracyjne wybranych systemów transportowych*, M. Michałowska (ed.), (Academic of Economy in Katowice, 263, 2007)

SHS Web of Conferences **58**, 01030 (2018) https://doi.org/10.1051/shsconf/20185801030
GLOBMAR 2018

Challenges for Polish ferry market development in Baltic Sea Region

Ilona Urbanyi-Popiołek[1,*]

[1] Department of Logistics and Transport Systems , Gdynia Maritime University, Poland

Abstract. Polish ferry market has become one of the prime segments of ferry shipping in Baltic Sea Region. It dominates in ferry industry in South Baltic. The ferry cargo and passenger traffic has been growing steadily since Poland's accession to the European Union. The estimates show that in 2017 the operators transported more than 1,2 million passengers and over 520 thousand cargo vehicles between ports in Poland and Sweden. The increase in trade turnover between Scandinavia and Central Europe as well as growth of tourism affect the demand for ferry transport. The aim of this paper is to explore the prime determinants influencing the development of Polish ferry shipping and to research the challenges that carriers operating the ferry services from Poland to Sweden must face such as internal competition between the operators in Polish market, competition from German and Lithuanian routes and low cost airlines as well as increasing trade volumes. The research hypothesis is: the growth of trade between Scandinavia and Central Europe as well as tourism traffic will increase the demand for ferry transport from Polish ports. Detailed research hypotheses are that: ferry services between Poland and Sweden constitute the primary market on the South Baltic and new tonnage investments to increase the capacity have to be taken.

1 Introduction

Ferry shipping is well developed in many regions of the world. Northern Europe and Mediterranean are regions of high concentration of ferry industry with Baltic Sea being one of the prime market [7].

Baltic ferry business is highly concentrated, at the beginning of January 2018, 16 major carriers operated on the primary international and domestic markets. Stena Line, Finnlines, Tallink, Scandlines and TT-Line dominate in Baltic Sea in terms of transport capacity and market share. The important carriers on the market include the Norwegian Color Line and Fjord Line as well as Polish companies operating in the South Baltic area. The network of ferry connections included 60 services with 116 ferries operated (ro-pax, cruise-roro i high-speed)[24].

It is not possible to estimate precisely the current volume of passenger and cargo ferry traffic in Baltic Sea Region. In previous years, the carriers published the annual and monthly volumes achieved on their routes, so it was possible to accurately estimate the

* Corresponding author: i.urbanyi@wpit.am.gdynia.pl

turnover. Since 2015, some operators have stopped publishing data. On the basis of previous empirical data as well as market trends, it was is assessed, that in 2017 around 49 million passengers, 9 million cars and 3 million cargo units were carried on international and main domestic Danish and Swedish ferry lines (e.g. Bornholm and Gotland) in Baltic Sea Region [24].

In the Baltic Sea Region three prime ferry markets are separated:

- West Baltic - comprises the lines between Denmark and Sweden, Norway and Germany, services from Sweden to Norway and Germany and line Germany - Norway; the west market accounts for about 54% of passenger and about 62% of freight traffic,
- East Baltic - covers services from Sweden to Finland and Estonia, lines from Estonia to Finland and Russia; the east market accounts to 37.5% of passenger and about 19% of freight turnover,
- South Baltic/ Central Baltic - includes services between Sweden and Poland, from Sweden to Latvia and Lithuania and the connections between Germany and Lithuania, Latvia and Finland; the central market constitutes the 8.5% of passenger transport and about 19% of freight.

Routes between Poland and Sweden constitute a prime ferry market on the latter Baltic region.

2 Methodology and literature review

In the literature and in practice, there is not a uniform and generally used definition of ferry shipping and ferry vessel. The reason for it is connected with permanent transformation of the this type of ships, changes in supply and demand at ferry market, as well as approach of the authors to the presented issues.

Stapford states that ferries transport people, goods and vehicles over short distances by sea [30]. Kotowska highlights the carriage by ferry shipping two segments: passengers traffic and cargo transportation [9]. Similarly ferry shipping describes Wergeland [34]. Most authors however focus either on cargo or passenger segment.

Some of researches describe the cargo traffic only. Musso, Paixao and Lynce, as well as Paixao and Marlow define ferry shipping as a segment of short sea shipping and emphasise that these connections are an element of road –sea transport system to and from continental Europe [16. 18]. Paixao and Marlow, Morales-Fusco, Sauri and Melo as well as Daduna and et al stress the importance of ferry services in multimodal transport chains [17, 15, 4]. Cargo segment of ferry shipping is regarded by e.g., Suarez-Aleman, Trujillo and Medda, as well as Baird [31, 1].

The others pinpoint the passenger service and tourism functions of ferry shipping, e.g. Peter, Heijveld and Grey, Pantouvakis, Wild and Dearing [20, 6, 19, 36]. Mańkowska,, as well as Kizielewicz, Haahti,, Lukovic and Gracan focus on passenger segmentation [13, 8].

Some researches consider geographical markets, as Mathisen and Solvoll, Polydoropoulou and Litinas, Graham and Russell [14, 21, 5]. The other papers focus on ferry service network and fleet, e.g. Lai and Lo, Lo, An and Lin, as well as Wang and Lo [11, 12, 35].

Ferry shipping is also considered as alternative for land transportation in context of sustainable transport development, e.g. Casteliss, Usabiaga and Martinez, Kotowska, Chen, Yang and Notteboom [2, 10, 3].

The Swedish ShipPax, the leading ferry consulting and information provider, define ferry as a ship larger than 1000 GT that sails on a regular line and has passenger accommodation and is using ro-ro technology for the transportation of cars and commercial vehicles (if any), having sufficient free height on car deck(s) for this [26]. This definition of ferry ship will be assumed in the paper.

Ferry shipping is a type of liner trade where passengers and cargo form one market. Both segments are carried by one ship, ferries ply the regular route.

The principal aim of this paper is to analyse the development of Polish ferry market and to research the prime factors and challenges affecting this market. The research hypothesis is: the growth of trade between Scandinavia and Central Europe as well as tourism traffic will increase the demand for ferry transport from Polish ports. Detailed research hypotheses are that: ferry services between Poland and Sweden constitute the primary market on the South Baltic and new tonnage investments to increase the capacity have to be taken.

For the purpose of this paper, the analysis of ferry shipping at the South Baltic has been conducted. The analyses embraces cargo and passenger turnover, services network, fleet and carriers. The research is based on data collected directly from Polish operators. Moreover the author has conducted in-depth interviews with chairmen of management boards as well as cargo and passenger operation departments. The paper is based also on data from ShipPax publications and official trade figures collected from Statistics Sweden. Upon the available data Polish ferry market has been analysed.

The article focuses on selected factors determining Polish ferry market. With reference to the demand, foreign trade, as prime factor for cargo flows has been discussed and further development has been predicted. The passenger segment is highly affected by low costs airlines activity. Moreover the competition from German and Lithuanian routes was presented.

Other factors affecting the ferry market, as environmental requirements, access infrastructure, fuels prices, competition of land modes of transport do not fall within the scope of the research due to the limited volume of the study.

3 Selected factors determining ferry traffic demand

The mutual economic connections between Nordic countries as well as between the Baltic States and the countries of West and Central Europe are essential for the development of trade in the Baltic Sea Region. Cargo traffic flows from Central Europe to Scandinavian countries and vice versa are complementary to mutual trade turnover.

Close commercial relations between Scandinavia and northern Continental states are strongly developed since these countries have acceded to the European Union. The mutual turnover however shows the differences in exports and imports among individual countries. The highest trading volume is observed with Sweden being the prime Scandinavian partner for Central European states, the next are Denmark and Norway. Among Central European countries, Poland has become a prime trading partner for Sweden. (Fig.1 and 2). In 2017 the Polish exports amounted to 48,86 million SEK whereas import from Sweden totalled 38,51 million SEK (respectively 2,36 million tons in export and 2,37 million tons in import). Polish share in total Sweden foreign trade amounted to 3% in exports as well as in imports. This applies to both value and volume turnover [28].

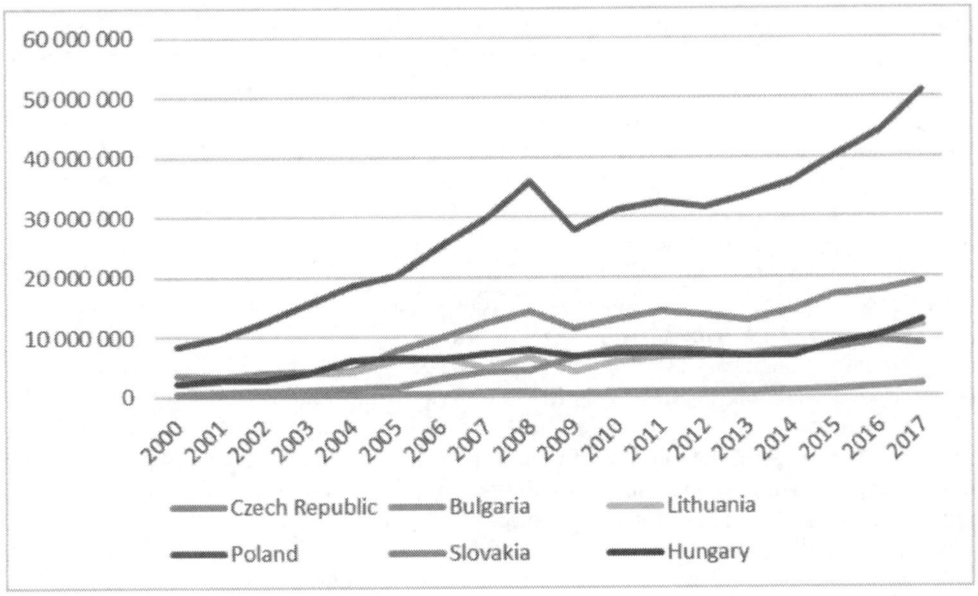

Fig. 1. Exports to Sweden from selected Continental countries in 2000-2017 (million SEK).Source: owns elaborations based on SCB Statistic Sweden.

The next trading partner are Czech Republic (1% in Swedish imports and exports), Hungary and Slovakia (both countries 1% in imports and less than 1% in exports). In comparison with Poland, these countries show significantly lower turnover, e.g. Czech Republic, Sweden's second-largest partner achieved a level more than 2.5 and 3.5 times lower in exports and imports respectively (Fig.1 and 2) [28].

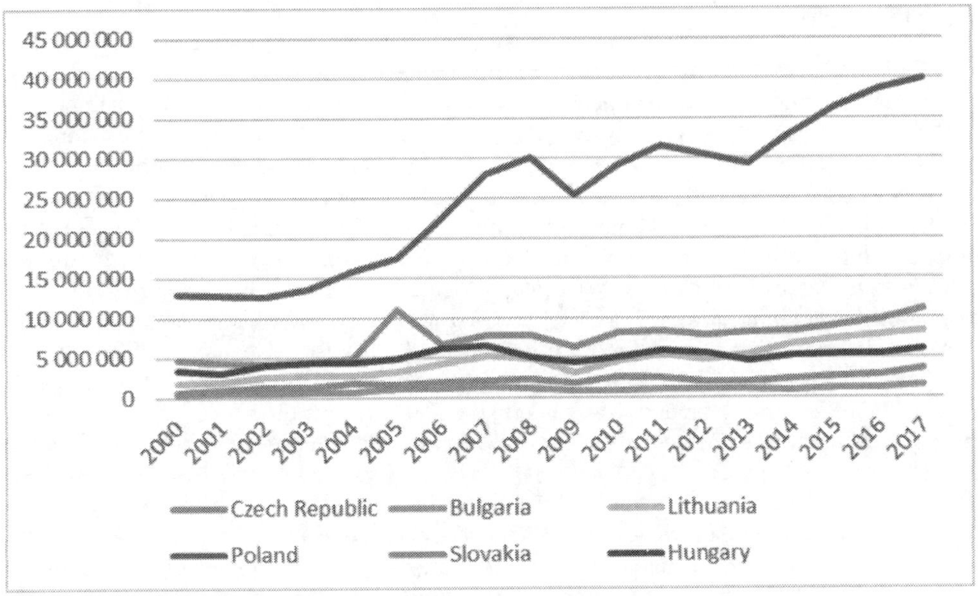

Fig. 2. Import from Sweden to selected Continental countries in 2000-2017 (million SEK). Source: owns elaborations based on SCB Statistic Sweden.

On the basis of data on trade turnover in the years 2000-20017 [28] and trend line the further growth of commercial traffic is forecasted. Polish export to Sweden is expected to rise by 5% till 2023 while Czech Republic and Slovakian is estimated to increase much higher - by 22% and 42% respectively. Import from Sweden to Central Europe is characterized by great instability, with large fluctuations in turnover, so the estimates and trend line are not reliable, however the volumes are expected to grow according estimates based on Statistics Sweden.

The commercial intercourse between Central European and two other Scandinavian countries is lower in value and volume. Poland is again the prime trading partner for Norway and Denmark with 3% share in foreign exchange of both countries. The other analysed states constitute around 1-2% in total turnover of Denmark and Norway [27, 29].

In mutual exchange dominate valuable commodities as electronic appliances, machinery, construction equipment, medical equipment, textiles, furniture, household appliance, food and agriculture commodities, as well as steel and semi-finished products.

Presented commercial intercourse, in regard to volumes and commodity structure, is the prime determinant developing the cargo flows from Central Europe to Scandinavia and creating the demand for ferry transport of cargo.

Passenger traffic is the second segment of ferry industry. The demand is created by people travelling in different purposes, e.g. visit chosen destination in connection with various needs, package trips, shopping, business. According to the prime factor generating the demand, the ferry passengers may be divided into categories:

- Liner passengers – travellers taking a trip in various reasons (visit an interesting destination, participation in culture or sport event, visit relations etc.), ferry crossing is only a part of whole trip,
- Cruise passengers – travellers participating in package trips of diverse nature, e.g. round trips for pleasure, trips with short stay in port of destination etc.,
- Conference passengers – participants in conferences and seminars organized on-board within liner trips,
- Shopping passengers – people who travel to purchase commodities on-board or in port of destination,
- Business passengers – passengers travelling in connection with their professional activities.

In practice, assigning the travellers to given category is not clear because passengers traveling e.g. as conference participants also do some shopping and enjoy entertainment. The assignment of traveller to a given segment results from the main transport need.

Polish traffic includes all mentioned categories, but over last ten years the structure of travellers has been changing. The number of liner and business passengers declined due to low cost airlines activity. The network of LCCs comprises flights from main Polish airports e.g. Warsaw, Gdansk, Cracow, Wroclaw, Poznan, Katowice to Swedish cities of Stockholm, Gothenburg, Malmoe, as well as Oslo and other Norwegian destinations. The frequency of Wizz Air and Ryanair flights are from 3 daily (e.g. Gdansk – Stockholm), 1 daily (e.g. Warsaw – Stockholm), to 2-3 weekly (e.g. Poznan – Stockholm, Poznan – Oslo, Cracow – Gothenburg, Cracow – Stockholm). The direct influence of LCCs on ferry markets is difficult to estimate, as the low cost airlines do not publish detailed figures for each destination. However in 2011 the total number of air passengers to or from Sweden amounted to 577,9 thousand, while in 2016 grew to 1 097 [32, 33]. The figure embrace LCCs traffic as well as flag carries, e.g. SAS and Lot as chief ones. So during 5 years the air traffic nearly doubled. In the same time the turnover of ferry passengers is stable and fluctuates around 1 000 thousand travellers yearly with slight increase in 1-2% y/y. One can assume, the increase in air traffic was caused primary by LCCs with cheap tariffs and some small destination, e.g. Swedish - Vaxjo, Malmoe, Polish - Wroclaw, Szczecin, Poznan, being not in traditional network.

However the Polish route Gdansk – Nynashamn is an example of LCCs` victims. In 2007 the ferries carried 154 thousand passengers, while in 2016 the traffic dropped below 90 thousand. The decline resulted from a number of factors, but the growing number of low costs flights to Stockholm destination has affected the ferry traffic chiefly [26, 25].

In the other hand the growing traffic of passengers travelling for shopping or taking parts in off season cruise trips is visible. Moreover in summer months more Czech and Slovakian motorized tourists are recorded on-board Polish ferries. The latter results from improved road infrastructure.

Another factor, competition from German and Lithuanian services is presented in the next paragraph.

4 The South Baltic and Polish ferry shipping market

The ferry market on the South Baltic comprises services from Germany to Denmark and Sweden, between Sweden and Poland as well as from Sweden to Lithuania. Routes from German ports in West Baltic - Puttgarden - Rodby and Travemunde - Trelleborg are less competing for Polish market, so there not studied. The traffic comprises cargo and passengers travelling between West Germany, France, Netherlands, Belgium, Switzerland, Austria and Denmark , Norway and Sweden. So the hinterland for Polish market is not disputable.

Competitive market for ferry traffic from Polish ports constitute services connecting eastern coast of Germany with Sweden and Denmark. The greatest potential is concentrated on Rostock – Trelleborg. The route is operated jointly by TT-Line and Stena Line, offering 3-5 day departures per day. The other is Rostock – Gedser, the short alternative for cargo transported from Central Europe via Germany to Denmark.

Table 1. Ferry sub-markets on South Baltic competitive for the Polish services

Market	Service	Operator	No. of ferries	No. of daily departures
Germany-Denmark	Puttgarden - Rodby	Scandlines	5	Up to 48
	Rostok - Gedser	Scandlines	2	8-10
Germany-Sweden	TravemundeTrelleborg	TT-Line	5	3-4
	Rostok –Trelleborg	TT-Line	5	3-5
	Rostok - Trelleborg	Stena Line	2	3-5
	Sassnitz - Trelleborg	Stena Line	1	1-2
Lithuania-Sweden	Kłajpeda - Karlshamn	DFDS Seaways	3	1-2

Source: owns elaborations based on Shippax Market 16, Shippax , Halmstad 2016, Shippax Guide 16, Shippax, Halmstad 2016.

Rostock-Trelleborg is the most competing in cargo segment for Polish operators. In the first decade of XXI century, the operators used to carry 280-290 thousand cargo units yearly. In 2009 the route revealed drop in volumes, the number of units decreased to 223 thousand trucks, to due to crisis. In subsequent years, the traffic was gradually recovering. In 2013 the joint service of TT-Line and Stena Line carried 246 thousand units, and during last 4 years the cargo traffic is stable and fluctuates around 250-270 thousand units yearly [24]. Sassnitz – Trelleborg is demonstrating fall in cargo volumes and now is not a competitor with 1 departure daily off season.

The Scandlines Rostock – Gedser, after decline in 2009, has been increasing the traffic since 2010. In 2016 the operator carried 86 thousand units. The route is convenient for traffic from Central Europe and Austria for hauliers driving to Denmark and to some extent to Norway. It is also a competitive alternative for cargo transported from Central Europe to Denmark, traditionally carried via Swinoujscie and fixed link – Oresundbroen.

The Klaipeda – Karlshamn is the most Eastern service on the Central Baltic, but is not so strong competitor for the Polish market. The latter transport trucks from Lithuania and some transit lorries from Belarus and Russia. Cargo from Ukraine and Central Europe are on marginal level due to the distance to Klaipeda and road connection. In 2014-2016 the turnover amounts to 130-140 thousand cargo units. [25, 24]. The growing demand on Lithuanian market caused addition sailings with third ship from the spring 2017.

Two presented connections from Rostock are the competitors for Polish routes from Swinoujscie, as both ports are based on the same hinterland.

Polish ferry market comprises links to Sweden from ports of Gdynia and Gdansk located at East Coast and from Swinoujscie at West Coast. The services are as follows:

- Gdynia – Karlskrona (Stena Line),
- Gdynia – Nynashamn (Stena Line),
- Gdansk – Nynashamn (Polish Baltic Shipping Company – Polferries),
- Swinoujscie – Ystad (Unity Line),
- Swinoujscie – Trelleborg (Unity Line),
- Swinoujscie – Ystad (Polish Baltic Shipping Company – Polferries),
- Swinoujscie – Trelleborg (TT-Line).

Altogether 18 ferries ply these routes (table 2). The largest capacity is concentrated on the Swinoujscie - Ystad and Swinoujscie - Trelleborg, which are operated by 12 ferries. Other services are Gdynia - Karlskrona (4 ferries) and Gdansk - Nynashamn with 1 ferry. The latter one will increase the capacity September 2018 with chartered ferry Nova Star.

Table 2. Supply on Polish ferry market

Service	Operator	No. of ferries	No. of daily departures
Swinoujscie – Ystad	Unity Line	4	4
Swinoujscie – Ystad	Polferries	3	3
Swinoujscie – Trelleborg	TT-Line	2	1-2
Świnoujscie - Trelleborg	Unity Line	3	3
Gdynia - Karlskrona	Stena Line	4	3-4
Gdynia – Nynashamn	Stena Line	1	3 weekly
Gdansk – Nynashamn	Pollferries	1	3-4 weekly

Source: owns elaborations based on Shippax Market 17, Shippax, Halmstad 2017, Shippax Guide 17, Shippax, Halmstad 2017.

Stena Line dominate in passenger capacity with 3,298 passenger space, Gdynia – Karlskrona deploys the largest on Polish market cruise-trailer ferries and offer the variety of package trips. In cargo segment Unity Line leads with 10,718 total line meter.

It is not possible to estimate precisely the current volume of passenger and cargo ferry traffic from the Polish ports. Since 2015 the operators has not published the results. Consider the previous volumes and trend line as well as ferry terminals data, the increase in cargo and passenger traffic is estimated by 8% and 2% respectively y/y.

In passenger segment Stena Line leads with 45% share. In 2016 the traffic amounted to 608 thousand travellers. Such high traffic volumes results from the Stena Line tourist offer, dedicated to both the Swedish and Polish market and the possibility that arise from the operation of the cruise ferries. In turn, services from Swinoujscie transported a total of 614,000 passengers (without truck drivers). It is estimated that Unity Line carried more than half of this number on the Swinoujscie - Ystad , where cruise ferries are deployed.

In the freight segment, the highest total turnover are observed from Swinoujscie to Ystad and Trelleborg. In 2016, over 420,000 freight units (lorries and trailers) were carried on that sub-market. The turnover in the truck segment increased by 8.9% and the semi-trailers - intermodal units by 10.5% y/y. Unity Line`s share amounted to 40-45% in total cargo turnover. The Gdynia-Karlskrona line handled some 180,000 freight units in 2016, which accounts for 29% of trade in goods. The share of PŻB and TT-Line in the discussed segment amounted to 25-30%.

5 Requirements for Polish ferry market

Growing demand for freight service generate the need for cargo capacity. Currently, operators are facing deficiency in load line. The use of the available capacity for individual operators in 2016 was at the level of 70-80% and rose to 85 – 87% in 2017, which means that there is no space reserve. To meet the growing demand, the carriers have to increase the transport ability.

Some actions were taken in 2017. Polferries introduced newly purchased ferry Cracovia on Swinoujscie – Ystad and increased daily crossings to 3 in both directions.

In turn Stena Line added a fourth vessel on the Gdynia - Karlskrona, chartered ferry Gute dedicated for cargo units. The operator`s next move was the opening a new connection from Gdynia to Nynashamn with chartered Elisabeth Russ, change now by Bore Bay. Currently, the operator offers 50 calls a week from Gdynia to Sweden.

Euroafrica bought from the Italian operator ro-pax which is a sister ship of Galileusz. The ferry will be renamed Copernicus and will ply Swinoujscie – Trelleborg under Unity Line brand. The newest acquisition on the Polish market is mentioned above Nova Star. Ferry is chartered by Polferries and will operate Gdansk – Nynashamn line as second vessel.

The activities of Polferries, Unity Line and Stena Line in terms of increasing transport capacity are response to the market demand. Ferry cargo segment is the most dynamically growing sector of maritime transport in the Baltic Sea Region and Polish services are the best example of this.

Most of ferries were built in 1980-1996, all of them need replacement. The potential renovations or conversions of ferry fleet particularly in the context of adaptation to SECA requirements is not profitable.

Hence, it is necessary to implement projects for the construction of new ferries for the Polish market. In March last year, Polferries ordered a ro-pax ferry which is scheduled for operation in 2019. A second vessel is planned for this operator as well as four more ropaxes are anticipated for Unity Line. Until the finalization of the extremely ambitious but necessary for the Polish market project, carriers will have to charter ferries or purchase vessels on the secondary market. This task is difficult due to the lack of used ferries on the market with appropriate parameters (load line) and acceptable age.

The next challenge for Polish ferry market focuses on terminal infrastructure. Ferry traffic is concentrated in 3 terminals.

The largest, both in terms of turnover and available infrastructure, is located in Swinoujscie. The terminal has 6 berths with ro-ro ramps. Nevertheless currently its traffic capacity is insufficient. In 2017 total 3576 calls were registered, up 9- 12 calls daily. At the terminal infrastructure investments are planned to adapt the potential to the growing volume and support for intermodal transport. As part of the modernization berth 5 and 6 will be jointed to serve the ship up to 265 m of LOA. For lorries three new parking spaces will be constructed. The new berth will be fitted with a gallery for passengers.

In Gdynia ferry terminal is located in Western part in the immediate vicinity of container terminals. The infrastructure limits the capacity of the terminal due to length of berths and parking space as well as number of calls 4-5 daily. The new public terminal is under construction and is planned for 2020. The terminal will berth ferries up to 245 m length and be fitted with double ramp enable the simultaneously loading and discharging.

The only terminal with currently sufficient infrastructure is operated in Gdansk, where ships to 225 LOA may berthed. Terminal is fitted with 3 ramps, the parking for approximately 125 lorries and 12 buses and is very well connected with the road network (the A1 motorway and the S 7 expressway being built). It also has access to the railway infrastructure.

Two presented challenges are the most important factors the ferry operators have to face. Moreover during last 5 years another phenomenon is visible. The cargo flows between Europe and Scandinavia have been shifting to East. The latter results from congestion in German motorways and improving road infrastructure in Poland. These impact the connection from Polish terminals. More Austrian, Czech, Slovakian and other European lorries are recorded on Swinoujscie services, before transported via Rostock and Sassnitz. The share of Central European origin hauliers amounted to 30-40% depends on service. Similar effect is registered on Gdynia and Gdansk routes, where share of foreign hauliers e.g. Czech, Slovakian, Hungarian or Romanian increase to about 35% of cargo segment.

6 Conclusions

Commercial intercourse between Central Europe and Scandinavia is showing an upward trend and a further increase in mutual exchange of goods is expected. Increasing trade will result in growing demand for transport of wheeled cargo from Polish ports. Moreover the shifting of cargo flows from West to central and East corridors is another factor influencing the demand. The passenger sector is stable primarily due to cruise and shipping travellers. Low cost airlines are the main competitor to ferry links.

Services from Polish ports constitute the most important ferry sub-market in South Baltic. The chief competitor is Rostock – Trelleborg with stable turnover, but investment in tonnage planned by TT-Line may result in another growth of cargo traffic. Current tonnage operated on Polish market is insufficient in capacity. Hence carriers to improve their competitiveness and adjust the transport possibilities to the increasing demand have to invest in tonnage.

References

1. A.J. Baird, MARIT. POL. MGMT, **32**, 4 (2007)
2. M. Castelis, J. Usabiaga, F. Martinez, J. Mar. Res. **9**, 3 (2012)
3. K. Chen, Z. Yang, T. Notteboom, Trans. Research E **61** (2014)
4. J.R. Daduna, K. Hunke, G. Gunnar Prause, Logistics, Journal of Shipping and Ocean Engineering **2** (2012)
5. M. Graham, V. Russell, J. of Vac. Mark., **7**, 3 (2001)
6. H. Heijveld, R. Gray, MARIT. POL. MGMT, **23**, 2 (1996)
7. J. Kizielewicz, I. Urbanyi-Popiołek *Rynek usług morskiej żeglugi wycieczkowej* (PWN, Warszawa, 2015)
8. J. Kizielewicz, A. Haahti, T. Lukovic, D. Gracan, Econ. Research. **30**, 1 (2017)
9. I. Kotowska, *Żegluga morska bliskiego zasięgu w świetle idei zrównoważonego rozwoju transportu*, (WNAM, Szczecin 2014)
10. I. Kotowska, Rev. Econ. Pers. **15**, 1 (2015)
11. M.F. Lai, H.K. Lo, Trans. Research A **38**, 4 (2004)
12. H.K. Lo, K. An, W.H. Lin, Trans. Research E **59** (2013)
13. M. Mańkowska, ICTS 2015, Conf. Proc. (2015)
14. T.A. Mathisen. G. Solvoll, EJTIR, **10**, 2 (2010)
15. P. Morales-Fusco, S.Sauri, G.D.Melo, J.T.R. **33**, 4 (2013)
16. E. Musso, A.C. Paixao Casaca, A.R Lynce, *The Handbook of Maritime Economics and Business*, (Lloyd`s List, Londyn ,2010)
17. A.C. Paixao Casaca, P.B. Marlow, MARIT. POL. MGMT, **36**, 1 (2009)
18. A.C. Paixao Casaca, P.B. Marlow, MARIT. POL. MGMT, **32**, 4 (2005)
19. A. Pantouvakis, MARIT. POL. MGMT., **34**, 6 (2007)
20. B. Peter, Baltic Ferries, (Ferry Publications 2009)
21. A. Polydoropoulou. L. Litinas, R.in Trans. Econ, **21** (2007)
22. *Shippax Guide 17*, (Halmstad: Shippax.2017)
23. *Shippax Guide 16*, (Halmstad: Shippax.2016)
24. *Shippax Market 17*, (Halmstad: Shippax 2017)
25. *Shippax Market 16*, (Halmstad: Shippax 2016)
26. *Shippax Market 12*, (Halmstad: Shippax 2012)
27. SBD Statistic Denmark.
28. SCB Statistic Sweden
29. SSB Statistic Norway.
30. M. Stapford, *Maritime economics*, (Routledge Taylor & Francis Group, London-New York 2009)
31. A .Suarez-Aleman, L. Trujillo, F. Medda, MARIT. POL. MGMT, **42**, 4 (2015)
32. *Transport, activity results in 2016*, (SCO, Warsaw 2017)
33. *Transport, activity results in 2011*, (SCO, Warsaw 2012)
34. T.Wergeland, *Ferry passenger markets, The Blackwell companion to maritime economics*, (Blackwell Publishing Ltd.2012)
35. D.Z. Wang, H.K. Lo, Trans. Research B **42**, 9 (2008)
36. P. Wild, J. Dearing, MARIT. POL. MGMT., **27**, 4 (2000)

Future CO_2 emissions from shipping: four-scenarios using a multi-level perspective – a proposed methodology

Peter Wells[1], *Stephen* Pettit[1], *Wessam* Abouarghoub[1], *Jane* Haider[1], and *Anthony* Beresford[1]

[1]Logistics and Operations Management, Cardiff Business School, Cardiff University, Colum Drive, Cardiff, CF10 3EU

Abstract. Previous work by Pettit et al (2017) drew on Socio-technical Transitions Theory (STT) to contextualise recent developments in the technological and operational eco-efficiency of ships which may ameliorate sustainability issues in shipping. Within STT the Multi-Level Perspective (MLP) is a framework used to explain the permeation of innovations into society (Geels, 2014). The framework has three levels: Landscape; Regime and Niche. In order to develop a better understanding of changes that may take place in the future, this paper uses the MLP to assess the broad range of challenges facing the shipping industry with respect to future CO_2 emissions at each of the three levels. The framework is applied to four scenarios for global shipping in order to understand the potential range of outcomes for CO_2 emissions. The four scenarios are: 'Business as Usual'; 'Managed Transition'; 'Chaotic Transition'; and 'Managed Degrowth'. It is contended that changes in the shipping industry alone are unlikely to lead to large-scale changes in its CO_2 emissions. Under 'Business as Usual' conditions, CO_2 emissions will continue to rise as the changes required at all three levels, but primarily at the regime level, will have limited impact. Under the other three scenarios, it is suggested that there is room for optimism that CO_2 emissions will decline over time.

Introduction

Previous work by Pettit et al (2017) drew on Socio-technical Transitions Theory (STT) to contextualise recent developments in the technological and operational eco-efficiency of ships which may ameliorate sustainability issues in shipping. Within STT the Multi-Level Perspective (MLP) is a framework which is used to describe and explain the permeation of innovations into society (Geels, 2014). The framework has three levels: Landscape; Regime and Niche. The 'Landscape' level pertains to over-arching trends and contextualising factors such as economic growth, long-run price movements, trade policies, and environmental conditions. The 'Regime' level is understood to be a dynamically stable ensemble of organisations, technologies, laws, social and employments policies practices, and related matters that essentially describe the world as we know it. Finally, the 'Niche' level pertains to protected spaces away from mainstream governance wherein new technologies or social practices can be experimentally initiated, and potentially nurtured to the point where they displace an incumbent regime.

As shipping is fundamentally a derived demand which arises out of, but also enables, the spatial separation of production and consumption the industry is a 'hostage to fortune' as it responds to, rather than leading, overarching economic changes. This has meant that historically shipping has been a facilitator of global economic growth, with CO_2 emissions growing steadily as the shipping industry has grown to meet need. Thus, in order to develop a better understanding of changes that may take place in the future, this paper uses the MLP to assess the broad range of challenges facing the shipping industry with respect to future CO_2 emissions at each of the three levels. The framework is applied to four scenarios with regards to the future of global shipping in order to understand the potential range of outcomes for CO_2 emissions. The four scenarios are: 'Business as Usual'; 'Managed Transition'; 'Chaotic Transition'; and 'Managed Degrowth'.

Socio-technical transitions, multi-level perspectives and futures thinking

The study of socio-technical transitions is concerned with understanding how, and to what effect, technologies permeate society. Theoretically, socio-technical transitions draws from multiple antecedents in evolutionary economics, the sociology of innovation, and institutional theory with a concern for understanding the relationship between technological innovation and economic growth. An initial impetus was long wave theory, as developed by the Russian economist Kondratiev, and elaborated more formally by Schumpeter. In a defining statement on the emerging research agenda for socio-technical transitions, Köhler et al. (2017) put emphasis on the multi-dimensional, multi-actor, and co-evolutionary character of change in which outcomes are uncertain. Hence while technologies may still be at the core of explaining change, socio-technical transitions scholars integrate an understanding of the contested (and hence political) character of change processes. The so-called Multi-Level Perspective (MLP) is one framework adopted in transitions studies (Geels, 2014), although others are also used (Smith et al., 2010). STT thereby seeks to embrace the complexity of social reality, but also to simplify that reality in order to make it susceptible to rigorous analysis. Under the MLP framework the initial understanding was that change and stability could be identified within a hierarchy of three levels.

The landscape level pertains to over-arching trends and contextualising factors such as economic growth, long-run price movements, trade policies, and environmental conditions. In principle landscape-level factors change relatively slowly and are systemic in impact. One example of a long-term landscape change may be the development of the 'One-Belt One-Road' (OBOR) initiative being promoted by China. This initiative seeks to further integrate Asia, Europe, and Africa through two interlinked components: the Silk Road Economic Belt, a land-based transport network connecting China to Europe and the Middle East through Central Asia and Russia, and a twenty first century Maritime Silk Road connecting ports in Asia, Africa and Europe. Development under this US$ 1.2 trillion project includes roads, rails, airports, seaports, energy pipelines, and other core projects in the region. The OBOR strategy (if successful) has far-reaching implications for the future volume and balance of shipping trade around the world. Although it will take many years to become a reality, the consequences may be profound for shipping if significant volumes of cargo are transferred from sea to alternative land routes (Ferdinand, 2016). With its claim to boost support for green and low-carbon development, there will be an impact on carbon emissions, and evaluating its contribution towards a low-carbon economy.

The regime level is understood as a dynamically stable ensemble of organisations, technologies, laws, social practices and related matters that essentially describe the world as we know it. Change certainly happens at this level, but with multiple feedback loops that act to dampen the pace and degree of change. In general then, regimes are self-perpetuating ensembles, conditioned from above by the factors at the landscape level. Emergent niches largely fail to displace incumbent regime structures, but sometimes do. Examples of 'regime' level organisations include the IMO and other representative organisations such as Tanker Owners (ITOPF), port representative organisations (ESPO), Labour organisations (ITF), ship owners, ship operators, naval architects, shipbuilders and suppliers, flag states, finance and investment organisations, and the customers of the shipping sector (both on the supply side and the demand side).

Finally, the niche level pertains to protected spaces away from the mainstream wherein new technologies or social practices can be experimentally initiated, and potentially nurtured to the point at which the niche can displace an incumbent regime. Niches tend to be conceptualised as emerging rapidly, but often disappearing equally rapidly. Niches may therefore expand and effectively displace one regime with another, as in the way that steam displaced sail in shipping (Geels, 2002). Such changes, or transitions, have the potential to follow several pathways variously combining new entrants with incumbents as defined by, for example, Berggren et al., (2015) and Hess (2013). Examples of 'niche' activities in the realm of shipping include: the development of battery electric power and other supplementary power through the use of sail and solar assistance technology, the further development of slow steaming and cold ironing, the 'greening' of port activities, and 'virtual' port arrival. The three levels are illustrated in Figure 1.

Transitions studies are concerned with future sustainability, but they tend to be empirically grounded in historical research. While transition pathways may be viewed as conceptual experiments to contemplate possible futures, in practice the tendency has been to explore observed empirical outcomes of the near or more distant past. Scenarios are an established method of contemplating possible futures, which may or may not be desired, with a view to informing corporate strategy or government policy (Dixon, 2011).

Socio-technical transitions and shipping

The shipping industry is in a state of transition as it adjusts to increasingly strict emissions standards promoted primarily at the regime level. There is an increasing awareness of shipping's contribution to global emissions of greenhouse gases; a low-carbon economy is identified as one of the Sustainable Development Goals from the UN. Landscape changes are also relevant however, as, for example, China's OBOR strategy will have profound implications in supporting this transition and acting upon long-term environmental change.

Discussion on the future of carbon emissions from shipping, and the future of the sector more generally, tend to be constrained by a focus on economic imperatives and rationality, along with an incremental view of future change possibilities. Shipping can be powerfully symbolic (and symbolic of the projection of power) as is the case with aircraft carriers for example, and as was the case with the vessels of the Cunard line. More generally, the future of shipping is not reducible to the economics of moving materials and products across the seas of the globe, but is also likely to be profoundly shaped by national and international policy measures, by shifts in geo-political possibilities, by technological and operational innovation, and indeed by climate change. Socio-technical transitions theory provides a framework in which the multi-dimensional and co-evolutionary aspects of change can be systematically captured. Shipping is more than just ships. Shipping is an increasingly complex system of inter-related parts including ports, land transport systems, satellites, and related matters but also extending all the way to consumption practices and norms that, for example, regard importing finished products from distant countries to be an acceptable means of achieving aspirational material lifestyles. In other words, cultural norms and expectations over things like work, employment, consumption, and globalisation are powerfully influential on existing shipping practices. Similarly, emergent expectations around sustainability in its broadest sense will come to bear on shipping.

Figure 1. The multi-level perspective of socio-technical transitions applied to shipping.

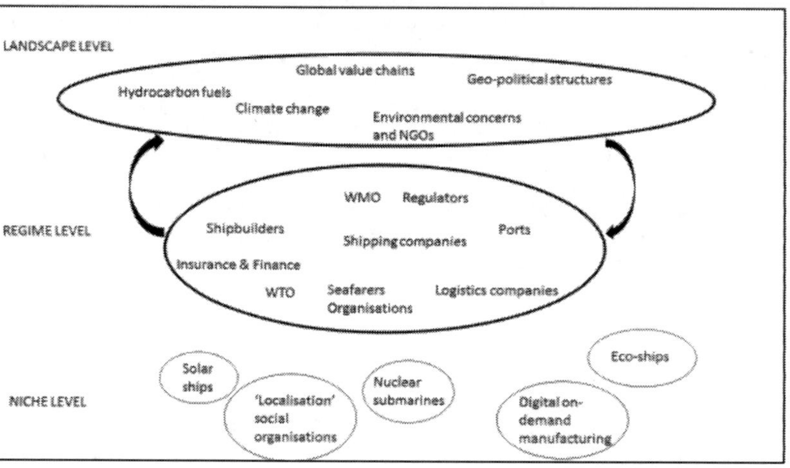

Source: Pettit et al, 2017

In principle then, STT can inform the construction of alternative scenarios, which can then be used to generate illustrative outcomes (see McDowall, 2014; Angheloiu et al., 2017). Historical studies may also be used to validate the assumptions underpinning expectations with regards to the pace and direction of transitions that might emerge from the scenarios (McDowall, 2016). Nonetheless, a concern with validation of this type is that unforeseen or unprecedented events may occur, and these may be precisely the 'regime changing' events that could have a significant impact on future outcomes. Scenarios are inevitably also concerned with multiple agents of possible change, and of interactions between agents.

Scenarios and future shipping carbon emissions

It is recognised in the transitions literature that there are competing visions for the future (Geels et al., 2015). In this paper we seek an explicit exploration of those competing visions through the use of scenarios as they apply to carbon emissions from shipping. Four scenarios are defined in this paper and summarised in Table 1. Two of the scenarios represent managed change within plausible or broadly accepted limits, albeit with rather different outcomes. These first two scenarios are 'business as usual' and 'managed transition'. The remaining two scenarios are much more dramatic in terms of impact but are included to illustrate alternative pathways to very low carbon emission outcomes: chaotic transition and managed degrowth.

It can be seen that the scenarios are described in terms of STT theory, with distinct characteristics at landscape, regime, and niche levels. The three main levels thus are interactive, with for example developments at landscape level pressuring change at regime and niche levels, but with say regime innovations having repercussions at landscape level. In this regard the scenarios can be understood as unfolding processes rather than simple destination states of a system.

Business as usual

The business as usual scenario does not mean 'no change'. Rather, it represents the idea that current trends will continue, and future changes of significance are readily able to be forecast. Hence it is anticipated in this scenario that global trade broadly continues, and that shipping continues to be a significant component of this trade. The regime in this respect continues to be self-regulating, while landscape-level pressures are not so profound as to demand drastic and enduring shifts in shipping. Technological innovation and 'normal' competition will continue to drive down carbon emissions per vessel or per tonne moved, but such efficiencies will be largely negated by continued growth in the overall volume of shipping. Thus, while shipping may continue to be considered to be environmentally friendly, as the level of emissions per unit moved will be low due to scale economies, the overall emissions burden created by shipping will remain a significant and increasing part of global emissions.

Managed transition

The idea behind the managed transition scenario is the beneficial restructuring of economies and the technologies embedded in many aspects of life can deliver more sustainable production and consumption. A likely outcome is an emphasis on the circular economy operating at a local or national scale, and in combination with reduced material and energy intensity in economies (i.e. a decoupling of economic growth from material and energy consumption) there is a concomitant reduction in the volume of shipping even as the activity becomes more efficient in carbon emissions terms. Hence the overall result is an accelerating process whereby carbon emissions are reduced, before stabilizing around a new regime structure. Carbon emissions in total are thus substantially below the business as usual scenario, but still end up higher than the latter two scenarios.

Chaotic transition

In 'futures' thinking, it is rarely acknowledged that there may be catastrophic change, by which we mean change so abrupt and profound that it destroys existing social structures. Yet historically many societies have witnessed cataclysmic collapse, and certainly in the contemporary world the prospect of nuclear Armageddon cannot be discounted. There are also current examples where individual countries have come close to collapse through the use of extreme measures to subdue a local population resulting in the largescale movement of people either within a country as 'internally displaced people or out of a country as 'refugees'. The consequent economic scenario for such countries is poor and will ultimately impact on a countries ability to trade internationally. The collapse of the global economic order does seem unimaginable, but a process of competitive disengagement is relatively easy to envisage – as is evident by the isolationism pursued by the USA since the 2016 election (see, for example, BBC, 2018a). Another current example is the political confrontation between Russia and the western bloc concerning cyber security and espionage (see, for example, BBC, 2018b). Chaotic transition may result from unbearable landscape pressures manifest across societies and their economies, including shipping. The resultant outcome would be a collapse in carbon emissions.

Table 1. Scenarios for shipping using the MLP framework

Baseline Features	Oil and other prices; GDP change; geopolitical events; trade and liberalisation policies; environmental changes; infrastructure developments; population size and structure; consumption behaviours and practices	Industry and companies; governance; representative organisations; social norms; legal structures and enforcement; markets; supply and demand; investment	Technological and material innovations; organisational, procedural and social (grassroots) innovations; micro-markets and specialist applications.	Current level of CO_2 emissions
Scenario	***Landscape Level***	***Regime Level***	***Niche Level***	***CO_2 trajectory***
Business as usual (Hickman and Banister, 2007; Meadows et al, 1972)	Steady global GDP growth; stable commodity prices; robust global trade agreements; reductions in CO_2 emissions in other sectors creating 'headroom' for shipping; conservative planning with no major transformation in the way transport is performed	Growth in shipping volumes; ship numbers; and in stock turnover rate. Further cost reductions achieved. Latent cyclical over-capacity remains. Actions of NGOs, IMO, investors, owners, etc. remain similar to today.	Some niche technology developments in specialist applications e.g. battery electric ferries; fuel cell ferries in harbours; solar panels and other renewables on some vessels.	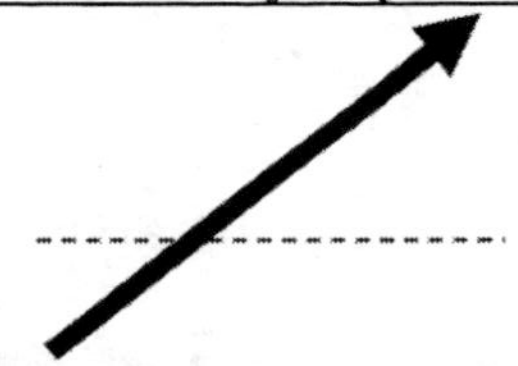
Managed transition (Geels, 2014; Geels et al., 2014)	Green growth strategy; de-couple resource consumption from GDP; reduced trade volumes in food; materials and products; policy targets towards the circular economy; global consensus politics.	Stronger regulation and policy towards shipping and ports; CO_2 trading in shipping; Eventual decline in scale of the sector; concentration of ownership; new routes open up.	Strong encouragement of emergent technologies and fuel efficiency measures; new operational practices e.g. on-board additive manufacturing in transit.	
Chaotic transition (Tainter, 1988)	Environmental, economic and political collapse on a global scale; return to isolationism; widespread loan defaults; severe climate change emerging; food and other shortages.	Collapse of intermediary governance organisations; widespread business failure; loss of investor confidence	No new technologies or operational practices; some innovation in 'scavenging' and retrofit of parts; some new uses found for vessels	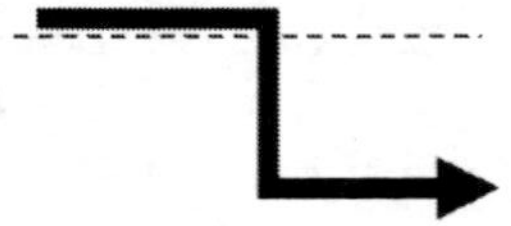
Managed degrowth (Kallis, 2011)	Chronic material shortages; abandoned use of GDP as a measure of progress; strong regulation at international level to ensure equity of outcomes; rapid restoration of some environmental degradation.	Sequential deconstruction of main fleets; industry atrophy; break-up of large MNCs. Strong role of regulatory agencies such as IMO in transition phase. Substantial pre-negotiation period before actions taken.	Potentially significant innovation in alternatives to shipping; reduced material consumption in any case; possible growth in passenger liner niches as alternative to flying.	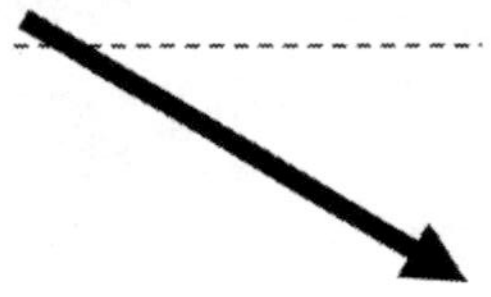

Managed degrowth

An emergent prescription is for managed degrowth, whereby the deliberate aim is to reduce an economy in scale. It is argued that landscape pressures can only be resolved by drastic de-consumption, and a rejection of materialism as a measure of economic success. In order to prevent chaos, and to protect the vulnerable (in a society; in a country) this process of change must be managed. Some aspects of shipping may benefit from these changes, for example with a renewed emphasis on leisure time perhaps the 'cruise' line sector will prosper more. Broadly, a combination of technological innovation and dramatic reductions in trade volumes will reduce carbon emissions to a very low level. One of the principal issues in this scenario is the balance between managed degrowth in developed economies and the need for economic growth in developing countries. Developing nations are likely to see degrowth as a first world problem, and such an approach will require a careful balancing or rebalancing of the global economy.

Scenario Modelling

In order to understand what the outcomes of the scenarios presented above might be, a methodology for modelling such changes is presented below. The modelling methodology will be explored in more detail in future analysis.

In terms of methodology there are variables identified that can influence outcomes at the regime level, such as economic growth rates or shifts in petroleum prices changing the economies of scale and hence economic structure of the shipping industry. Such relationships may vary according to the type of shipping concerned (bulker, container, etc.) or indeed by the route, by source or by destination. The initial step is to understand how far the landscape descriptors explain the current structure and recent past changes in the global shipping industry as the most aggregated levels of analysis. This 'top down' approach can then be cascaded into types of shipping, specific commodity flows, etc., as desired. Given that we anticipate landscape level changes to be the most important determinant of the future of shipping, the focus is on how this overarching level influences the regime and the niche levels.

Note also, however, that the more extreme scenarios are less amenable to traditional econometric modelling. There may be distinct de-coupling of outcomes from variables that were previously explanatory. Indeed a central thrust of much policy with regards to sustainability is to de-couple economic growth from material and energy consumption. In shipping terms, this may mean de-coupling growth in shipping from economic growth. However, the scenarios should enable a statement on the likely character of the shipping industry under each scenario case, and for the implications of that character in terms of carbon emissions to be expressed. In so far as shifts in the industry at regime or niche level occur, there may thus be outcomes in terms of changes at a landscape level (i.e. as measured by CO_2 emissions to the atmosphere).

The arrows in Table 1 are indicative of expected outcomes under the scenarios. However the landscape, regime and niche descriptors are also, in effect, expected outcomes or internally consistent constructs of different 'worlds' of shipping under different conditions. A key question in these outcomes is that of time. In other words, at what rate will change occur or, alternatively, by what date would an outcome be observed? Generally scenarios are useful for long-term projections of the order of 20 to 30 years maybe. The chaotic transition and managed de-growth scenarios both imply drastic change over a short period of time consistent with virtually halting carbon emissions. The time required for the managed transition scenario is the least predictable, but elements are known. For example, it should be possible to forecast a rate of technology penetration and shift in shipping volumes consistent with a steady decline in global shipping emissions. Equally, the business as usual scenario largely implies a continuation of current trends: more efficient shipping offset by volume growth leading to gradual growth in total emissions.

Proposed methodology

In order to model what is described above, a multi-step approach to estimate and forecast shipping CO_2 emissions at the three levels of the MLP model, and to determine the effect of external and internal economic variables on shipping emissions structural change, under the four scenarios, is proposed.

First, an annual time series for shipping emissions in tonnes will be structured using a fuel-based (top-down) approach, as described in Pettit et al 2017, and using equation (1):

$$SE_{annual_t} = \sum_{i=1}^{n} \left(\frac{Bf_{i,daily}}{AC_i} \times ST_{i,annual_t} \times EF_{CO_2} \right)_{annual_t} \quad (1)$$

where SE_{annual_t} is annual shipping emissions at time t, $Bf_{i,daily}$ is daily burned fuel for trade i, AC_i is average transported cargo for trade i and at any given day, $ST_{i,annual_t}$ is annual seaborne trade for trade i at time t, EF_{CO_2} is the CO_2 emission factor, and n is total number of trades (e.g. tankers and dry bulk). The result is a figure for total CO_2 emissions for the shipping industry in the form of an annual time series.

The structured time series of shipping emissions in equation (1) is then used to determine the significance of the MLP framework in explaining historical structural changes in shipping emissions and second, to determine the drivers of shipping emissions under the proposed four scenarios. Thus, shipping emissions data is disaggregated, and hierarchically structured in the form of a top-down approach by landscape, regime and niche, as depicted in Figure 2.

Figure 2. Hierarchical structure of bulk shipping emissions categorised according to landscape, regime and niche levels within the MLP

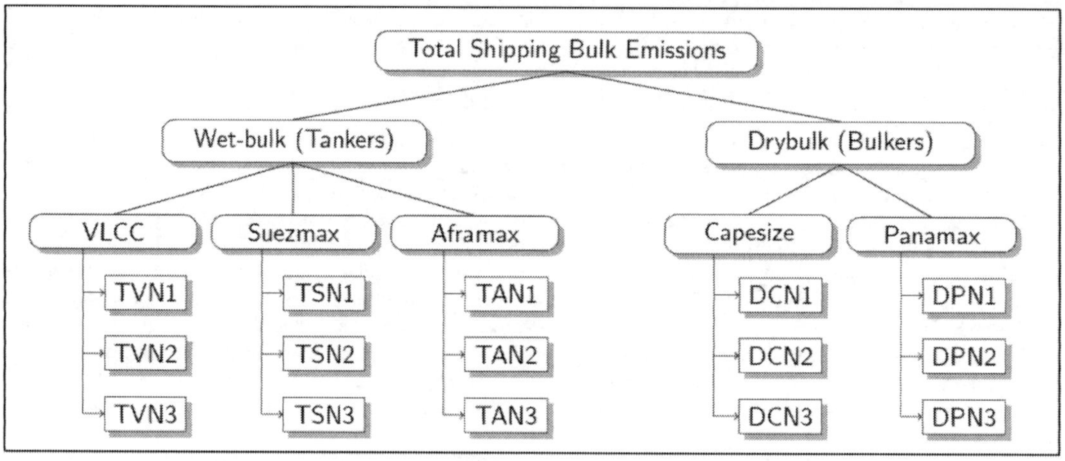

In the hierarchical structure shown in Figure 2, a structured time series of shipping emissions estimated by equation (1) donate the completely aggregated level "Total Shipping Bulk Emissions" as level 0. This level is then further disaggregated to lower levels: "Wet-bulk (Tankers)" and "Drybulk (Bulkers)" as level 1, "VLCC", "Suezmax", "Aframax", "Capesize" and "Panamax" as level 2 and the most disaggregated level, by vessel type, as level 3. Consequently, time-series data aggregated at level 0 represents estimated shipping emissions at the landscape-level of the MLP model, data disaggregated at levels 1 and 2 represents changes in shipping emissions at the regime-level and data disaggregated at level 3 represents developments in shipping emissions at the Niche-level. Thus, a 3-level hierarchy structure to study structural changes in shipping emissions at the three-levels of the MLP framework is proposed. At level 3 in Figure 2, T, V, S, A, D, C, P and N represent Tankers, VLCC, Suezmax, Aframax, Capesize, Panamax and Niche. For example, TVN1 refers to emissions from a VLCC tanker that is categorised in this study as Niche 1 (e.g. modern tanker), TVN2 as Niche 2 (e.g. old tanker), (other categories would cover short-haul and long-haul trades as appropriate).
Second, a multilevel model is suggested to estimate the macroeconomic and microeconomic drivers of structure change in shipping emissions under the assumptions of the MLP framework. Finally, the estimated assumptions are adjusted according to the four scenarios and used for forecasting shipping emissions for the next 20 – 30 years. To this end, we implement a vector autoregression (VAR) model that has proven to be useful for capturing and forecasting the dynamic behaviour between economic variables. The VAR model is used to analyse, at a multilevel within the MLP framework, the dynamic interaction between exogenous and endogenous economic variables. Our bivariate VAR(p) model equation has the form:

$$Y_t = C + A_1 Y_{t-1} + A_2 Y_{t-2} + \cdots + A_p Y_{t-p} + \varepsilon_t, \quad t = 1, 2, \ldots, T \qquad (2)$$

where $Y_t = (y_{1,t}, y_{2,t}, \ldots, y_{n,t})'$ donate an $(n \times 1)$ vector of time series variables, A_i are $(n \times n)$ coefficient matrices and ε_t is an $(n \times 1)$ unobserved zero mean white noise vector process with time invariant covariance matrix Σ.

The overall outcome of this modelling process will therefore be a disaggregation of CO_2 emissions from shipping across the range of shipping types at the three MLP levels and across the four scenarios over a medium-term time horizon.

Conclusions

It is contended that changes in the shipping industry alone are unlikely to lead to large-scale changes in its CO_2 emissions. Under 'Business as Usual' conditions, CO_2 emissions will continue to rise as the changes required at all three levels, but primarily at the regime level, will have some impact. However, the overall large-scale changes required will not be significant enough to lead to overall carbon emissions reductions and are therefore unlikely to have a significant impact on the overall carbon burden (see, for example, Jonkeren et al, 2009). Under the other three scenarios, it is suggested that there is room for optimism that CO_2 emissions will decline over time. In the 'Managed Transition' scenario, de-coupling resource consumption from GDP, greater global political consensus, stronger regulation and policy towards shipping, and encouragement for emergent technologies are likely to lead to a net decline in CO_2 emissions (Geels et al, 2015). In the 'Chaotic Transition' scenario, which was first proposed by Meadows (1970) in the context of global resource depletion, environmental, economic and political systems collapse at a global scale. As a consequence, this would bring about a return to isolationism, and widespread business failure, with little or no innovation which would in turn result in significant CO_2 emissions reductions as there would be significantly less need for shipping (Tainter, 1988). Finally, in the 'Managed Degrowth' scenario, global material shortages, stronger regulation at an international level, and innovation in alternatives to shipping would lead to a rapid decline in CO_2 emissions before a steady state at a lower level occurs (Kallis, 2011).

The technologies and operational innovations that could reduce the environmental burdens of shipping, while useful, do not represent the socio-technical system 'regime' shifts that the international maritime sector requires in order to contribute to reduced CO_2 emissions, which would enhance sustainability. It is the underlying 'landscape' shifts in production and consumption that are likely to bring about a reduction in the demand for shipping, and which will be more significant over the longer term.

References

Angheloiu, C.; Chaudhuri, G. and Sheldrick, L. (2017) Future Tense: Alternative Futures as a Design Method for Sustainability Transitions, **The Design Journal**, 20:sup1, S3213-S3225, DOI: 10.1080/14606925.2017.1352827.

BBC (2018a) US steel and aluminium imports face big tariffs, Available at < http://www.bbc.co.uk/news/world-us-canada-43249614>. accessed 27th March 2018

BBC (2018b) Spy poisoning: Russia accuses US of 'expulsion blackmail', Available at http://www.bbc.co.uk/news/world-asia-43550938#. accessed 27th March 2018

Berggren, C., Magnusson, T., Sushandoyo, D. (2015) Transition pathways revisited: Established firms as multi-level actors in the heavy vehicle industry, **Research Policy**, 44(5), 1017-1028.

Dixon, T. (2011) 'Low Carbon' Scenarios, Roadmaps, Transitions and Pathways: an Overview and Discussion, Retrofit 2050 Working Paper, October 2011. Available at: http://www.retrofit2050.org.uk/sites/default/files/resources/WP20116.pdf. accessed 2nd March 2018

Ferdinand P (2016) Westward ho—the China dream and 'one belt, one road': Chinese foreign policy under Xi Jinping, **International Affairs,** 92 (4), 941–957

Geels, F.W. (2002) Technological transitions as evolutionary reconfiguration processes: A multi-level perspective and a case-study, **Research Policy**, 31(8-9), 1257-1274.

Geels, F. W. (2014). Regime Resistance against Low-Carbon Transitions: Introducing Politics and Power into the Multi-Level Perspective. **Theory, Culture & Society**, 31, 21–40.

Geels, F.W., McMeekin A. Mylan J., Southerton D. (2015). A critical appraisal of Sustainable Consumption and Production research: The reformist, revolutionary and reconfiguration positions. **Global Environmental Change**, 34, 1–12.

Hess, D.J. (2013) Industrial fields and countervailing power: The transformation of distributed solar energy in the United States, **Global Environmental Change**, 23(5), 847-855.

Hickman R, Banister D (2007) Looking over the horizon: Transport and reduced CO2emissions in the UK by 2030, **Transport Policy**, 14 (5) 377-387

Jonkeren O, Jourquin B, Rietveld P (2011) Modal-split effects of climate change: The effect of low water levels on the competitive position of inland waterway transport in the river Rhine area, **Transportation Research Part A: Policy and Practice,** 45 (10) 1007-1019

Kallis, G. (2011). In defence of degrowth, **Ecological Economics**, 70(5) 873-880.

Köhler, J.; Geels, F.; Kern, F.; Onsongo, E. and Wieczorek, A. (2017) **A research agenda for the Sustainability Transitions Research Network**, Paper obtained from The Sustainability Transitions Research Network (STRN).

McDowall, W. (2014) Exploring possible transition pathways for hydrogen energy: A hybrid approach using socio-technical scenarios and energy system modelling, **Futures**, 63, 1-14.

McDowall, W. (2016) Are scenarios of hydrogen vehicle adoption optimistic? A comparison with historical analogies, **Environmental Innovation and Societal Transitions,** 20, 48-61.

Meadows DH, Meadows DL, Randers J, Behrens III WW (1972) **The Limits to Growth**, Club of Rome, New York, 205 pp.

Pettit S, Wells P, Haider J, Abouarghoub W (2017) Revisiting history: Can shipping achieve a second sociotechnical transition for carbon emissions reduction?, **Transportation Research Part D**, http://dx.doi.org/10.1016/ j.trd.2017.05.001

Smith, A., Voβ, J.-P., and Grin, J. (2010) Innovation studies and sustainability transitions: The allure of a multi-level perspective and its challenges, **Research Policy**, 39(4), 435-448.

Tainter, J.A. (1988). **The collapse of complex societies**, Cambridge: Cambridge University Press.

Assessing the Economical Efficiency of Intermodal Chains in Global Trade

Bogusz Wiśnicki[1,*] and *Dariusz* Milewski[2]

[1]Transport Engineering and Economics Faculty, Maritime University of Szczecin, Waly Chrobrego 1-2, 70-500 Szczecin, Poland
[2]Faculty of Management and Economics of Services, University of Szczecin, Cukrowa 8, 70-001 Szczecin, Poland

Abstract. The paper relates to the development of intermodal technologies used in international trade, together with factors which will influence that development at present and in the future. The aim of the paper is to assess impact of the integration of intermodal chains on logistics costs and thus on global trade. The special emphasize is laid on the problem of the diversity of standards of containers and transport technology in different parts of the world. This research is based on the available statistical data and information obtained from leading logistics operators. The authors conducted comparative analysis of the economic efficiency of usage of different sizes of cargo units, both standard and non-standard, relating to representative intermodal chain between the USA and Europe. In order to assess different scenarios a model was elaborated for calculating the total costs of such a chain, and transport efficiency analysis with the use of this model were conducted. The results of research presented in the paper confirm that integration of intermodal chains thanks to standardization of loading units in a global scale could be profitable. The results of logistics cost analysis prove measurable benefits for logistics operators and as well as their customers.

1 Introduction

In order to draw a comparison between present state and technological development prospects of intermodal transport, one should start with differences in understanding of the notion of intermodal transport on a global scale. The main two areas of development and the standardisation of intermodal transport to Europe and the United States. Hence, these are the basic areas of further research in the context of this publication.

In Europe a point of reference is the definition of Economic Commission for Europe (UN/ECE), which says that intermodal transport is "the movement of goods in one and the same loading unit or road vehicle, which uses successively two or more modes of transport without handling the goods themselves in changing modes" [1]. In the USA the following explanation of the Intermodal Association of North America (IANA) which is commonly used: "Intermodal transport - the movement of freight, in a container or on a trailer, by

* Corresponding author: b.wisnicki@am.szczecin.pl

more than one mode of transportation; the movement can be made from rail to truck to ship in any order". The difference in the phrase „without handling the goods themselves" brings about consequences in transport technologies used on both continents. In the USA so-called container transloading, which is rare in Europe, is a common practice. It consists in dismantling loading units in order to set up new, differently sized container units. This situation takes place in sea ports where the cargo is loaded from ISO maritime containers (20' and 40') to larger domestic 53' containers. The most important advantages of transloading are: a possibility of sorting out loading batches in the port, and the use of a smaller number of containers in activities between the port and the hinterland (three 40' containers can correspond to two 53' container).

Among many differences between the intermodal transport in the EU and USA the most important ones from the viewpoint of a further analysis are as follows:

- fewer infrastructure limitations in the USA make it possible to implement cost-efficient technologies in rail (longer trains, heavier railcars, double stack loads) and road transport (longer road trains);
- a considerable difference in transport organization which is affected by greater liberalization of the market in the USA as well as conditions of tough competition in the EU often depending on factors beyond the market;
- the EU is characterized by a much denser network of terminals belonging to different operators and short-distance transits;
- the use of transport technologies typical for continents, that is Ro-La in the EU and bimodal modes in the USA
- transloading in the USA which has broken the European definition of intermodal transport;
- ports play their major role in both cases in port-to-door distribution, but in the USA long-distance landbridges and trade corridors are typical, whereas in Europe there is rather a network of intermodal corridors with mixed logistic solutions.

By analysing the generic structure of intermodal units in the USA, it can be stated that according to the IANA data from the year 2014, 50% of the carried units were ISO containers, 40% 53' domestic containers, and 10% share belonged to semitrailers. The available statistical data from 2011 about Europe show an important participation of 50% ISO containers, 42% of swap bodies, 8% share of road trains and 3% share of semitrailers (fig. 1). A 50% share of containers which handle cargo on sea routes is common for both markets. A kind of counterpart for swap bodies which are used in the continental transport are 53'containers in the USA. Both types of transport units are domestic standards and as such are better adapted to the market needs of the two analysed continents.

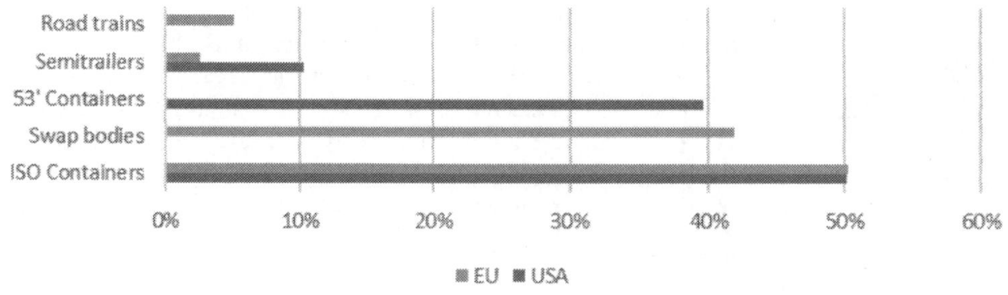

Fig. 1. Structure of ITU in Europe and USA according to transport volume.
Source: [2] and UIRR Reports (www.uirr.com), 2012 and IANA Statistics (intermodal.org).

The phenomenon of economies of scale and associated cost advantages is common in any business. In transport lower unit costs can be obtained due to greater performance of transport operations measured by distance and volume. The loading capacities of transport type and transport equipment have been increasing for many years. In some regions of the world the global standards are being extended through the introduction of the use of larger domestic transport units. In the USA, large intermodal loading units are widely used, including containers 53', which are not a worldwide ISO standard. Interestingly, the use of such units on a local market is inconsistent with the idea of technological integration of the door-to-door supply chains and involves additional manipulations. Despite this, it is cost-efficient. This shows that the aforementioned phenomenon of scale actually occurs in this case, so one could assume that the cost efficiency would be even higher if such larger units were used on the whole door-to-door route.

The authors' intention was to examine under what conditions the admission of containers larger than 40' (45' and 53') in the transport links between the United States and Europe could be economically viable. The aim is to assess impact of the integration of intermodal chains on logistics costs and thus on global trade. To realize this purpose the authors used a cost model of a transport chain. With the help of this model economic calculations of proposed solutions were made for a number of examples of different sizes of containers and different intercontinental transport links.

2 Literature review

The problems of the usage of intermodal technology are quite often described in the literature – both in scientific and the professional publications. Topics presented refer to a wide range of issues: optimization of intermodal chains, the choice of transport technologies, the transport services market and transport policy [3]. The mafjority of research concerns transport links between the EU and the Far East. It is much more difficult to find studies relating to the transport between the USA and Europe. This is understandable, as the turnover between the two continents is much smaller than with China.

One of the most important problems connected with intermodal transport is its economical efficiency measured primarily by its costs. The share of investment costs is very high and, according to some authors, can even reach 50% of the total intermodal terminal cost [4]. Although the cost of intermodal transport and structure of these costs are given in different ways in different publications [5-8], we can agree that the share of fixed costs is relatively higher and requires a high efficiency of operations resulting from utilization of transport types, their capacity and scale of operations.

The efficiency of one process usually depends not only on the performance of that process but also others connected with it. Transport and logistics are very good examples of such system effects, because the transport and logistics processes are strongly interrelated. Some authors point to organizational matters, strategy and the relations between transport technology with other logistics processes such as warehousing [9], allocation of resources in an intermodal terminal or the assignment of trucks and the fleet size in intermodal drayage operations [10]. These relations can be conditions for achieving economies of scale and they can be found in the maritime intermodal transport. Big container ships allow economies of scale, but economies of container ship operations are dependent on port productivity. Hence, there are also disadvantages of scale and there is often an optimum scale by which the benefits are maximal. The same problem refers to the whole logistics chain (understood here as being wider than the pure transport chain), in which striving for economies of scale in transport can result in higher costs for other logistics processes. Numerous models have been developed to prove that such relations actually exist [11].

Many transport models have been developed to allow decision-making processes [12]. Among these there are also many intermodal transport models, which relate to a wide range of issues such as operational and organizational problems [13], the optimal level of investment [14, 15] and the market for transport services and transport policy [16].

Models presented in the literature have been designed in many cases to solve specific problems in individual parts of the intermodal chain. An example of such a model is the PLATFORM Model [17]. Another problem concerns optimization criteria, which are sometimes of a non–economic character. There are models that take into account the system approach and even cover an entire intermodal chain. An example of such a model is the Heuristics Intermodal Transport Model (HIT-model) [18, 19]. But there is a lack of information of the verification of these models in scientific publications. Another problem linked with these is a too high level of aggregation, and data which are quite often averaged. It should be noted, however, that if economic criteria are used in a model this means that external costs are also being taken into account [20].

So far, the most developed model for calculating total costs of a planned intermodal connection is that made by Janic [21]. Three types of costs were included in his model: the internal costs - borne by transport operators, the time costs of goods tied in transit and the external costs - the costs of the impacts on society and the environment. The model concerns the problem of choice of a transport solution for the transport of containerised cargo, i.e. for cargo which comes through sea ports.

None of the existing models presented in the literature concern the problem of the different sizes of intermodal units in different regions. None can be directly used for analysis of containerized supply chains in which goods are reloaded from one cargo unit to another. The authors in the present study undertook this research challenge by adapting for this research the above-mentioned model of Janic.

3 Research methodology

The purpose of this research was to assess the cost-efficiency of the use of various sizes of container transport units carried in sea-land intermodal chains. The authors conducted a comparative analysis of the economic efficiency of the transport chain using cargo units of different sizes, including standard and non-standard. For assessing different scenarios of transport, the mathematical model was elaborated and calculations with the use of this model were conducted. The obtained results were used for the assessment of the influence of the use of different combinations of containers on the economic efficiency of global trade by comparing the costs of deliveries of goods with the value of these goods.

The model reflects the economic trade-off relation between the costs of handling and the carriage costs. The decision problem occurs when there is non-integrated intermodal transport forced by differences in standards in transport infrastructure or means of transport. As a result, in the supply chain the differences in standards dictate a change in the means of transport or loading unit. Such a change from a smaller unit to a bigger one can allow a reduction of total costs if savings in transport costs offset increased transloading costs.

The construction of the proposed calculation model is based on the Janic's model [21] and is presented as a mathematical formula, enabling a calculation of the total cost of the transport chain (tab. 1). The total costs include four components: transport (carriage) costs, time costs, handling costs, external costs. Carriage costs are prices paid by transport users.

Table 1. Mathematical formula for calculating total costs of an intermodal chain

Carriage cost	$C_{I/k} = \sum_{i=1}^{n} Fr_{I/k/i} \times N_{TU/k/i}$
Handling costs	$C_{H/k} = \sum_{i=1}^{n} Hr_{k/i} \times N_{TU/k/i}$
Transport External costs	$C_{E/k} = \sum_{i=1}^{n} Q_k \times d_{k/i} \times c_{E/k/i}$
Total costs	$C_{FULL/k} = C_{I/k} + C_{T/k} + C_{H/k} + C_{E/k}$

$Fr_{I/k/i}$ - transport freight rate per unit per leg

$N_{TU/k/i}$ - number of transport units on a transportation leg

Q_k - total weight of a cargo

$Hr_{k/i}$ - handling rate for transhipment (including transloading) a transport unit

$d_{k/i}$ - number of km of a transport leg

$c_{E/k/i}$ - unit external costs of a transport mode

Source: Authors' calculations.

Such a model could be used as a tool for making decisions concerning the choice of type and size of containers used in an analysed transport chain. The model could also provide the base for wider macroeconomic calculations and thus could be useful for policy makers. Firstly, because various technologies affect external costs. Secondly, because the integration of the various technologies would probably require investment, e.g. roads.

The transport relations chosen for this study are sea-land connections between the USA and Europe. This choice stems from the fact that in the supply chains between these regions different cargo units are used, including standard sea containers as well as domestic units. What is important, in some cases transloading container-to-container operations are observed, which an unusual procedure for integrated intermodal transport is.

It is not possible at this stage of research to put forward a concrete hypothesis because in order to provide proof extensive research would be needed. Therefore, the authors conducted here only limited calculations with the use of the developed model.

4 Analysis of the economic efficiency of the use of large containers in US-Europe intermodal transport

With the use of the presented above model a simulation of economical efficiency of the international trade with the use of big containers has been conducted. The calculations were made for the following variants transportation of cargo from the USA to Europe:

- transportation of container 40 'on the entire route,
- transport of container 53 'to the port in the US and transloading of goods to container 40',
- transport of 45 ' containers on the entire route,
- transport of 53 'containers to the port in the US and transloading of goods to container 45',
- transportation of 53 'containers on the entire route.

The model considers the transport and handling costs at land terminals and seaports, which vary depending on the length of land routes, the total cargo weight to be transported, cargo density, freight and transload rates. In a case of a change of a transportation unit during transport, a transloading fee is additionally charged. The capacity of containers depends on their technical capabilities (tonnage, volume in m3 and the number of pallets) and the permissible payloads for road transport in different countries. The limitations are regulations in the countries with the lowest permissible gross weight (36 tons in the USA).

Data used in the simulation are presented in tables 2-3. Transport rates are estimated based on freight tables of transport operators in the USA and Europe.

Table 2. Capacity of containers in USA and EU

Container size	40' pallet wide	45' pallet wide	53' pallet wide
Number of europallets (two layers)	60	66	80
Space [m3]	79	89,2	112
Tare weight [t]	15,8	16,3	17,1
Netto weight on road EU [t]	28,2	27,7	26,9
Netto weight on road USA [t]	20,2	19,7	18,9
Netto weight on rail or inland transport [t]	31	32	34
Weight limit EU [t]	44	44	44
Weight limit USA [t]	36	36	36

Source: Authors' calculations based on market data and regulations.

Table 3. Freight rates for 40' on individual sections of the door-to-door route

USA								EU	
Dist. road [km]	Road rate [$/cont.]	Terminal rate [$/cont.]	Rail dist. [km]	Rail rate [$/cont.]	Port rate [$/cont.]	Sea voyage [$/cont.]	Port rate [$/cont.]	Land distance [km]	Intermodal rate [$/cont.]
10	40	324	450	800	293	2 130	293	200	220
110	100	324	600	900				300	300
210	150	324	750	1 000				400	370
310	200	324	900	1 100				500	440
410	230	324	1 050	1 200				600	480
510	250	324	1 200	1 250				700	515
610	260	324	1 350	1 280				800	550
710	265	324	1 500	1 300				900	570
810	270	324	1 650	1 350				1000	590
910	280	324	1 800	1 400				1100	600

Source: Authors' calculations based on freight rates of transport operators: worldfreightrates.com and imtis.contargo.net.

One of the most important factors is the density of a cargo. Hence, the calculations were carried out for various cargo density options assuming that the total load weight is 200 tons (tab. 4). Containers 53' are the most efficient for lighter cargo of density up to 0,25 t/m3. For heavier goods option 53'-45' is the best option, what confirms high efficiency of the 53', which are profitable even, when they are not used on the whole route and an additional transloading takes place.

Table 4. Impact of a cargo density of a commodity on internal costs of a delivery

Cargo Density [t/m3]	Combination of containers				
	40'-40'	53'-40'	45'-45'	53'-45'	53'-53'
0,10	179 600	173 600	157 589	154 085	132 677
0,21	89 800	86 800	78 794	77 042	66 338
0,25	78 575	77 161	68 288	68 223	60 810
0,36	56 125	62 727	57 783	55 429	60 810
0,50	56 125	59 119	57 783	55 429	60 810
0,63	56 125	59 119	57 783	55 429	60 810
0,80	56 125	59 119	57 783	55 429	60 810
0,89	56 125	59 119	57 783	55 429	60 810
1,13	56 125	59 119	57 783	55 429	60 810
1,33	56 125	59 119	57 783	55 429	60 810

Note: Calculations are made for the following scenario: distance on land legs – 3810 km, total weight of delivery – 200 t, difference in freight rates – 45'/40' increase 103%, 53'/40' increase 110%.
Source: Authors' calculations.

The next simulation concerned the impact of differences between rates for transporting containers, assuming the increase for larger containers in comparison to standard 40' unit. Option 53'-53 'is the cheapest one, provided that the freight rates for these containers will not be 50% higher than 40' (tab. 5). Option 45'-45 ' is more profitable from option 40'-40' when freight rates for 45' containers are not higher by 24% compared to 40 'container rates.

Table 5. Impact of diversification of rates on internal costs of the delivery

Average changes of rates		Combination of containers				
45'/40'	53'/40'	40'-40'	53'-40'	45'-45'	53'-45'	53'-53'
3%	10%	179 600	173 600	157 589	154 085	132 677
6%	20%	179 600	177 632	161 128	160 144	142 114
9%	30%	179 600	181 664	164 666	166 202	151 550
12%	40%	179 600	185 696	168 205	172 261	160 987
15%	50%	179 600	189 728	171 744	178 320	170 424
18%	60%	179 600	193 760	175 283	184 379	179 861
21%	70%	179 600	197 792	178 822	190 438	189 298
24%	80%	179 600	201 824	182 360	196 496	198 734
27%	90%	179 600	205 856	185 899	202 555	208 171
30%	100%	179 600	209 888	189 438	208 614	217 608

Source: Authors' calculations.

The benefits of the analysed combinations for two factors: load density and the increase in freight rates. Large containers are relatively very profitable for transporting loads with a density of 0.10 t/m3 and 0.21 t/m3, even with considerable increases in freight rates. With heavier loads of density from 0.25 t/m3, there is a significant drop in efficiency, and the differences in freight rates have a big influence.

The model has also been used to estimate the possible benefits of using larger containers in the entire USA-Europe trade. Not only the costs of the delivery were analysed, but also costs of purchase. Based on previously presented data on trade in goods between the USA and the EU it can be estimated that the average weight of cargo carried in containers is 7.1 t/TEU (based on Eurostat). This corresponds to the load density of 0.25 t/m3. Assuming also that transport costs together with transport insurance amount to 15% of the value of the goods [22], it was assumed that value of a cargo is 600 $/t. Assuming also that the total turnover between the US and the EU is 109 million tons, simulations have been carried out for 1% of this potential turnover.

Simulation results showing possible cost savings in percentage terms, including additional external costs, are presented in table 6. External costs slightly, i.e. by 3%-4%, increase the total costs, which can be explained by the fact that the analysis assumed the use of rail transport in the USA and only the inland waterway in the EU to a large extent. These branches, especially sea transport (the longest segment) generate relatively the lowest external costs. The simulation proved that the biggest savings relate to the 53'-53' combination and increase with distance.

Table 6. Total costs of procurement of goods (incl. external costs)

Total land distance [km]	**Combination of containers**				
	40'-40'	53'-40'	45'-45'	53'-45'	53'-53'
660	0,0%	0,0%	-3,3%	-2,7%	-6,9%
1 010	1,9%	-0,2%	-3,9%	-3,4%	-7,5%
1 360	3,6%	-0,4%	-4,4%	-4,0%	-8,0%
1 710	5,4%	-0,6%	-4,9%	-4,6%	-8,6%
2 060	6,7%	-0,7%	-5,1%	-5,0%	-8,9%
2 410	7,6%	-0,8%	-5,4%	-5,3%	-9,2%
2 760	8,3%	-0,9%	-5,6%	-5,6%	-9,4%
3 110	8,8%	-0,9%	-5,7%	-5,7%	-9,6%
3 460	9,5%	-1,0%	-5,9%	-5,9%	-9,7%
3 810	10,1%	-1,1%	-5,9%	-6,0%	-9,8%

Note: Calculations are made for the following scenario: cargo density – 0,25 t/m3, total weight of delivery – 10900 t/year, total value of delivery - 6540 m$/year, difference in freight rates – 45'/40' increase 103%, 53'/40' increase 110%.
Source: Authors' calculations.

Conclusions

The results of research presented in the paper confirm that integration of intermodal chains thanks to standardization of loading units in a global scale could be profitable. The results of logistics cost analysis prove measurable benefits for logistics operators and as well as their customers. The important factor of integration of intermodal chains is a unification and increasing the dimensions of loading units, what in turn will allow reduce logistics costs and accelerate global trade.

The conducted analysis relating to representative intermodal chain between the USA and Europe allows to conclude that the use of 53' or 45' containers instead of standard 40' units in door-to-door relation could have an impact on lowering the cost of trade between America and Europe. However, it should be noticed that introduction of larger units needs to overcome some important technological limitations, both on maritime section (construction of containerships) and in the land infrastructure in Europe. The authors are of an opinion that it is in American and European interest to overcome this barrier.

The following phenomena and processes which take place in the global trade allow the use of the above conclusions in the world scale, taking into account the specifics of each transport relationship:

- there is a great potential of the containerization development;
- more and more light and very light commodities are carried in containers which tend to be transported in larger loading units;
- transport of large containers, i.e. 45' and 53', by the water and rail means of transport does not require significant changes in law, but only design (construction) adjustment.

The calculations have been made with an aim to use the most credible data sources and avoid approximate data. Therefore, the authors' aim was to avoid subjectivism. It is advisable to conduct further research based on a thorough analysis and extender models. Authors postulate extension the scope of economic analysis of intermodal global chains considering other factors of their costs. Further research should identify the relations between logistics costs and volume of global trade.

References

1. UN ECE, *Terminology on Combined Transport*, www.unece.org, (2001)
2. UIC Report, *Report on Combined Transport in Europe*, International Union of Railways - UIC, Paris, (2012)
3. D. Milewski, B. Wisnicki, *Modelling intermodal transport systems - directions for scientific research*, Research Journal of the University of Gdańsk Transport Economics and Logistics, Vol. 66, 35-42 (2017)
4. A. Ballis, J. Golias, *Towards the improvement of a combined transport chain performance*, European Journal of Operational Research, 152, 420-436, (2003)
5. J. Wronka, *Transport intermodalny/kombinowany – teoria i praktyka*, Wydawnictwo Naukowe Uniwersytetu Szczecińskiego, Szczecin 2008, 114, (2008)
6. S. Dărăbanț, P. Ștefănescu, R. Crișan, *Economic benefits of developing intermodal transport in the European Union*, Annals of the University of Oradea, Economic Science Series, Vol. 21, Issue 2, 81-87, (2012)
7. J. Stokłosa, Š. Liščák, M. Jaśkiewicz, K. Ludwinek, *Systemy transportu intermodalnego - kierunki rozwoju w świetle europejskich doświadczeń*, Logistyka, 6, 10070-10074, (2014)
8. K. Liljestrand, *Intermodal transportation from a haulier's perspective. An analysis on how to increase the usage of intermodal road-rail transportation for hauliers in Sweden*, Chalmers University of Technology. Göteborg, Sweden, Report No. E., (2010)
9. B. Guszczak, *Intermodal transport as an attractive mode of transport*, Research in Logistics and Production, 4(3), 209-219, (2014)
10. A. Caris, C. Macharis, G.K. Janssens, *Planning Problems in Intermodal Freight Transport: Accomplishments and Prospects*, Transportation Planning and Technology, 31(3), 277-302, (2008)
11. K.Baumgartnera, A. Fuettererb, U.W. Thonemannb, *Supply chain design considering economies of scale and transport frequencies*, European Journal of Operational Research, 218(3), 789–800, (2012).
12. G. Jong de, I. Vierth, L. Tavasszy, M. Ben-Akiva, *Recent developments in national and international freight transport models within Europe*, Transportation,. 40, s. 347–371, (2013)
13. L.M.Gambardella, M. Mastrolilli, A.E. Riz Zoli, M. Zaffalon, *An optimization methodology for intermodal terminal management*, Journal of Intelligent Manufacturing, 2001, Vol. 12, No. 5/6, 521–534, (2000)

14. M.B. Pedersen, *Optimization models and solution methods for intermodal transportation*, PhD thesis. Centre for Traffic and Transport Technical University of Denmark, 171, (2005)
15. E. Kozan, *Optimising container transfers at multimodal terminals*, Mathematical and Computer Modelling, Vol. 31, 235–243, (2000)
16. C.Macharis, E. Van Hoeck, E. Pekin., T. Van Lier, *A decision analysis framework for intermodal transport: comparing fuel price increases and the internalization of external costs*, Transportation Research, Part A 44, 550–561, (2010)
17. A.E. Rizzoli, N. Fornara, L.M. Gambardella, *A simulation tool for combined rail/road transport in intermodal terminals*, Mathematics and Computers in Simulation, No. 59, 57–71, (2002)
18. J. Flodén, *Modelling Intermodal Freight Transport - The Potential of Combined Transport in Sweden*, Department of Business Administration. Gøteborg University, (2007)
19. B. Groothedde, C.Ruijgrok, L. Tavasszy, *Towards collaborative, intermodal hub networks. A case study in the fast moving consumer goods market*, Transportation Research Part E: Logistics and Transportation Review, 41, 567-583, (2005)
20. B. Sahin, H. Yilmaz, Y. Ust, A.F. Guneri, B. Gulsun, E. Turan, *An Approach for Economic Analysis of Intermodal Transportation*, The Scientific World Journal, 10 p., (2014)
21. M. Janic, *Modelling the full costs of an intermodal and road freight transport network*, Transportation Research D, 12(1), 33–44, (2007)
22. UNCTAD *Review of Maritime Transport 2017*, UNCTAD/RMT/2017, United Nations Publication, 55, (2017)

Role of maritime ports and shipping in the creation of the economic value of the sea areas

Jacek Zaucha[1] , and *Maciej* Matczak[2]

[1] University of Gdańsk, Faculty of Economics, 119/121Armii Krajowej str., 81-824 Sopot, Poland
[2] Gdynia Maritime University, Faculty of Entrepreneurship and Quality Science, 81-87 Morska Str., 81-225 Gdynia, Poland

Abstract. The sea coast location of a country has been regarded as a critical value for its development and prosperity for years. It relates to economic, military, and environmental issues. A wide scope of economic activities, like shipping and maritime ports, fishing and aquaculture, underwater mining (e.g., oil, gas, sand) and sea coast tourist activities all require access to the sea. The sea area should also be regarded as a limited resource thus the activities need to compete for maritime space. Today, comprehensive economic valuation of selected activities, both from the private and public point of view, became significantly important. The aim of this article is an attempt to evaluate the economic value of maritime space, which is generated by ports and shipping. Initially, review of the literature regarding the valuation of economic effects of maritime port and shipping was carried out. The analysis has been based on a model of the interrelation between ports and shipping and its socio-economic environment. Then, redefinition and extension of identified values to the sea areas were provided. Finally, a calculation of the unit value of sea space [PLN per sq. km] as well as identification of most valuable spatial parts of the sea was completed. The application part of the study was based on the example of Polish seaports as well as Polish maritime areas on the Baltic Sea.

1. Methodology approach

The aim of this article is an attempt to evaluate the economic value of maritime space which is created by ports and shipping sector. For that purpose, the following stages of research were implemented:

- Investigation of current achievements regarding valuation of the impact of the maritime sector on the socio-economic development by review of relevant literature.
- Identification of the key types and sources of spatial rents from the maritime sector.
- Development of the model of the interrelation between the seaport and shipping sector and its socio-economic environment.
- Investigation of the value created by the sector with a focus on gross added value and gross financial results.
- Recalculation of the unit value of maritime space in Poland [PLN per sq km].

- Transfer of results to maps illustrating the space distribution of values as well as identification of most valuable spatial parts of the Baltic Sea (the Polish internal waters part).

The empirical part of the study was based on the case of the Polish maritime sector, which resulted mainly from the availability of detailed information necessary to carry out the study.

2. Socio-economic impact of the shipping and port industry development in research

The impact of shipping and seaport sector on the socio-economic activity is undeniable. The history of human development and civilisation growth is directly connected with the access to the seas. Thus, development of studies concerning interrelations between maritime transport (shipping and ports) and its surroundings, including socio-economic development, could be noted. Considering the key areas of research, the following levels of analyses can be distinguished:

- port-city relations & policies,
- maritime ports versus regional and countries economy development,
- ports and shipping in the global economy.

Referring to the first scope of analysis, spatial development should be regarded as the main area of the researcher's interest [1-5]. Despite the relevant identification of interactions, this type of research does not valuate the impact on the socio-economic level in the monetary terms.

The interrelation between regional and national economic system constitutes the second level of investigation. In that respect, two issues are considered – the role of maritime ports in economic development (defined as an engine for gross domestic product or gross added value increase) at regional and country level [6-10] and its place in the creation of the labour market [11]. Commonly discussed issue, in that respect, is the direction of influence. On the one hand, the ports are regarded as an accelerator of regional and country economy, on the other they are treated only as a response to an economic demand from a socio-economic system [12]. For the needs of this study, the first approach is adopted.

The role of maritime transport and seaport in economic development is also researched on the global scale. Considering the main area of research, the problem of transport cost is dominating. In this case, the transport is regarded as an issue restricting the volume of external trade [13]. Similarly, international trade, a vital element of GDP calculation (net export within the expenditure approach), is also considered as a necessary element for further development of the national economy. Thus effective transport can be treated as an accelerator of growth.

All in all, investigation of value created by sea areas is strictly connected with the second level of analysis, so further research is conducted according to that approach.

3. Economic valuation of the sea space – issue of maritime ports and shipping

Valuation of benefits deriving from the use of sea space and the costs of obtaining them is important for knowing the spatial rental, which was considered one of the most important mechanisms for shaping the maritime space.

As part of market processes, the spatial rent can be considered as private (private spatial rent). Such approach can be understood as the tendency of the user of the space to pay for its use (the so-called bid-rent), assuming that the level of utility/profit remains constant.

Otherwise, in public selection processes, the rent can have a public nature. It can be equated with the difference between social benefits (private benefits and external benefits) and social costs (private costs and social loss). Both rentals, private and social (public), can vary significantly. For the progress and development of spatial economics, analyzing these differences may have a high cognitive significance, but from the point of view of the development of the maritime space, this is not so important [14].

A private pension remains an important mechanism where the legislator has allowed the market advantage (e.g., maritime transport, tourist or fishing), and public rent where public choice is of primary importance (environmental protection, wind energy, mineral extraction, etc.). Differences between both types of rent for a given method of benefiting from the sea space, serve to verify the validity of these decisions. For example, decisions limiting the uncontrolled development of coastal tourism prove that the legislator sees the difference between its public and private rents. These differences are also an important decision-making parameter when the public administration makes administrative decisions on changing the use of sea areas (in a situation where new ways exclude pre-existing ones). For example, the occupation of fishery areas for wind farms should be preceded by a comparison of the public pension from both uses of sea areas.

The scale of difficulty in obtaining a reliable annuity index increases with the decrease in the possibility of using a market valuation for this purpose. This is due to the problem of spatial attribution of economic data. Considering the main areas of socio-economic activity provided within (or on base) the sea areas, the following sectors could be distinguished:

- fishing,
- coastal tourism,
- seaports and shipping,
- denitrification services of marine sediments,
- offshore wind energy [14].

In that respect, the possibility of calculating the level of spatial rent for listed sectors differs significantly. As per the initial assumption, the study is focused on the ports and shipping activities. There is a possibility of market pricing of private benefits and costs but to a limited extent. Their spatial attribution is complicated; private rent is more important, but public rent also seems important due to the considerable support of public authorities for the port sector. Further identification of spatial rent from those sectors required identification and valuation of their socio-economic benefits on the national level.

4. Economic value of maritime ports and shipping industry in Poland

A comprehensive valuation of socio-economic benefits from maritime transport development requires definition and quantification of the basic relations occurring between seaports and the regions or countries of their location. The implementation of this goal requires multi-faceted and a multidimensional view of the functioning of modern seaports. The Oxford Economics research is an example of a similar relationship [15, 16]. The authors divide the impact of port sector between three levels: a direct impact, indirect impact and induced effects. The results of port activity were evaluated from labor market, national accounts as well as exchequer revenues. Worth noting is the fact that such a method allows the determination of multiplier effects. Because of lack of access to relevant sources of information, a modified approach has been adopted in this study.

Analysis of the functional structure of Polish seaports, as well as its verification from the point of view of the availability of information, allow to define a four-layer model (Fig. 1) of port sector [17]:

- First level: *MARITIME PORT*,
- Second level: *SEAPORT SECTOR* (incl. maritime ports),
- Third level: *DIRECT PORT ENVIRONMENT* (incl. port sector),
- Fourth level: *SOCIO-ECONOMIC PORT ENVIRONMENT* (incl. direct port environment).

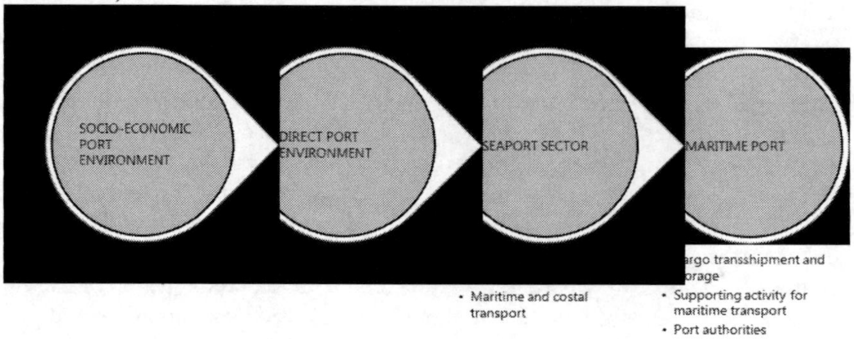

Fig. 1. Model of interrelation between ports and socio-economic environment
Source: Matczak 2017

Such an approach allows taking into account the fact that the real beneficiaries of sea transport and transshipment are Polish companies importing and exporting goods by sea. Their private (monetary) net benefit from the existence of sea space is the difference between their profits obtained by using maritime transport and profits that would be possible using other modes of transport that do not use ports, anchorages and sea routes, and those using foreign ports. In practice, this figure can't be empirically calculated, or it would require very extensive and often time-consuming research, hence the simplified method of GVA estimation has been used. Comprehensive investigation based on the data from the Central Statistical Office in Poland has been completed within the study [18].

The total GVA of PLN 1 619.7 million (an increase of 2.7% compared to the previous year) was generated in Poland in 2016. The regions of the main seaport locations in Poland- Pomerania and West Pomerania were responsible for 5.7% and 3.7% of country GVA respectively.

Based on the parameters of the labor market in the port sector and its environment as well as the average GAV generated in specific port regions by one employee [19], the level of gross value added in individual areas of the analysis was estimated.

Taking into account the broadest view of the effects of seaport activities (Level IV) covering both seaports, port sector, and the immediate and economic environment of ports, the value of PLN 32,155.02 million was generated in 2016. When comparing the results to the national GVA, it can be said that 1.99% of the value in the country is created by port operations, which should be considered significant.

The results obtained can be directly related to the share of the port sector and its environment in generating gross domestic product.

Worth noting here is the fact that results obtained are based on the average values assigned to particular areas of analysis, which in the light of the above-average level of value creation in the transport sector allows us to suppose that the real impact is greater.

Another method of assessing the value of maritime space implemented in the research is an investigation of the gross financial results of companies engaged in the seaport and shipping sector. Information on the financial results of shipping and ports is also available in the annual statistics of the maritime economy [18]. They only include, unfortunately, maritime enterprises registered in Poland, with accounting books in which the number of employees exceeds 9 persons. Information on the gross financial result is available by type of activity, and the following should be included in those related to maritime transport: reloading, storage and storage of goods in seaports, other activities supporting maritime transport, operations of maritime transport agencies, seaports, maritime and coastal water transport.

In 2016, the gross financial result of enterprises conducting the abovementioned type of activities amounted to PLN 874,1 million. This approach is conservative. It does not include multiplier effects, such as indicating how ports and shipping stimulate their economic environment.

5. Valuation of the maritime space – Poland's case

As was previously touched, free navigation on the seas is the main reason for leading importance of private rent in maritime transport. The public rent differs from it because its value is greater by very significant external benefits (multiplier effects) and costs. Social losses from shipping include those related to the use and degradation of the environment (e.g., oil spills, worsening of the welfare of marine fauna, atmospheric pollution, introduction of alien species). Even if their monetary value could be obtained, their spatial attribution will always be subjective. For example, atmospheric pollution in ports may be lower than on the high seas if the port uses electric assistance for ships entering it.

Also, the pricing of private rent is not easy. At sea, there is also a complication related to the fact that the law of the sea (UNCLOS) guarantees (with certain exceptions, e.g., traffic separation zones) free access to sea areas to ships of all flags. Therefore, foreign shipping companies benefit from maritime areas for free, and their activity does not benefit the national economy if these ships do not call at those relevant ports. Only environmental costs appear. The rent from transit shipping routes is therefore of an international nature. It is difficult to estimate it using the national economic information available in the country. However, if foreign shipping companies support Polish exchange of goods, it is true that their profits do not create the wealth for their country, but they contribute to the development of the Polish economy.

All this makes it very difficult to calculate the spatial rent from the existence of ports and sea routes connecting them. What is more, a different construction of categories: gross financial result (PLN 740.3 million) and gross value added (PLN 29,295.29 million) causes a significant difference between the presented values. In case of gross financial result,, a difference between revenues and costs is estimated, thus in case of GVA, a difference between global output and intermediate consumption is used.

The above-mentioned value of GVA and gross financial result was assigned to individual ports on the basis of the average size of their reloading in the years 2005-2015 [20]*.

The methods of assigning the value of gross financial result in their spatial attribution are shown in Table 1. The assumption was made that the port cannot function without the necessary space of sea areas, which includes sea basins, roads, anchorages, and flaps serving a given port. All elements of this space are equally valuable. It was also assumed that the Szczecin-Police-Świnoujście port complex is served by the common sea area. On the basis of existing documents, it was not possible to distinguish such areas for small ports, with the exception of Kołobrzeg. It was also assumed that the ship must use both the space serving the port directly and the shipping lanes. Therefore, 50% of GVA and the gross financial result of maritime transport were assigned to these routes, and 50% to marine areas serving the ports directly.

* Presented approach should be considered simplificd, as it can be assumed that the containerized cargo provides a higher gross financial result per unit weight of the load. The proposed division does not take this fact into account

Table 2. Value of spatial rent for sea areas serving ports based on GVA and gross financial result in 2015.

No	PORT	Port share in total turnover in% (average 2005-2015)	Gross financial result assigned to a port in PLN million in 2015	Gross value added attributed to a given port in PLN million in 2015	The sea area that serves the port directly [sq. km]	Gross financial result per sq km of Polish sea areas serving a given port in PLN	Gross added value per sq km of Polish sea areas serving a given port in PLN
Total		100.00	740.3	29,295.29	15,435.00		
1	Gdańsk	39.68	146.9	5,812.19	166.4	882,665.38	34,928,999.62
2	Gdynia	24.64	91.2	3,609.18	35.41	2,575,683.71	101,925,437.11
3	Szczecin-Świnoujście	32.08	118.7	4,698.96	223.33	580,923.19	22,988,400.90
5	Police	2.97	11.0	435.04			
6	Kołobrzeg	0.25	0.9	36.62	17.11	54,083.87	2,140,216.98
7	Other ports	0.38	1.4	55.66	-	-	-

Source: Own calculation based on the data & information from above presented sections and the Central Statistical Office in Poland [18]

Table 1 presents the spatial rent for ports based on GVA and gross financial result. The latter approach underestimates the significance of sea areas used for port purposes, since it omits other port operations, aside from transshipments. In GVA approach, additional types of maritime related activities are taken into account, e.g., scientific research, sea mining (oil), construction of water engineering facilities for maritime economy, architecture, engineering for maritime economy, renting water transport, consultancy, research and analysis technical, recruitment of employees and other activities for maritime economy. Unfortunately, this is not recognized by the GUS financial result indicator.

The level of the annuity with the use of the gross financial result varies from over PLN 2.58 million for the port in Gdynia to PLN 54.08 thousand for the port in Kołobrzeg. For the GVA, this spatial rent is PLN 101.93 million, respectively, up to PLN 2.14 million.

The calculation of the spatial rent of shipping lanes in Polish sea areas was made on the basis of information on the intensity of ship traffic in these areas. AIS information was used, a system which allows to indicate the frequency of passage of the ship through sea squares with an area of approx. 1 square km. Spatial rent is calculated as the product of the number of vessels passing through a square and GVA during the year, or the gross financial result of one crossing calculated as 50% of their total value in 2015 for Polish sea areas and the sum of all transitions through all AIS squares in these areas. In 2015, there were a total of 5,108,559 such transfers for all squares, and the value of one passage through the square was PLN 72.4 for the gross financial result and PLN 2.87 thousand PLN for GVA.

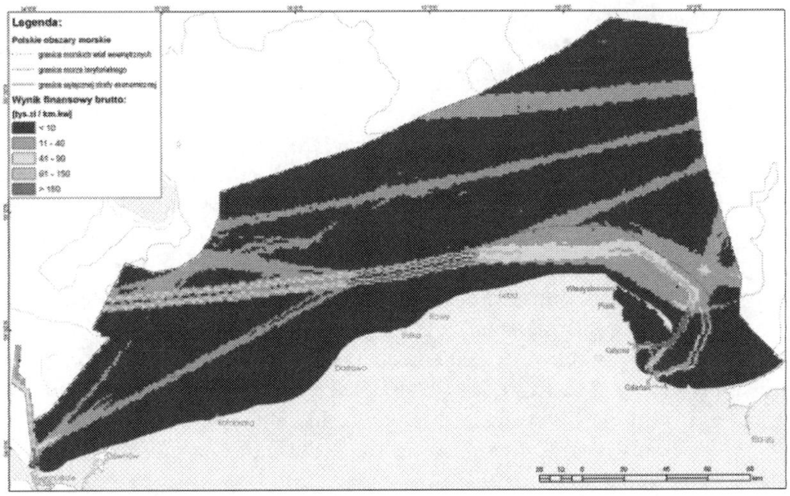

Fig. 4. Spatial rent from shipping routes for 2015 using the gross financial result of shipping and ports [14]

The highest value of the estimated shipping rent is about 281 thousand PLN per sq. km for the gross financial result and PLN 11,604 thousand per sq. km for GVA. The average value is PLN 11 thousand and 472 thousand PLN respectively. Figures 4 and 5 contain a graphic image of a spatial rent from commercial shipping.

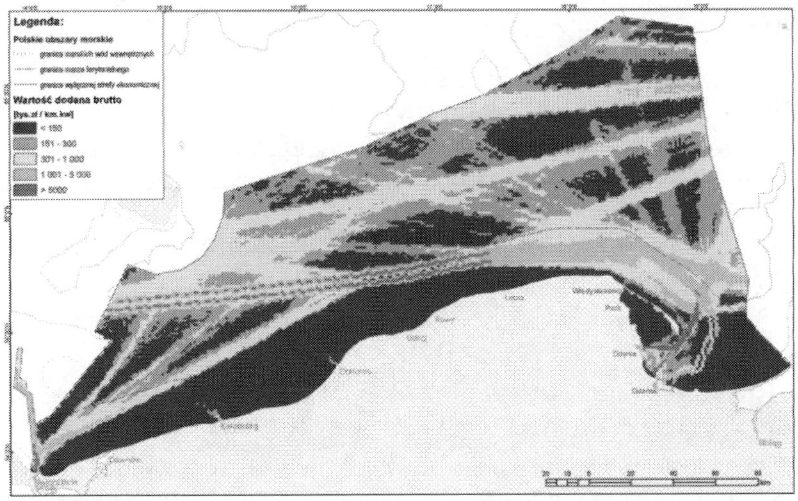

Fig. 5. Rent from shipping routes for 2015 using WDB for port sector and its business environment [14]

The intensity of shipping in both engravings is the same, and the apparent differences result from the adoption of different limit values for rents illustrated in both drawings. The estimates presented above overstate the rent, as they assume that ships only use Polish sea areas and Polish ports. In fact, depending on the distance that a ship has to travel, this type of disbursement should be spread over the entire shipping route. Such calculations based on information from AIS would, however, be significantly complicated and costly.

Conclusions

Estimation of the value of access to the sea areas is an important element of modern maritime policy, because the need to make rational choices related to the management of the sea is very important. Maritime areas are used to implement a number of economic and social activities, which create benefits and costs for society, economy and environment. Reliable assessment of the impact of such activities allows for better use of marine space and thus contributes to the development of the economy and prosperity.

This article presents an attempt at a valuation of the sea areas pertaining to transport activities. The results obtained indicate a number of problems and challenges that limit the analytical possibilities. On the one hand, it is an access to relevant data, as well as their scope and quality. The second is the different ways of assessment of the value created by the maritime transport sector, as indicated by discrepancies in the results obtained. For instance, the calculation based on the GVA created by ports and shipping in Poland reached level of PLN 29,295.29 million in 2015. At the same time, the value based on the gross financial results of the sector companies reach PLN 740.3 million. A significant gap is also identified in the case of particular maritime ports. The level of the annuity with the use of the gross financial result varies from over PLN 2.58 million per sq. meter for the port in Gdynia to PLN 54.08 thousand per sq. meter for the port in Kołobrzeg.

An important issue, deliberately omitted in the study, is the comparability of results from various areas of activity (e.g. tourism, fishing).

There is no doubt that the valuation of access to sea areas requires further research and analysis, which should bring further approximations to their final, reliable determination.

References

1. T. Palmowski, M. Pacuk, T. Michalski, Przemiany przestrzeni miast portowych na przykładach Gdańska i Gdyni, XIV Konwersatorium Wiedzy o Mieście (2001) https://depot.ceon.pl/bitstream/handle/123456789/5469/Przemiany_przestrzeni_miejskiej_miast_portowych_na_przyk%C5%82adach_Gda%C5%84ska_i_Gdyni.pdf?sequence=1&isAllowed=y
2. P. Lorens, Obszary Portowe – problemy rewitalizacji, ISR, Szczecin (2013) https://arch.pg.edu.pl/documents/174968/51761959/poportowe.pdf
3. K. Krośnicka, Wpływ działalności portów na strukturę funkcjonalno-przestrzenną miast, Inż. Mor. i Geotech. **5**, (2012)
4. K. Krośnicka, *Ewolucja relacji port-miasto na tle rozwoju technologii żeglugi*, Publishing House of Gdynia Maritime University (2005)
5. Współzależność funkcjonowania portu i miasta portowego (w warunkach gospodarki rynkowej), Edit. B. Szwankowska, Publishing House of Maritime Institute, No. **4946**, Gdańsk (1994).
6. Ferrari, C., Merk, O., Bottasso, A., Conti, M., Tei, A. (2012), *Ports and Regional Development: a European Perspective*, OECD Regional Development Working Papers, (2012), OECD Publishing http://dx.doi.org/10.1787/5k92z71jsrs6-enA.
7. C. Ferrari, *Ports and regional economic development,* Global Ports and Urban Development: Challenhes and Opportunities, OECD, Paris (2011), OECD Publishing www.oecd.org/gov/regional-policy/49456330.pdf

8. G.s. Dwarakish, Akhli Muhamman Salim, *Rewiev on the Role of Ports in the Development of a Nation,* Aquatic Procedia, **4** (2015), pp. 295-301.
9. Q. Sakalayen, The Strategic Role of Australian Ports in Regional Development, Univ. of Tasmania (2014) https://eprints.utas.edu.au/18706/5/Whole-Sakalayen-thesis_ex_pub_mat.pdf
10. L. Berkoz, D. Tekba, The Role of Ports in the Economic Development of Turkey, 39th European Congress of the regional Science Association (1999) https://pdfs.semanticscholar.org/9abe/14d3487645c81729506f0d8f73a19b9cb0f1.pdf
11. Ports: job generation in the context of regional development, Department of Infrastructure and regional Development, Australian Governance, Information sheet 56 (2014)
12. B-m. Jung, *Economic Contribution of Ports to the Local Economies in Korea,* The Asian Journal of Shipping and Logistics, Vol. **27**, No. 1, pp. 001-030 (2011)
13. P.R. Krugman, M. Obstfeld, M.J. Melitz, *International Economics Theort & Policy,* Addison-Wesley (2012),
14. J. Zaucha, *Gospodarowanie przestrzenią morską,* Academic Publishing House Sedno, Warsaw (2018).port
15. *The economic impact of the UK Maritime Service Sector,* A Report for Maritime UK, Oxford Economics (2012)
16. The economic impact of the UK Maritime Service Sector: Ports, A Report for Maritime UK, Oxford Economics (2012)
17. M. Matczak, *Polskie porty morskie jako biegun rozwoju gospodarczego kraju I regionów lokalizacji,* Raport: Rada Interesantów Portu Gdynia/Baltic Container Terminal Gdynia/Actia Forum (2017)
18. Statistical Yearbook of Maritime Economy 2017, Central Statistical Office, Statistical Office in Szczecin (2017)
19. Local Data Bank, Centra Statictical Office, Warsaw (2017) https://bdl.stat.gov.pl/BDL/start
20. M. Matczak, Polish seaports in year 2015. Summary and future prospects, Actia Forum (2016)

AUTHOR INDEX